Efficient and Enhanced Management of Plastic Waste in India

Paromita Chakraborty • Girija K. Bharat
Eirik H. Steindal • Smita Mohanty
Rachel Hurley • Marianne Olsen • Suneel Pandey
Editors

Efficient and Enhanced Management of Plastic Waste in India

Case Studies from Vapi and Surat

Springer

Editors
Paromita Chakraborty
Environmental Science and Technology Laboratory, Centre for Research in Environment, Sustainability Advocacy and Climate Change (REACH)
Directorate of Research
SRM Institute of Science and Technology
Kattankulathur
Chengalpattu district, Tamil Nadu, India

Eirik H. Steindal
Norwegian Institute for Water Research
Oslo, Norway

Rachel Hurley
Norwegian Institute for Water Research
Oslo, Norway

Suneel Pandey
The Energy and Resources Institute (TERI)
New Delhi, Delhi, India

Girija K. Bharat
Mu Gamma Consultants
Gurugram, India

Smita Mohanty
CIPET
Bhubaneswar, Odisha, India

Marianne Olsen
Norwegian Institute for Water Research
Oslo, Norway

ISBN 978-3-031-89397-1 ISBN 978-3-031-89398-8 (eBook)
https://doi.org/10.1007/978-3-031-89398-8

This Springer imprint is published by the registered company Springer Nature Switzerland AG
The registered company address is: Gewerbestrasse 11, 6330 Cham, Switzerland

Foreword

The issue of plastic waste management in India is a complex challenge, deeply interwoven with rapid economic growth, urbanization, and evolving consumption patterns. This book delves into the persistent gaps and hurdles in managing plastic waste, offering a comprehensive analysis rooted in on-the-ground assessments and emerging international discussions.

The book highlights the critical need for accurate and representative baseline data on waste generation, management, and leakages. Current data collection methods often fall short by not capturing inadequately disposed waste or the significant role of the informal sector. The lack of standardized methodologies for measuring plastic waste makes it difficult to compare results across different geographical contexts.

While India has made significant strides in policy with the plastic waste management (PWM) rules and the concept of extended producer responsibility (EPR), implementation remains a challenge. The book explores how policies, often focused on cleanliness and consumer behavior, need to shift to address the entire life cycle of plastics. It also examines the complexities of banning single-use plastics (SUPs) and the need for viable alternatives. The rapidly evolving policy environment presents challenges for industries, especially micro, small, and medium-sized enterprises (MSMEs).

Existing waste management structures lack the capacity to handle the sheer volume of plastic waste. This book emphasizes the need for improved waste segregation and collection infrastructure, as well as access to technology that supports environmentally sound management of waste. The complexities of waste-to-energy (WTE) solutions are also explored, noting their potential limitations and the need for more data and evidence on their wider implications.

The book sheds light on the socio-economic impacts of plastic pollution and reduction strategies, stressing the importance of a just transition. The vulnerabilities of informal waste workers and the need to integrate them into formal waste management systems are key considerations. The book also acknowledges that waste reduction strategies can have unintended consequences for MSMEs and marginalized communities. The COVID-19 pandemic has exacerbated challenges, impacting both the waste sector and the livelihoods of informal workers.

This book argues that addressing the complex problem of plastic pollution requires a systemic approach with collaboration among stakeholders across the plastic's lifecycle. Key priorities include investment in organized waste management facilities, capacity building, awareness raising, and ensuring inclusivity and justice in waste management initiatives. Additionally, there needs to be a focus on understanding the interlinkages between plastics and persistent organic pollutants (POPs), as well as creating financial incentives for consumers to segregate waste. It also highlights that the integration of the informal sector is essential.

Ultimately, this book makes a strong argument that a holistic approach, which incorporates the informal sector, tackles the lack of standardization in measuring plastic waste, and addresses the whole life cycle of plastic is essential for a more sustainable future. The information contained in this book aims to drive the necessary political momentum and public awareness required to achieve real change.

City Net Secretariat, Seoul, Republic of Korea Vijay Jagannathan
World Resources Institute, Washington, DC, USA

Contents

About the Editors

Paromita Chakraborty is Professor and Head of the Centre for Research in Environment, Sustainability Advocacy and Climate Change (REACH) under Directorate of Research in SRM Institute of Science and Technology, Kattankulathur, India. Prof. Chakraborty is an expert scientific member for UNESCO Chair on Ecohydrology and Applied Ecology, Faculty of Biology and Environmental Protection, University of Lodz, Poland.

Prof. Chakraborty leads the Environmental Science and Technology research group and works extensively on multimedia movement, risk assessment, and cost-effective sustainable remediation techniques for micro-organic pollutants including several plastic-associated chemicals. For more than a decade she has been conducting research by working closely with industrial partners on recycling/upcycling of waste plastics and developing new circular economy technologies. She is the Indo-US Science and Technology Forum's Water Advanced Research Innovation Fellow 2016, and recipient of the American Chemical Society Early Career Chemist Award 2015, Hiyoshi Environmental Award 2020, and Dr. Ajmal Khan Award 2024. She has published 100+ scientific articles in peer-reviewed international journals and is editor of 4 books of Springer/Elsevier. She is the Associate Editor for the Elsevier journal, *Science of the Total Environment*. She has been enlisted by Elsevier and Stanford University as one of the top 2% global scientists for the year 2024 in Earth and Environmental Sciences.

Girija K. Bharat is the Managing Director of Mu Gamma Consultants, Gurgaon, and is an international expert with 34 years of experience in plastic pollution, chemical waste management, and water resource management. She has a PhD in Chemistry and was a Presidential Scholar at George Mason University, USA. She has over 130 publications on water resource management, plastic pollution, persistent organic pollution, endocrine-disrupting chemicals, and the Plastic Waste Strategy Report (Gujarat). She has co-developed the National Policy Framework on Safe Reuse of Treated Water (India), and the Global Sustainable Development Reports for 2015 and 2016. She has been a Research Council Member for CSIR-NEERI and Technical Member of NITI Aayog. Dr. Bharat has worked on over 50 projects funded by the Norwegian Embassy, Research Council of Norway, The World Bank, GIZ, UNEP, UNICEF, Swiss Development Cooperation, and ADB. She is the recipient of several awards including the Women Transforming India Award (2021) from NITI Aayog and UNEP.

Eirik H. Steindal is the Head of Business Policy and Public Relations at Grønt Punkt Norge and a PhD fellow in Environmental and Development Studies at the Norwegian University of Life Sciences. With over 20 years of international experience in environmental and development fields, he has worked in research, civil society, as well as the private and public sectors.

As a trained biologist, Steindal focuses on sustainability, business, knowledge, and policy, particularly in regulatory development, environmental issues, and corporate social responsibility (ESG). Previously, he was a senior researcher at the Norwegian Institute for Water Research (NIVA), where he led large international research and capacity-building projects in China and India, aiming to address severe environmental challenges through interdisciplinary approaches in collaboration with governments, industry, and research institutions.

Steindal has built an extensive international network in the environmental field through his work in the Norwegian Government and with Norwegian and international environmental organizations and the UN. For seven years, he served as a Norwegian negotiator for international environmental agreements and has been a government-appointed expert under the Arctic Council and the UN.

Smita Mohanty is the Pr. Director (Sr. Principal Scientist) and Head of the CIPET:SARP-LARPM, Bhubaneswar, Department of Chemicals and Petrochemicals, Ministry of Chemicals and Fertilizers, Government of India. Dr. Mohanty has more than 20 years of experience in teaching, research, and technical consultancy. She has been the key to industry connect, bridging academics and industry. She has been involved in the development and establishment of novel technologies, scaling-up from lab to bench pilot scale, industrial tie-ups and pathway for commercialization in the areas of biopolymers, plastics waste and recycling, additive manufacturing, and composites. Dr. Mohanty has also been serving as the Convener for the Standardization Committee of BIS—PCD 21 for Biodegradation.

She has mentored groups of faculties, scientists, and PhD scholars in their ventures. She has been the author of more than 200 international publications, with current citations received = 19,956, h-index 68, and i10-index 315. She is editor/author of 16 books/book chapters with leading international publishers and has 16 patents to her credit. She is also the recipient of several prestigious awards like: "National Award" by Government of India in appreciation toward Technology Innovations in Petrochemicals and Downstream Plastics Processing Industry, Outstanding Research Faculty Award by Scopus Careers 365, etc. She has also been represented among top 2% scientists for the years: 2020–21, 2021–22, 2022–23, and 2023–24 by Stanford University, USA.

Rachel Hurley is a Key Researcher at the Norwegian Institute for Water Research (NIVA). She has been active in the field of (micro)plastic research for the past decade, with a background in geography and fluvial geomorphology and a PhD in Physical Geography from University of Manchester. Today, she focuses on developing a process-based understanding of macro- and microplastic pollution, specializing in plastic pollution dynamics in freshwater and soil environments. Important areas in Dr. Hurley's research include: (1) the fate of microplastics in agricultural soils and their impact on soil resilience, and (2) understanding the fate and transport of riverine plastic pollution, developing and harmonizing monitoring methods, and capacity building related to macroplastics in rivers across Asia as part of several large projects in India and the ASEAN nations. She has produced seminal work on the role of flooding in redistributing plastic pollution in river catchments, within this context.

Dr. Hurley has garnered an extensive international scientific network. She serves as Deputy Coordinator of a large EU Horizon 2020 consortium, leading scientific progress in the long-term sustainability of agricultural plastic use. She has been invited as an expert in policy processes related to the European Chemicals Agency and the European Commission's Scientific Advice Mechanism.

Marianne Olsen is Research Director (Chief Operating Officer) at NIVA with operational and scientific strategic responsibilities within NIVA and takes part in NIVA's leader group. Dr. Marianne holds an MSc in Marine Biology and a PhD in Ecology, addressing ecological effects of contaminated sediment.

As a researcher, she has worked with a wide range of pollution challenges including environmental contaminants, oil spills, and plastic pollution, and she has extensive experience with monitoring, environmental risk assessments, mapping of pollution, tracking of sources, mitigation measures, and stakeholder involvement.

Dr. Marianne has work experience from industry and private sector running projects related to the environmental effects and mitigation measures to prevent negative consequences of offshore oil and gas extraction and of various types of land-based process industries. She has worked as a special advisor to the Norwegian Environmental Authorities related to polluted seabed and remediation actions. In recent years, Dr. Marianne has been heavily involved in the development of NIVA's research and project portfolio of international aid-funded capacity development projects on chemical and plastic pollution, on ocean accounting and on marine spatial planning, particularly in Asia.

Suneel Pandey is Senior Fellow and Director, Circular Economy and Waste Management Division, TERI. In addition, he is also working as *Adjunct Faculty* at TERI School of Advanced Studies (TERI SAS). He has more than 30 years of consultancy/research experience in the areas of municipal, industrial, and hospital waste management; their linkage with climate change aspects and circular economy and resource efficiency; waste-to-energy issues; impact assessment; air, water, and soil quality monitoring; site assessments; performance evaluation of ETP; and institutional strengthening and capacity building. He has obtained his PhD in Hazardous Waste Characterization from Nagpur University while working as *Project Fellow* at NEERI, Nagpur. Prior to joining TERI, Suneel has worked for ERM India and at the Hong Kong University of Science and Technology on development of landfill liner for containment of land-disposed hazardous waste and deriving material balances and associated environmental pollution for Hong Kong region as part of his *postdoctoral research*.

Status of Litter and Plastic Pollution Globally and in India

Girija Prasad, Himalaya S. Vardikar, Abdullah Atiq, Smita Mohanty, Twinkle Dev, Avanti Roy Basu, Suneel Pandey, Girija K. Bharat, and S. K. Nayak

1 Existing Monitoring Data

Polymers, owing to their wide range of properties and applications, have become an inevitable part of modern society. Among the widely used polymers, commodity plastics like polyethylene (PE), polyvinyl chloride (PVC), polypropylene (PP), polystyrene (PS and GPPS general-purpose polystyrene), poly(methyl methacrylate) (PMMA), and engineered plastics like acrylonitrile butadiene styrene (ABS), polycarbonate (PC), and polyamides (nylons) play an important part in many technological areas (Table 1). Due to their widespread use, inadequate management including inefficient collection and the end-of-life, and lack of proper recycling techniques, plastic wastes are gradually accumulating, which will have various hazardous effects on land and water. Average global plastic production crosses 150 million tonnes per annum with uses that span packaging, films, wrapping materials, shopping and garbage bags, fluid containers, clothing, toys, household and industrial items, as well as construction materials (Swachh Bharat Mission (Urban), 2019) (Fig. 1).

A significant proportion of pollution comes from carelessly trashed plastics. When throwaway plastics (ranging from micrometre to meter in size) enter the

G. Prasad (✉) · H. S. Vardikar · S. Mohanty
Central Institute of Petrochemicals Engineering and Technology (CIPET): School for Advanced Research in Petrochemicals (SARP)-Laboratory for Advanced Research in Polymeric Materials (LARPM), Bhubaneswar, Odisha, India

A. Atiq · T. Dev · S. Pandey
Environment & Waste Management Division, TERI, New Delhi, India

A. R. Basu · G. K. Bharat
Mu Gamma Consultants, Gurugram, India

S. K. Nayak
Ravenshaw University, Cuttack, Odisha, India

P. Chakraborty et al. (eds.), *Efficient and Enhanced Management of Plastic Waste in India*, https://doi.org/10.1007/978-3-031-89398-8_1

Table 1 Different types of commodity plastics (Swachh Bharat Mission (Urban), 2019)

Types of Plastic	Identification code	Types of Plastic	Identification code
Polyethylene terephthalate (PET or PETE)	1 PETE	Polypropylene (PP)	5 PP
High-density polyethylene (HDPE)	2 HDPE	Polystyrene or Styrofoam (PS)	6 PS
Polyvinyl chloride (PVC)	3 PVC	Others (polycarbonate (PC), Polylactide, acrylic, acrylonitrile butadiene styrene (ABS), fiberglass, and nylon)	7 Other
Low-density polyethylene (LDPE)	4 LDPE		

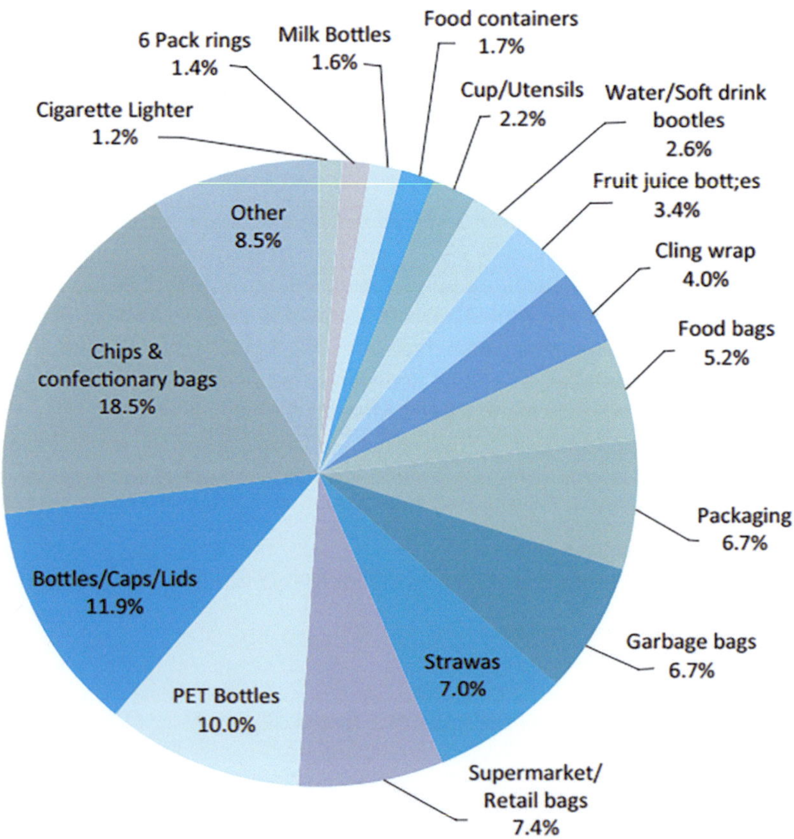

Fig. 1 Contribution of different daily-use commodities in a plastic waste generation (Source: Kurniawan et al., 2017)

aquatic environment, it can pose a threat to the environment. A majority of such trashed waste comprises of household goods, personal care items, commercial commodities, industrial debris from plastic industries (classified as a primary source) and/or from secondary sources such as remnants resulting from the breakdown of bigger plastic goods. The rising consumption of single-use plastic items represents a significant portion of the waste responsible for pollution in rivers and beaches globally. Marine activities, such as fishing, account for approximately 20–30% of ocean plastic, whereas land-based sources are responsible for a substantial 70–80%. (Li et al., 2016).

1.1 Plastic Waste and Its Management in Environmental Matrices

Plastic is used across several sectors. Plastic waste from industries is produced in the form of products and residues during the pre-processing stage. Industries use plastics to improve the quality aspects of products. The agricultural sector uses plastics in vast quantities as bags, mulch or protective films for protecting their crops from unfavorable environment and, also for the transport and storage of fertilizers and food. The hospital sector uses plastics for single-use medical items like syringes, saline bottles, blood bags, and personal protective equipment (PPE) kits. Construction sector uses plastics for packaging as well as for piping, which is mostly PVC at present. Plastic packaging accounted to about 39.9% in 2018, which was the largest plastic end-usage categories in Europe (Plastic Europe, 2019). Figure 2 illustrates the per capita consumption of plastics in different countries during 2014–2015, out of which India's consumption was 11 kg, which is expected to reach 20 kg by 2022 (Swachh Bharat Mission (Urban), 2019 & TERI, 2018).

Plastic leakage occurs when the macro and microplastics are not kept in a circular loop or are improperly managed at the end of their life. Mismanaged plastic waste materials pose great risk of entering the ocean via wind or tidal transport or being carried to coastlines from inland waterways. Inadequately disposed waste is not formally managed and includes disposal in dumps or open, uncontrolled landfills, where it is not fully contained. Rivers carry plastic waste from deep inland to the sea, making them major contributors to ocean pollution. A lot of waste comes from the world's rivers, which serve as direct conduits of trash from the cities to the marine environment (UN Environment, 2018). Leakages can occur at all stages of the production and consumption system, particularly when systems for capturing the waste are not adequately established and/or functioning. Plastic leakage from land-based sources occurs along the following four paths:

- Inadequate waste management that loses control over the waste generated, with waste entering the environment and eventually reaching water bodies (e.g., during storms, floods, windblown).

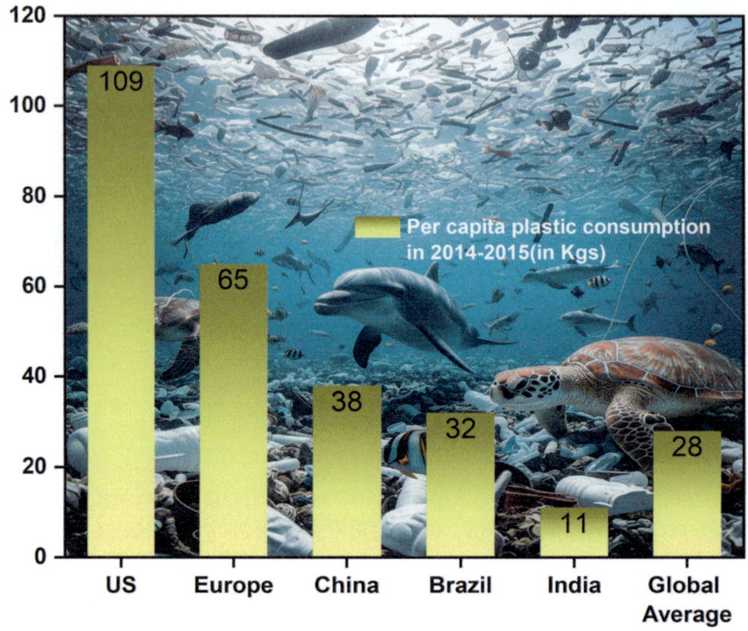

Fig. 2 Schematic showing the per capita consumption of plastics (in kg) of countries during 2014–2015 (Source: Swachh Bharat Mission (Urban), 2019)

- Direct littering or dumping of waste into water bodies (e.g., at beaches or along riverfronts).
- Illegally discharged waste entering waterbodies through surface water drainage systems (stormwater drains).
- Plastics or waste is discharged into sewage or wastewater treatment systems (e.g., microplastics from laundry or cosmetics) and not retained during the treatment stages (Mathews & Stretz, 2019).

Inadequate waste management on land, lack of awareness, and indifferent behavior in the absence of an adequate disposal system, resulting in plastic being dumped or abandoned wherever it is convenient. This contributes to the uncontrolled release of plastics into the environment. About 83% of plastic waste that is inadequately managed poses a high risk of polluting rivers and oceans, whereas the remaining 17% originates from ocean-based sources like discarded fishing nets and fishing vessels, oil rigs, piers and boats transporting freight or passengers (Jambeck et al., 2015a, b).

Land-based sources of plastic marine litter are numerous and include both direct littering and dumping of waste (by householders, members of the public and businesses), as well as by the release of waste plastics via wastewater and effluents. Other potential sources of marine litter include: the release of plastics from agricultural and horticultural activities, particularly polyethylene films which are used extensively in both the livestock and horticultural sectors; leakage of plastic particles from industries in the form of leakage of pellets from plastic manufacturing, and leakages from the waste management itself, particularly where illegal and unregulated waste management activities are taking place in poorly regulated systems.

Individual behaviors play a crucial role in preventing plastic waste from being transported into rivers, waterways, and oceans. Cleaning up litter poses a huge cost to local governments and communities. A key intervention is needed to address the issue of marine litter at the source and also to raise public awareness. One such example is Keep America Beautiful (KAB), a nonprofit organization in the United States is dedicated to community improvement through litter prevention, waste recycling and beautification (Keep America Beautiful, 2021). In 2008 and 2009, KAB funded a series of studies focusing on two broad topics: litter and littering behavior. Concerning litter, the composition of litter across America was explored: its volume, locations, and costs to local communities and businesses. With regard to littering behavior, the factors that were explored included the frequency of littering, the individual and contextual variables that contribute to littering, and the effectiveness of various approaches to reducing littering rates. In 2009, the Curbside Value Partnership (CVP) was established under KAB to enhance participation in curbside recycling programs across the country (Keep America Beautiful, 2021).

Uncontrolled dumping of waste is still a commonplace in many municipalities. Dumpsites are a key source of marine litter, and in many cases are intentionally located near rivers or near coasts for the waste to be carried away by heavy rains or currents, fortifying the capacity of the dump to receive more waste. One such example is the Saida landfill in Lebanon, which was established in 1982 and is managed by the Municipality of Saida (Anon, 2013). The site was located at the beach and not specifically engineered as a landfill. It had a surface area of over 6 hectares and rose to 58 m above sea level, with extremely steep-sided slopes. The site was estimated to contain demolition debris, municipal solid waste, hazardous waste, and slaughterhouse waste from municipalities of the Saida district. It was a result of waste collection vehicles dumping directly into the sea. To deal with the menace, rehabilitation started in 2009 with the closure of the site and the installation of a seawall around the site. The city's waste started going to a newly opened facility further south. By mid-2013, construction works began for the treatment of some of the waste at the old site and land reclamation. The rest of the dump was turned into a sanitary landfill, lined with protective material and gas pipes. In 2016, a park was opened on the reclaimed land (Anon, 2013).

It can therefore be learned that improved waste management systems, infrastructural investments, and awareness raising activities resulting in a positive behavioral change in people can contribute to controlling the leakages of plastics into the environment, and has a strong potential for controlling marine litter.

1.2 *Environmental Hazards of Plastics*

Plastics are lightweight, flexible, moisture resistant, strong, and relatively cheap; these attractive qualities have drawn the world's attention to making plastics a part of everyday life. But certain advantages, such as slow degradation and durability have become a disadvantage upon meeting the end of life of these plastic-based products. Our prodigious attraction to plastic with an irresistible behavior propensity of increasingly over-consuming, discarding, littering and thus polluting, has become a combination of lethal nature. Careless disposal and littering of plastics have the most alarming ecological impacts. Natural organisms can find it difficult to break the synthetic chemical bonds in plastic making it nonbiodegradable, which in turn increases its potent hazard to the environment and organisms. Out of the total amount of plastic waste generated, a very negligible amount is efficiently recycled globally and the remaining is usually sent to landfills or dumpsites where it lays undecomposed for hundreds of years or to incinerators where the toxic compounds are released into the atmosphere creating a potential danger to the ecosystem. Current plastic recycling rates at the global level are about 14–18% (OECD, 2018).

Human health: The burning of plastics causes the evolution of halogens and other toxic gases which can lead to severe health problems such as cancer, endometriosis, neurological damage, birth defects, and multiple organ damage.

Groundwater and soil pollution: Plastics, unlike other materials are not capable of biodegradation but break down to smaller and smaller pieces that can stay in the environment forever as microplastics, thereby contaminating soil and sediments. When plastic is buried in a landfill, over the years, the toxic chemicals from plastics drain out and leach into groundwater or flow into the lakes and rivers.

Pollution in oceans: Plastic pollution in the oceans arises from a complex interplay of both land-based and sea-based activities (see Fig. 3), with a strikingly high

Fig. 3 Schematic diagram showing the sources of plastic waste entering the oceans

share linked to fishing gear, shipping operations, and discarded plastic waste entering marine environments through inadequate waste management and riverine input (Ferraro and Failler 2020). Once in the ocean, these plastics, ranging from large items such as nets and packaging to microscopic particles, pose a significant threat. Macroplastics, such as abandoned nets, ropes, and larger debris, cause entanglement and suffocation of marine life, while also fragmenting over time into microplastics that infiltrate food webs at every level (Aminur Rahman et al., 2023). Microplastics have become ubiquitously distributed throughout pelagic and benthic zones, readily ingested by plankton, fish, seabirds, and mammals, resulting in adverse effects such as impaired feeding, reduced growth, inflammation, and the bioaccumulation of toxic compounds in tissues (Prasad et al., 2023). These plastics act not only as physical hazards but also as chemical vectors, adsorbing and gradually releasing toxins, including persistent organic pollutants and heavy metals, into the marine ecosystem, which can potentially disrupt microbial communities, carbon cycling, and the oxygen balance in ocean sediments. The accumulation of plastic on the seafloor further limits gas exchange and light penetration, transforming once vibrant marine habitats into inhospitable zones. The persistent presence and degradation of plastics sustain and magnify the threat to ocean biodiversity and human health via the seafood chain (Smith et al. 2018).

In addition to suffocation and ingestion (Fig. 3), macro-plastics cause the death of birds, fish, mammals, reptiles, microbes, and bio-accumulates in larger concentrations up the food chains to humans at the top. In the ocean bed, the accumulation of plastic waste suffocates the seabed and prevents gas exchange turning it into an inhabitable hard surface for marine plants and living beings (Gregory, 2009; Haward, 2018 & Luo et al., 2018). Over the years, the accumulated plastics are broken down to small fragments resulting in the formation of microplastics, which may be ingested even by the tiniest organisms that absorb these hazardous chemicals.

2 Plastic Waste Generating Industries in the Catchment

Tapi River channels through central India between the Godavari and Narmada rivers, flowing westwards before draining into the Arabian Sea. The river has a length of 724 km and flows through the states of Maharashtra, the plain of Surat (Gujarat), and Madhya Pradesh. Similarly, Daman Ganga (131 kms or 81 miles) River divides the township into Nani Daman and Moti Daman. It borders the southern end of the industrial town of Vapi before it meets the Arabian Sea.

Plastic waste from various sources, including paper mills, synthetic textile manufacturing, chemical and petrochemical plants, rubber and plastic industries, as well as domestic leakage, has been evaluated in the cities of Surat and Vapi, which fall within the catchment areas of the Daman Ganga and Tapi rivers. The detailed overview is given below.

2.1 Preliminary Assessment of Plastic Pollution Sources in Surat City

Of the total amount of waste generated by the state of Gujarat, the city of Surat contributes massively by producing 1,50,000 kilos of plastic waste/day. Surat's key industries encompass textiles (responsible for 40% of the nation's synthetic fabric output), chemicals and petrochemicals, rubber and plastics, as well as paper mills. These industries significantly contribute to the city's plastic waste generation (Figs. 4 and 5) (Surat District Profile, 2011). The total daily plastic waste generated in Surat is approximately 149.62 metric tonnes/day (CPCB, 2015).

Fig. 4 Map showing major industries in Surat district (Source: Surat District Profile, 2011)

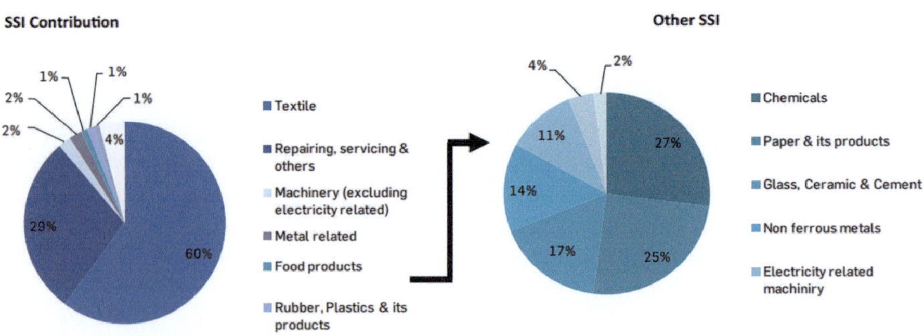

Fig. 5 Contributions of small-scale industries (SSI) in Surat district (Source: Surat District Profile, 2011)] (It shows SSI contribution to almost 40% to gross industrial value added in Indian economy)

2.2 *Preliminary Assessment of Plastic Pollution Sources in Vapi City*

As a designated chemical estate, Vapi's industrial landscape comprises 70% pharmaceuticals and fine chemicals, while the remaining 30% includes paper mills, packaging, engineering, plastics, textiles, and various small-scale industries (Goswami & Patel, 2015). The region generates approximately 400 metric tonnes of plastic waste (Shah, 2018), with paper mills alone contributing around 180–200 MT per day, totalling 3500–4000 MT per month (GCPC ENVIS, 2018). Common disposal practices for this waste include landfilling, incineration, or illegal dumping into water bodies, leading to environmental contamination (UNEP, 2018).

In recent years, the Gujarat State Pollution Control Board has initiated efforts to collect non-recyclable plastic waste generated by paper mills and redirect it to cement plants for co-processing (UNEP, 2018). In Gujarat, around 86,553.62 metric tons of plastic waste have been utilized in cement plants for co-processing (CPCB Annual Report, 2018–2019). Co-processing involves converting waste into energy and materials. For example, one ton of plastic can replace 1.2 tons of coal, making cement production more cost-effective, with plastic waste priced at Rs. 3–4 per kg compared to coal at Rs. 8–9 per kg (Prabhune, 2018). In Vapi, plastic waste from paper mills was sent to Ambuja Cements Ltd.'s Kodinar unit for use in their cement kilns. From 2012 to 2016, a total of 1,54,018 metric tons of plastic waste was co-processed from paper mills in Vapi. (CPCB, 2017).

2.3 *Study of Daman Ganga*

According to the study conducted by CPCB in the city of Daman, the plastic waste quantified is given in Table 2. The total amount of plastic waste obtained during the survey was ca. 46.37 kg/MT where the MSW produced is 25 MT/day (CPCB, 2015). The most serious issue associated with the plastic waste littered in the city is the unavailability of collection points and no proper dumping ground. The HDPE/LDPE plastics from packaging contributed to about 75% of the collected waste from Daman city (CPCB, 2015).

Table 2 Study of Daman City: PW (kg/MT) (CPCB, 2015)

Serial of Days	PET	HDFE/LDPE	PVC	PP	PS	Other	Total
Day 1	0	27.28	0.112	0	13.44	9.856	50.688
Day 2	0.664	34.12	1.112	6.152	0.372	2.816	45.236
Day 3	1.168	39.2	0	0.84	1.56	0.424	43.192
Average	0.611	33.53	0.408	2.331	5.124	4.365	46.37

2.4 Study of Tapi

Tapi houses many ports, chemical plants, petrochemical industries (like Reliance, ONGC), textile mills, and paper mills that release industrial effluents into the Tapi River (Kamal et al., 2014; Dubey & Ujjania, 2013; Surat & Tapi District Profiles, 2010). In addition to industrial effluents, domestic littering is a major concern that affects the quality of river water. To improve the water quality and save the river from plastic pollution, the Government of Gujarat has introduced the "Save River Tapi" initiative through which public awareness was created and strict restrictions were imposed against littering plastics in water bodies (Deshpande, 2019; Mehta, 2019).

2.5 Other Sources of Plastic Waste Generation

As aforementioned plastic waste primarily originates from various sources, including synthetic textile and paper industries, rubber and plastic manufacturing, chemical and petrochemical sectors, and the widespread use of single-use plastics in households. In addition to these primary contributors, other activities such as tourism, shipbreaking, and e-waste disposal also play significant roles in plastic waste generation. The Tapi catchment areas, which are popular tourist destinations, experience a heightened accumulation of single-use plastics like food wrappers, straws, diapers, PET bottles, carry bags, and bottle caps. The shipping industry is another source of pollution, producing solid, liquid, and gaseous waste, including materials like asbestos, insulation products, scrap paint chips, and sealants. Around 40–50% of the waste from shipbreaking consists of plastics, which often end up in the marine ecosystem (Reddy et al., 2006). Once introduced into marine environments, plastics degrade into smaller particles, remaining suspended in the water column or settling in sediments, posing significant risks to marine life and ecosystems. In the Alang-Sosiya shipbreaking yard (Fig. 6), located northwest of the Gulf of Cambay, India (Reddy et al., 2006), approximately 180 ships are dismantled annually, leading to a considerable increase in plastic waste. Four common polymers—polyurethane, nylon, polystyrene, and polyester—were found in the sediments, largely used for insulation, fibers, and packaging. Additionally, small plastic fragments were identified, contributing to the accumulation of plastic waste in coastal areas and exacerbating pollution in the surrounding marine environment.

Another source of plastic waste is electronic waste (E-waste), which is a growing concern and a major addition to the waste generation stream. In E-waste recycling, the non-plastic parts are often recycled while the plastic parts are discarded or burnt (Chakraborty et al., 2016). This results in either the accumulation of plastics or the release of toxic gases into the environment (Chakraborty et al., 2021). To control the amount of plastic waste generated from E-waste, the Gujarat Pollution Control Board has taken controlled measures to ensure a "Clean e-Waste Channel" which ensures the proper recycling of residual waste and educates recyclers about the hazardous effects of careless dumping (Gujarat Pollution Control Board, 2008). This initiative ensures

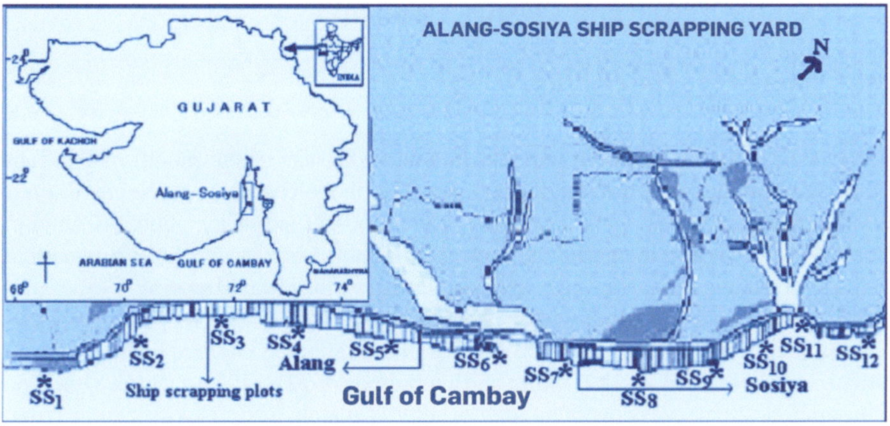

Fig. 6 Location of scrapping plots labeled SS1 to SS12 along coastline and the Alang-Sosiya ship-breaking yard along the Gulf of Cambay, Gujarat (Source: Reddy et al., 2006)

proper collection of E-waste generated from industries and domestic areas, dismantling, and processing every component separately and generating commodities.

2.6 *Plastic and Biomedical Waste Related to the COVID-19 Pandemic*

In this chapter we focused on enhancing knowledge to address plastic and chemical pollution from key sources across various industries, public sectors, and civil society. In response to the significant societal changes brought on by the COVID-19 pandemic, the team recognized the importance of shifting attention to how the pandemic specifically affected marginalized groups engaged in plastic waste management. Between March and July 2020, a study to explore the situation of plastic and biomedical waste management in Delhi and Surat during the pandemic (NIVA, Toxics Link, & MGC, 2020). The study aimed to evaluate the effects of COVID-19 on waste management systems in select Indian cities, considering the sudden shifts in context. Key findings revealed that while both cities had sufficient infrastructure to manage biomedical and COVID-19 waste, they struggled with nonsegregated waste due to CPCB guidelines prioritizing incineration and the heightened concerns among waste handlers and generators. Additionally, it was noted that cities with better overall waste management systems were more successful in adapting to the new waste stream and changes in waste types and volumes. Further details of the study can be found in the INOPOL Special Report titled "Pilot Study for Plastic and Biomedical Waste Management during the COVID-19 Pandemic in Surat and Delhi." (NIVA, Toxics Link, & MGC, 2020).

3 Potential Litter and Plastics Hotspots

3.1 Surat City Description

Surat, located in the southern part of Gujarat, India, sits at the mouth of the Tapti (Tapi) River, which runs through the city. The Mindhola River defines the city's southern boundary, and both rivers ultimately drain into the Gulf of Khambhat (Cambay). Surat serves as a key industrial and commercial hub, showcasing India's historical, cultural, and navigational heritage, and is governed by the Surat Municipal Corporation (SMC). The city's industrial sectors encompass activities such as port operations, shipbuilding, textiles, and the trade of gold and diamonds. Geographically, Surat is situated at a latitude of 21.112° North and a longitude of 72.814° East (Surat Municipal Corporation, 2018) and covers an area of approximately 326.515 km^2 (Surat Urban Development Authority, 2017). The city experienced rapid growth after India's independence, with its industrial expansion contributing significantly to its development.

The Surat Urban Development Authority (SUDA) was created under the Gujarat Town Planning and Urban Development Act of 1976 to oversee SMC and 195 nearby villages. SUDA spans an area of 715 km^2 between latitudes 21°03′ and 21°19' North and longitudes 72°41′ and 73°00′ East, at an elevation of 13 meters above sea level. In addition to SMC, SUDA encompasses four urban centers or Talukas: Olpad, Chorasi, Palsana, and Kamrej (Surat Urban Development Authority, 2017). According to the 2035 Development Plan Report, "Extended SUDA" will incorporate the industrial zone of Hazira, one of India's prominent seaports. This expansion will increase SUDA's area from 715 km^2 to 985.16 km^2 by 2035 (Surat Urban Development Authority, 2017).

Surat experiences a tropical climate with temperatures in summer ranging from 37.78 °C to 44.44 °C and 10 °C to 15.5 °C in winter. The average annual rainfall of Surat is approximately 60 inches (i.e. 1525 mm) with about 90% of the rainfall occurring in the monsoon season from June to September. The highest temperature recorded is around 45 °C during the months of May and June. The months of October and November experience retreating monsoon and the winter season runs from November to February. January is the coldest month with the lowest recorded temperature of 4.4 °C (Surat Urban Development Authority, 2017). The maximum reported humidity in Surat is around 80%. The annual average wind speed of Surat is around 18 knots (i.e. 9.26 m/s) in northwestern direction (Surat Municipal Corporation, 2014). Surat experiences wind flowing from southwest to northeast, and the breeze is flowing from the sea to the main city area.

The area commanded by SUDA has a gradual slope at the western and southern sides of the city (Figs. 7 and 8). The southwestern side of the coastal area slopes towards the Mindhola River. The city is divided into two parts by River Tapi. Urban and industrial development has happened on both sides of the river. The areas of Hazira, Mora, Damka, Limla, Dumas, Bhimpore, Abhawa, Gavier, Sarsana, and Vesu lie at a lower elevation between the river Tapi and Mindhola. The city is located on a low-lying flood plain and estuarial region having a number of creeks that lie in the southwestern part. Experts suggest that Surat is prone to climate change related

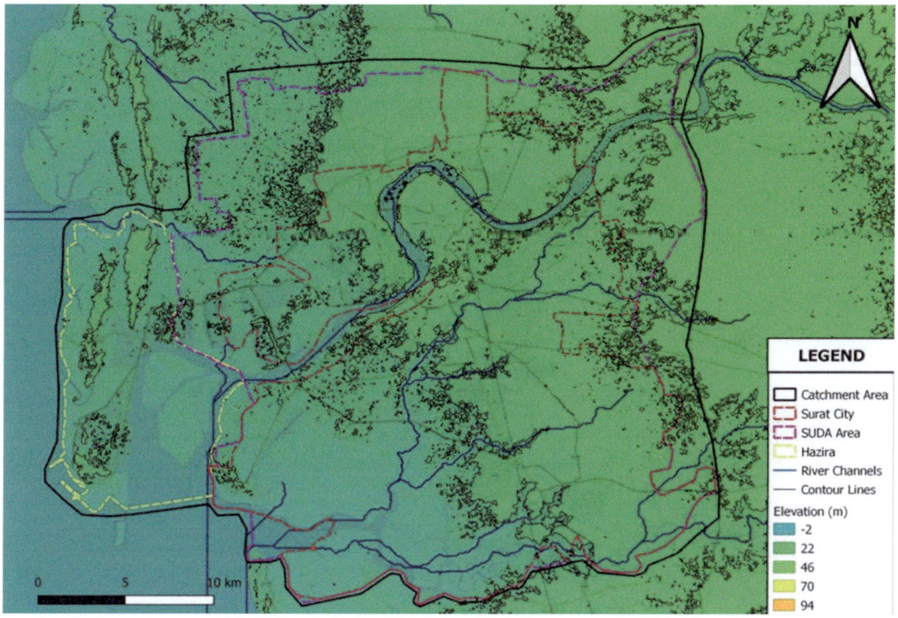

Fig. 7 Topographic map of Surat

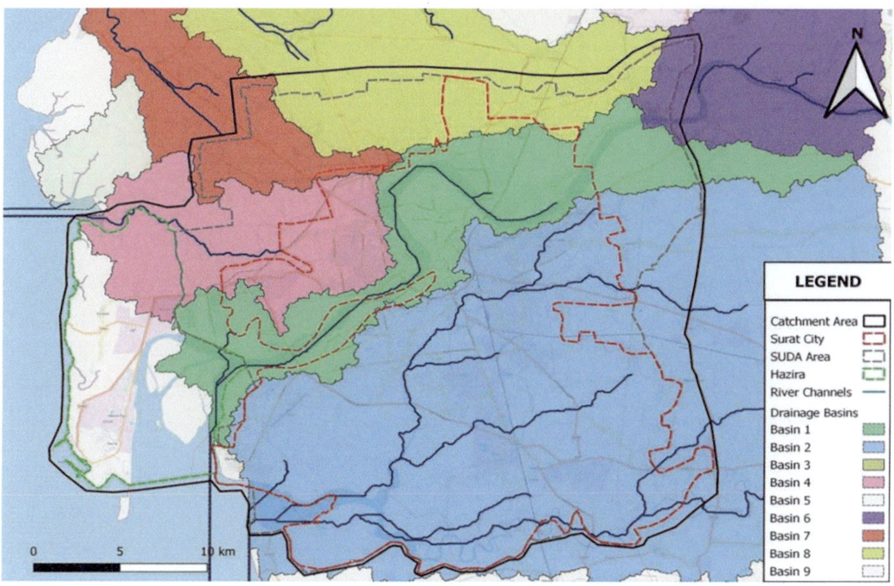

Fig. 8 Flow direction map for Surat

disasters such as flooding, owing to factors including unplanned rapid urban development, lack of surface runoff, and predictable heavy monsoons (Bouissou, 2014). Having experienced several floods since 1869, the 2006 floods have been the most devastating affecting almost two-thirds of Surat's population (Waghwala & Agnihotri, 2019). Two expert committees were set up by the state government to examine the cause of the 2006 flood situation. Key causes included inadequate urban development practices like illegal encroachment by industries, housing colonies, and buildings for development purposes. It was also understood that mismanagement of the release of Ukai dam water had made the situation worse. (Jamwal, 2015). The drainage system continues to be inadequate and being on a lower plane, some areas endure waterlogging issues. Flood-like situations arise in the Tapi River every year during the rainy season (Fig. 9). This results in the closing of the floodgates, which further leads to stormwater backflow from drains. This stormwater drainage issue creates problems for the general public (Bhadiyadra, 2015). In 2020, Surat experienced a similar flood-like situation, with overflowing stormwater drains submerging close to 6000 houses in low-lying areas (Times of India, 2020).

3.1.1 Socioeconomic Profile

Surat has an estimated GDP of $59.8 billion, which is the ninth largest among Indian cities. The city is well known for its diamonds and textile industry, and hence goes by the names "The Diamond City" and "The Silk City." Surat is the largest

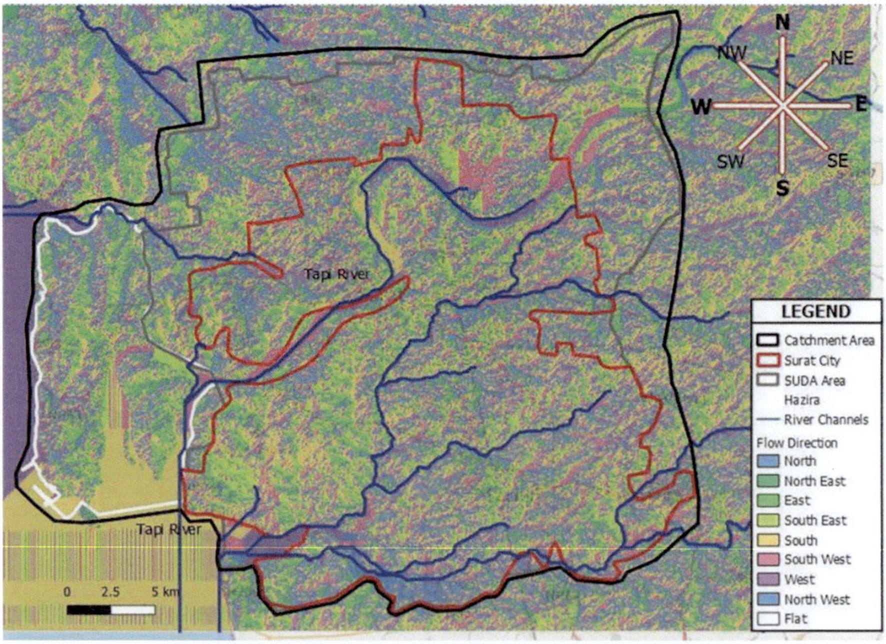

Fig. 9 Drainage basin map of Surat

diamond-cutting and polishing hub in the world and accounts for about 90% of the world's diamond-cutting and polishing exports (World Economic Forum and PWC, 2017). Major companies like GAIL, ONGC, ESSAR, Ultratech Cement, Ambuja Cement, Reliance Industries, and Shell have their presence in this city (Business World, 2017). As per GPCB data, Surat has a total of registered plastic waste recyclers that produce granules, fibers, and other materials from plastics (GPCB). Surat is expected to further grow economically in the coming years due to its selection in the "Smart City Mission" launched by the Ministry of Urban Development. Under the Smart City Mission, Surat will be developing in various sectors including IT, renewable energy, solid waste management along with town planning and development (Surat Smart City, n.d.). Post the outbreak of the plague in 1994, the city of Surat prioritized cleanliness, hygiene, and sanitation to make significant improvements in these sectors. In the Swachh Survekshan-2020 (annual cleanliness survey of urban areas) Surat ranked the second cleanest city in India. The city has also won recognition for introducing a smart underground waste collection system in 2018.

The population of Surat grew from about 2.8 million in 2001 to nearly 4.6 million in 2011, with a population density of 13,680 people per km^2 (Census of India, 2011a, b). Official population censuses are only carried out every 10 years in India, with new data expected in 2021. It is observed that the estimated population of Surat in 2019 was close to 6.3 million, representing more than a doubling since 2001 (Surat Municipal Corporation, n.d.). Table 3 below represents a decadal population of Surat city as per the 2011 census.

3.1.2 Waste Generation in Surat

Surat is a metro city and one of the largest cities in India. SMC is the authorized body designated to manage municipal solid waste (MSW) in the city (Fig. 10). According to SMC, the city generated a total of 2200 tons per day (TPD) of MSW

Table 3 Decadal population of Surat city. (Christian, 2016)

Census	Decadal Population
1901	119,300
1911	114,900
1921	117,400
1931	98,900
1941	171,400
1951	223,200
1961	288,000
1971	492,700
1981	912,600
1991	1,519,000
2001	2,811,614
2011	4,591,246
2021	7,651,367[a]

[a]Estimated data for 2021 as 2021 census was postponed.

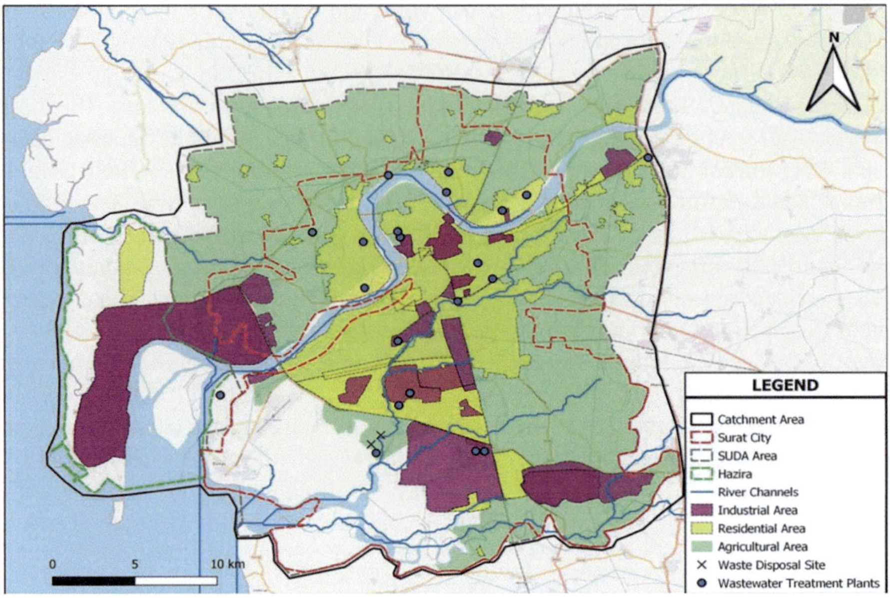

Fig. 10 Land use map and catchment of Surat

at a rate of 0.45 kg per capita per day in 2018–2019. In comparison, the global average was 0.74 kg per capita, reported in a World Bank, 2018 report. The highest levels of waste generation were between 2 kg and 4.5 kg per capita per day with the highest per capita waste generation of 4.54 kg being witnessed in Bermuda (World Bank, 2018). The plastic waste generation in Surat and India are given in Figs. 11 and 12 respectively. Following the requirements of the Solid Waste Management Rules, 2016 of the Ministry of Environment, Forest & Climate Change (MoEFCC), the duty of every municipal corporation is to annually report the quantity, quality, and handling measures of solid waste generated within their jurisdiction to their concerned State Pollution Control Board (SPCB), who pass on this information to the Central Pollution Control Board (CPCB) (MoEFCC., 2016). SMC further lays down Public Health Bye-laws for the city wherein local regulations of solid and liquid waste management, air and noise pollution along with other sanitation and public health aspects are covered (Surat Municipal Corporation, 2015).

The waste generation trend of Surat has been developed based on data generated by SMC in recent years, along with other waste quantification and characterization studies conducted in Surat City by CPCB and Central Institute of Petrochemicals Engineering & Technology (CIPET), Federation of Indian Chambers of Commerce & Industry (FICCI) National Environmental Engineering Research Institute (NEERI), and Environment Protection Training and Research Institute (EPTRI), and in the years 2010–2011, 2008–2009, 2004–2005, and 1999–2000 respectively. The above information has been tabulated in Table 4 below.

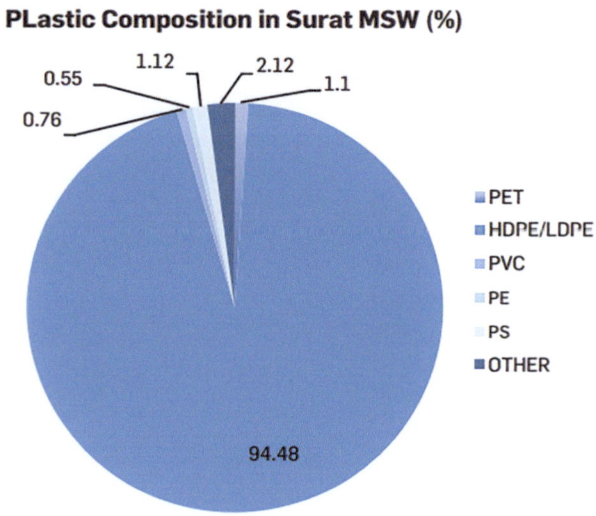

Fig. 11 Plastic waste composition in Surat. (Resilient Cities Network; Ocean Conservancy; The Circulate Initiative; Centre for Environment and Education; in collaboration with Surat Municipal Corporation, 2024).

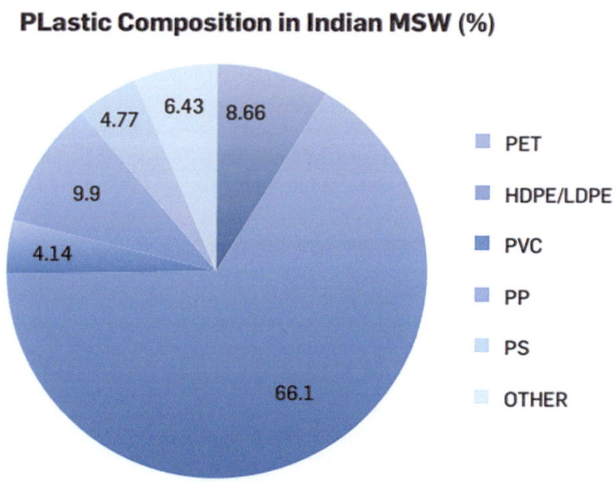

Fig. 12 Plastic waste composition in India. (Central Pollution Control Board 2015)

The data in Table 4 has been gathered through various studies that have been carried out in Surat City. A brief write-up about these studies (in timeline sequence) and their outputs are as follows:

- **Year 1999–2000:** A study was conducted by CPCB through the Environment Protection Training and Research Institute (EPTRI) in 1999–2000 in 210 Class I cities and 113 Class II towns indicated that Class I cities generated 48,134 TPD of MSW while Class II towns generated 3401 TPD of MSW. Surat (a class I city) documented a waste generation of 900 TPD at a rate of 0.6 kg per capita per day (CPCB ENVIS, n.d.).
- **Year 2004–2005:** The National Environmental Engineering Research Institute (NEERI) conducted a study titled "Assessment of Status of Municipal Solid Waste Management in Metro Cities and State Capitals" wherein waste quantifi-

Table 4 Reported waste generation in Surat over the years

Year	MSW (TPD)	Per Capita MSW (kg per day)	Plastic% in MSW	Plastic Waste (TPD)
1999–2000	900[a]	0.6[b]		
2004–2005	10,00[c]	0.41[c]	11.21[c]	112.1[a]
2008–2009	12,25[d]			
2010–2011	12,00[e]		12.47[e]	149.62[e]
2011–2012	12,50[f]			
2012–2013	13,50[f]			
2013–2014	15,00[f]			
2014–2015	1680.5[g]			
2015–2016	16,80[h]			
2016–2017	18,20[f]			
2017–2018	20,50[f]			
2018–2019	22,00[f]	0.450[f]		
2019–2020	23,20[f, i]			

Assessment & Quantification of Plastics Waste Generation in Major Indian Cities	
About study	The study for determining plastic waste generation in 60 Indian cities was conducted by Central Pollution Control Board (CPCB) in collabouration with CIPET in the year 2010–11. The 2015 report extrapolated data based on the data received from the cities. The study mentioned that 15,342 tonnes waste was being generated daily (five million tonnes per year). CPCB then extrapolated to the rest of country and further estimated that 25,940 TPD of plastic waste was generated (5.5 MT per annum). The study found that about 70 percent of plastic packaging of the products turned to waste
Methodology	The study methodology was to estimate the plastic waste from municipal dumping sites. ASTM method was used for assessment and quantification of plastic at dumpsites. The waste reaching the dumpsites mainly came from households and community waste bins kept at streets and institutions. About 1 tonne of waste was collected from the municipal dumpsite and was analyzed. This was done to check quantum of plastic waste present in the one tonne of waste collected and also to see what ends up in dumpsites
Findings	1. The study found that on average, plastic waste represents 7% of the total mass of waste in municipal dumpsites. In some cities, such as Delhi, Kolkata, Surat, and Raipur, plastic waste represents up to 10% of the total mass of dumpsite waste. 2. The study found that 66% of plastic waste is usually mixed waste, which includes polybags, multilayer packaging, and pouches. 3. The study suggests out of the total plastic waste, thermoplastic is about 94% (recyclable) while 6% is thermoset plastic (nonrecyclable) in nature. On a country level, the study classifies plastics waste into different categories, namely PET (66.10%), HDPE/LDPE (9.9%), PVC (8.66%), PP, (6.43%), polystyrene (4.77%), and unclassified or nonrecyclable plastic and rubbers (4.14%), as illustrated in Fig. 12

[a] Calculated value
[b] (CPCB ENVIS)
[c] (CPCB)
[d] (FICCI, 2009)
[e] (CPCB, 2015)
[f] (Surat Municipal Corporation, n.d.)
[g] (Surat Urban Development Authority, 2017)
[h] (CPCB, 2018)
[i] The MSW, which is taken from SMC is an estimate based on the aggregation of monthly data of waste collected and transported (Surat Municipal Corporation)

cation and waste characterization were carried out in the selected 59 cities. In this study, the total MSW generated in Surat was reported as 1000 TPD and there was 11.21% of recyclables in this, the majority of which were plastics (CPCB, n.d.).

- **Year 2008–2009:** Another study titled "Survey on the Current Status of Municipal Solid Waste Management in Indian Cities and the Potential of Landfill Gas to Energy Projects in India" was conducted in this region by the Federation of Indian Chambers of Commerce & Industry (FICCI). In this study, surveys were conducted among Municipal Corporations of 22 cities with the aim to determine the quantity, quality, and handling of the waste generated within the corporation's jurisdiction (FICCI, 2009).
- **Year 2010–2011:** The most prominent document to determine the plastic waste generation was a waste characterization study conducted by CIPET and CPCB in 60 cities in the year 2010–2011. The methodology of this study included assessing samples of 1000 kg of MSW arriving at dumpsites on different days from different sources and regions of the city. The study conducted in Surat shows approximately 12.47% of plastic waste in MSW, which was the highest in percentage amongst the 60 cities. The highest constituent was HDPE/LDPE, which was approximately 94.5% of the total plastic waste reaching the Surat dumpsite (CPCB, 2015).
- **Year 2019–2020:** The Surat Municipal Corporation reported 2320 TPD as the net waste collected in the city accounting for all waste generation sources (not a study).

3.1.3 Case Study Area in Surat

For this case study, the catchment area includes the existing SUDA region and the industrial town of Hazira, chosen to encompass the river system's flow to the sea. This allows for an in-depth analysis of the current marine litter situation in the area. Hazira is located within the expanded SUDA region (Surat Urban Development Authority, 2017).

Surat catchment consists of three components, the Surat Municipal Corporation Area, the existing SUDA area and the adjoining town of Hazira (extended SUDA area). Although the SUDA area is predominantly agricultural, it has been included in the catchment because its urban development is promoted under the 2035 Surat City Master Plan. The Hazira mangrove has been included in this catchment.

Mangroves acting as carbon sinks
Mangrove forests, most commonly found in tropical regions are home to animals and birds. It covers an estimated area of 140,000 km2. Most of its roots are submerged under water and become a protective nursery to juvenile fish and sanctuaries to mammals. Mangroves are useful for coastal communities as they act as stormwater barriers, which protect the inland areas from flooding and erosion. The excess of sediments and river pollutants get washed off by the mangroves before they reach the oceans Mangroves act as carbon sinks as they can store large amounts of carbon. Mangroves have the ability to sequester four times more carbon than a rainforest. The storage of carbon particularly occurs in the soil present beneath the mangrove (Erickson-Davis, 2018). Mangroves are the most carbon rich habitat on our planets and further act as natural carbon scrubbers which take carbon from the atmosphere and pack it tight in its rich soil (Spalding, 2013). When these coastal habitats are destroyed, a large amount of carbon is emitted back to the atmosphere that further leads to climate change (National Oceanic and Atmospheric Administration, 2019) Mangroves are part of coastal ecosystems, which contribute to sustain life and livelihoods. The roots of the mangrove trees form a barrier against the violent storm surge and floods. River and land sediment are trapped under the large roots that protect coastline areas and prevents erosion. The mangroves act as natural filters that prevent any harmful sediment reaching the coral reefs and sea grass meadows. Mangroves also protect the coastal communities from sea level rise and other catastrophic events caused by climate change. Mangrove trees provide livelihood for the coastal communities as they are one of the reliable sources for wood. The plant extracts are used by the people for its medicinal quality and leaves of trees are used for animal fodder (Wood, 2019). Mangroves have proved to be beneficial for our marine environment as they provide a basis for complex marine food chains, create breeding habitat, establishe restrictive impounds that contribute to protect maturing offspring, filter out the pollutants that come from upstream, stabilize the sediments that are present at the bottom, improve water quality, and protect shorelines from erosion (mangrove.org, n.d.). Mangroves trap the plastic litter in their roots, which is accumulated in the marine environment. One of the studies about Mangrove forests conducted earlier on the coasts of Saudi Arabia highlighted that those plastics made upto 90% of the litter found along the coast. This included plastic bottles, bags, and food wrappers. This plastic debris impacts the mangroves and can be ingested by the marine organism residing in the mangrove nursery (Martin, 2019). Mumbai, India, lost about 40% of its mangrove forest over the recent years. These mangrove forests have been destroyed and encroached by housing societies and associated municipal waste and litter. The city experiences climatic situations like sea level rise and extreme weather events that make the coastal communities more vulnerable, as there has been disappearance of mangroves. The sea level rise in the city results in flood-like situations. The local government of Mumbai created a mangrove cell at the forest department, responsible for patrolling the land and restricting people from dumping their waste (DW, n.d.)

Source: Based on CPCB, 2015

It is a heavy industrial area having a huge impact on the pollution level of the region, especially in marine pollution as it is located near the Gulf of Khambhat. Two rivers traverse the catchment area: the Tapi River, which flows through the center of Surat, and the Mindhola River, which marks the southern boundary of the catchment. This catchment area also has Hazira Mangroves.

Since one of the primary objectives of this project includes assessing plastic waste in marine litter, the information box added below links land-based plastic leakage to marine pollution.

Plastic waste leakage from land-based sources contributing to marine pollution
A Jambeck et al. study from 2015a, b quantifies the plastic waste that flows from land to oceans and provides a roadmap that contributes to solutions to curb the problem of marine plastic pollution. Key findings included that the amount of plastic waste entering oceans from land exceeds each year, equaling to approximately 4.8 MT and may be as high as 12.7 MT. plastic waste discharge is coming from 192 coastal countries, but largest amount and quantities are coming from middle-income and developing countries. It is evaluated that a 50% reduction of mismanaged plastic waste by the countries ranked as the top 20 producers would have the potential to decrease the mass of mismanaged waste by 40% by 2025. This strategy would require infrastructure investment in low-and middle-income countries. (Jambeck et al., 2015a, b) National Productivity Council's Plastic Leakage Pathways report assesses land plastic waste management in four cities; namely Haridwar, Agra, Prayagraj, and Mumbai. The report identifies major plastic hotspots and describes plastic leakage scenarios from land-based sources to rivers and marine environments. Prominent plastic categories observed during the leakage across the four cities include polythene bags, multilayer plastic bags, plastic pouches, monolayer plastics, multilayer bags, disposable cups and glasses, and woven bags. It was found that littered waste was not effectively collected and would eventually leak into the water bodies through drainage systems and runoff pathways. Waste being dumped by tourists and pilgrims is another direct source of marine pollution. Furthermore, there are major plastic waste hotspots, which include plastic processing factories and densely populated residential areas, particularly slums (National Productivity Council and UN Environment Program, 2019) Keeping these two studies in mind, the key conditions contributing to plastic pollution from land-based sources can be summarized as follows: 1. The concentration of plastic packaging and manufacturing industries along coastal regions contributes significantly to marine pollution 2. Presence of residential, commercial, and institutional areas in close proximity to the coast also add to plastic marine pollution 3. Littered plastic waste accumulated in drains is a major concern since this leads to flood-like situation through their blockage. This also results in large movements of plastic waste into the water bodies.

The details of Surat catchment are tabulated in Table 5 below:

Figure 7 presents a topographic map of Surat, depicting the elevation of the catchment area. This map indicates that Surat is situated at a relatively low elevation, particularly in the town of Hazira.

Table 5 Surat catchment details

Details	Surat catchment
Name of river flowing	Tapi
Length of river system	724 km[a]
Area of catchment	715 sq.km[b]
Number of basins within catchment	1 (lower Tapi Basin)[c]
Number of inhabitants	4,805,101[b]
Number of urban clusters	5[b]
Number of industries	41,300[d] (approx.)
Number of WWTPs/CETPs	5[b]

[a] (Central Water Commission, 2014),
[b] (Surat Urban Development Authority, 2017),
[c] (Gundaliya, 2016),
[d] (MSME)

Figure 9 distinctly illustrates the various river drainage basins present in the catchment area, as determined by hydrogeological mapping. The map indicates that there are six primary basins in the catchment area, corresponding to the different river streams.

The drainage basins illustrated in the Figure 16 represent a portion of the Tapi River basin, which encompasses approximately 2% of India's total land area, covering about 65,145 km^2. The Tapi River itself stretches for 214 km through Gujarat before it empties into the Arabian Sea at the Gulf of Cambay, passing through the city of Surat. It is fed by around 14 significant tributaries, five of which flow within Surat district (Central Water Commission, 2016). Additionally, the Mindhola Basin, which is entirely situated in Gujarat, constitutes the second largest drainage area within this catchment, measuring 1180 km^2. The Mindhola Basin is included in the Lower Tapi basin (National Water Development Agency).

Figure 8 presents a hydrological map that illustrates the flow direction of water within the catchment area. This map for Surat indicates the pathways along which water travels and utilizes distinct colors to represent various flow directions. It serves as a valuable tool for understanding not only the river's flow patterns but also the dynamics of surface runoff in the study region.

The flow direction map indicates that surface runoff in Surat primarily moves toward the Tapi River. Additionally, Hazira, an area mainly characterized by industrial activity, also channels its runoff toward the Tapi River, which subsequently flows into the Arabian Sea.

Figure 10, created using QGIS 3.18, depicts the land use within the catchment, highlighting areas designated for industrial, agricultural, and commercial/residential purposes. This map also identifies locations for waste disposal and wastewater treatment facilities situated along the banks of the Mindhola River. It indicates that the primary settlement lies within Surat City's boundaries, while the other SUDA clusters predominantly consist of agricultural land. Additionally, it is evident that Hazira is primarily an industrial zone.

3.2 Vapi City Description

Vapi City is the second-largest urban center in Southern Gujarat, situated in the Valsad district within the Pardi Tehsil, along the banks of the Daman Ganga River. It covers an area of 22 km^2, positioned at a latitude of 20.385° north and a longitude of 72.91° east. The city stands at an elevation of 27.33 meters (89.67 feet) above sea level and is bordered by the Union Territories of Dadra & Nagar Haveli and Daman & Diu. Adjacent to Vapi, Daman city is divided by the Daman Ganga River into two sections known as "Nani" Daman (Small Daman) and "Moti" Daman (Big Daman). This river flows onwards, eventually emptying into the Arabian Sea near Daman (refer to Fig. 2.14). Vapi is characterized by its industrial significance and is home to one of Asia's largest Common Effluent Treatment Plants (CETP), managed by the Vapi Municipality.

The climate in Vapi is tropical, featuring three distinct seasons: a mild winter, a moderate summer, and a heavy monsoon. The average annual temperature is approximately 26.8 °C, with summer temperatures ranging from 22 °C to 41.2 °C, while winter temperatures vary between 9.9 °C and 36.5 °C. The peak temperature, recorded in April, reaches 42 °C, while January can see lows dropping below 10 °C (Central Ground Water Board, 2015). Rainfall during the summer months exceeds that of winter, with an average annual precipitation of 90 inches (or 2279 mm) and average humidity levels peaking at 88.5% and dipping to a minimum of 43%. Wind speeds in Vapi, as measured at the Valsad observatory, range from a maximum of 6.7 knots (9.26 m/s) in June to a minimum of 2.3 knots (1.18 m/s) in December (National Water Development Agency, 2014).

3.2.1 Socio-Economic Profile

Vapi City is recognized as the most industrialized city in Gujarat, following Ahmedabad, Surat, Vadodara, and Rajkot (Industrial Scenario in Gujarat). The original town was located in the western part, while the eastern section is primarily characterized by industrial activity and newer residential developments. Vapi is renowned as a central hub for chemical-based industries (MSME, 2013). The establishment of the Gujarat Industrial Development Corporation (GIDC) Industrial Estate served as a significant catalyst for Vapi's growth. The industrial landscape of Vapi is predominantly occupied by chemical and paper manufacturing facilities. Over 300 medium and large-scale industries are concentrated in the area, driving the economic development of the district (MSME-DI n.d.). The industrial estate spans an area of 1163.77 hectares (11.64 km^2) (Central Ground Water Board, 2015). As per GPCB data, Vapi hosts 18 registered plastic waste recyclers that produce a variety of plastic products ranging from plastic bags, granules, pipes, packaging materials, and sheets (GPCB).

The Vapi Waste and Effluent Management Company has made a pivotal contribution by establishing the largest Common Effluent Treatment Plant (CETP) to date. This facility is notable for its capacity to process industrial pollutants before they are discharged into the river (Grotal, n.d.). Spanning 22 acres, the CETP further handles the treatment of chemical and biological waste originating from the Daman Ganga River. Operated by the privately owned Vapi Waste and Effluent Management Company, the plant manages both residential and industrial wastewater. It employs secondary and tertiary treatment processes for effective waste management. Approximately 98% of the local chemical, dyeing, and related industries utilize this CETP, which can process up to 54 million tons of waste each day (Sreenivas Janyala, 2006). Additionally, the facility has the capability to treat 82,000 m^3 of wastewater daily (Jay Chemicals, n.d.). In the city, there are 10,716 small and medium enterprises (SMEs) engaged in various sectors, including chemicals, textiles, engineering, and paper industries (Government of Gujarat, n.d.).

The city is organized into 14 wards and had a population of 163,630 in 2011. This reflects an increase from approximately 71,406 in 2001, resulting in a population density of 7292 individuals per square kilometer (Census of India, 2011a, b). Table 6 illustrates the population changes in Vapi city over the decade. A new census is scheduled for 2021.

3.2.2 Waste Generation in Vapi

Vapi Municipality is the authorized body designated to manage municipal solid waste in the city. According to the municipality, the city generated a total of 50 tons of "municipal waste" per day (TPD). (Valsad District Committee, 2019). The MSW dumpsite spanning 4046 m^2 in Vapi is located beside the river Daman Ganga at Chandor Village where the waste is dumped directly without scientific processing. Table 7 has been developed based on the waste data for Vapi available publicly. The waste generation data of Vapi is not as widely available and as comprehensive as that of Surat.

Vapi catchment details are tabularized in Table 8.

According to Valsad District Environmental Action Plan report of 2019, Vapi generated a total of 1 TPD of plastic waste (out of 50 TPD of waste generated), which corresponds to 2% of plastics in Vapi MSW (Valsad District Committee, 2019). However, understanding the characteristics of this plastic waste generated in Vapi is a challenge due to the lack of waste characterization studies in the city. To predict the characteristics of plastic waste in Vapi, one can reference the characteristics of Daman due to their close proximity (11 km apart) and comparable population sizes (Vapi's population was 163,630, while Daman's was 191,173 according

Table 6 Population of Vapi in decades

Year	Decadal Population
1981	19,620
1991	31,533
2001	71,406
2011	163,630
2021	336,145

Table 7 Waste generation data for Vapi

Year	Waste generated (TPD)
2011–2012	*39.6*[a]
2012–2013	*40.94*[b]
2013–2014	*42.33*[b]
2014–2015	*43.76*[b]
2015–2016	*45.24*[b]
2016–2017	*46.78*[b]
2017–2018	*48.36*[b]
2018–2019	*50*[c]

[a] (GPCB, 2012)

[b]Not official data. These figures have been calculated keeping 2011–2012 data as base and increasing it annually at a rate of 3.387% to reach 50 TPD in 2018–2019

[c] (Valsad District Committee, 2019)

Table 8 Waste generation data for Vapi

Details	Vapi catchment
Name of river flowing	Daman ganga
Length of river system	131.3 km[a], [b]
Area of catchment	22 sq.km[c], [d]
Number of basins within catchment	2 (daman ganga basin and Kolak basin)[e]
Number of inhabitants	163,630[c], [f]
Number of urban clusters	2
Number of industries	12,000[g, h] (approx.)
Number of WWTPs/CETPs	11 (approx.)

[a] This is the length of the entire river of Daman Ganga
[b] (U.T. Administration of Daman & Diu, 2019)
[c] Information only for Vapi City and not the Nani Daman area
[d] (Valsad District, n.d.)
[e] (Central Ground Water Board, 2013),
[f] (Census of India, 2011a, b)
[g] (MSME, 2013)
[h] (MSME Development Institute, n.d.)

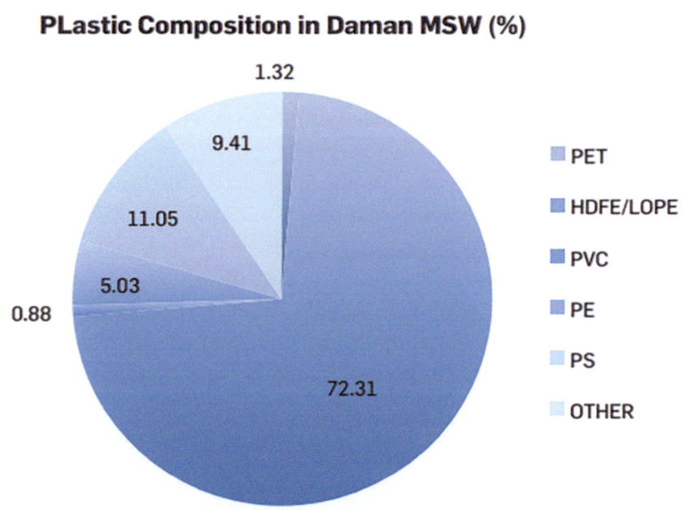

Fig. 13 Plastic waste composition in Daman. (Central Pollution Control Board (CPCB) & Central Institute of Plastics Engineering & Technology (CIPET), 2015)

to the 2011 census). In the CPCB-CIPET study of 2011, Daman was one of the sixty cities to be assessed and the city MSW consisted of 4.64% plastic waste (CPCB, 2015). It is important to note that Daman's plastic waste generation only determines the types of Vapi's plastic waste (and not quantification). Most authentic and recent data is used to quantify and estimate the plastic waste generated in Vapi. The composition of plastic waste in Daman as reported by CPCB in their 2015 report is shown in Fig. 13 below.

3.2.3 Case Study Area in Vapi

For this case study, the catchment area majorly encompasses the industrialized regions of Vapi and Daman and Dadra Nagar Haveli, given their closeness to the coastline where the Daman Ganga River flows into the Arabian Sea. The lower limit of the catchment is defined by the boundaries of the Vapi Municipal Corporation, while the upper limit is marked by the coast. Additionally, the catchment features the Daman Ganga Dam, often referred to as the Madhuban Dam. This dam, constructed by the Government of Gujarat in 1998, spans the Daman Ganga River downstream from the confluence with the Vagh Rivulet, a tributary of the Daman Ganga (National Water Development Agency, 2014).

To better understand the Vapi catchment area and visualize its features, GIS technology has been employed to create four distinct maps, similar to those produced for the Surat catchment area.

Figure 14 presents a land use map that depicts the various industrial, agricultural, and commercial/residential zones within the catchment area. This map highlights locations of waste disposal facilities and wastewater treatment plants in the specified catchment region. Additionally, the Daman Ganga Dam is marked on the map. It indicates that the catchment is primarily characterized by industrial and residential/urban development, with limited agricultural land.

Figure 15 presents a topographic map detailing the elevation of the Vapi catchment area, highlighting that Vapi is a low-lying region with an elevation of approximately less than 45 m.

Figure 16 illustrates the unique river drainage basins within the Vapi catchment area, as determined through hydrogeological mapping conducted using GIS. The map reveals the presence of two primary basins, associated with the Daman Ganga and Kolak river streams, along with four additional basins located in and around the catchment area.

The Daman Ganga basin covers a total area of 2331 km^2, with 1438 km^2 located in the state of Maharashtra, 430 km^2 in Gujarat (entirely within Valsad district), 393 km^2 in the Union Territory of Dadra & Nagar Haveli, and 70 km^2 in the Union Territory of Daman & Diu (National Water Development Agency, 2014).

The Kolak River represents the next significant basin within the catchment area. This perennial river, located in Gujarat, western India, originates from the Saputara hills and flows into the Arabian Sea (Central Ground Water Board, 2013). The basin extends up to 50 km in length and covers an area of 533 km^2, flowing along the northern edge of Vapi city in Gujarat (Gujarat Pollution Control Board, 2019).

Figure 17 presents a hydrological map of Vapi. The flow direction map indicates that surface runoff travels toward the two rivers, Kolak and Daman Ganga, suggesting that any waste or chemical pollutants entering these rivers will be carried along their course.

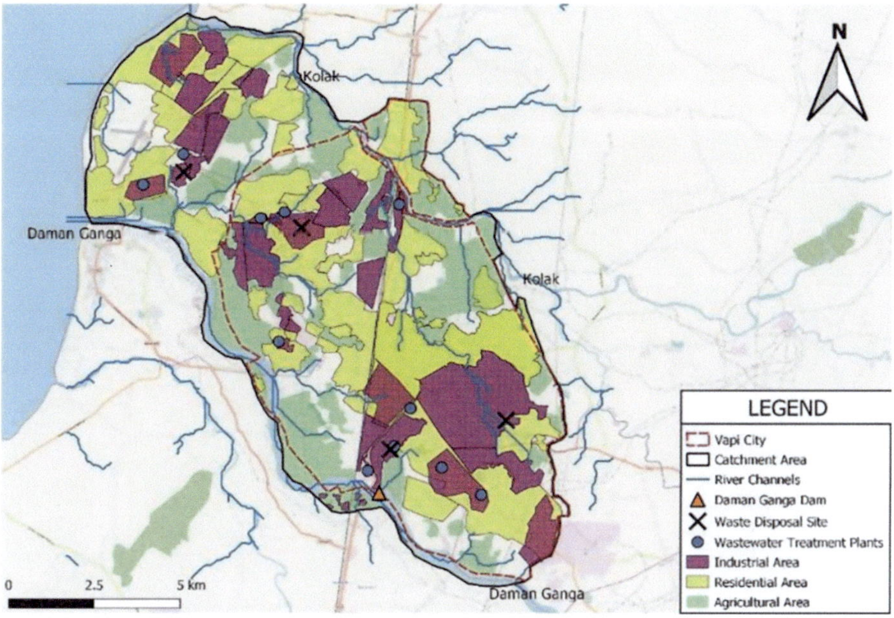

Fig. 14 Land use map and catchment of Vapi

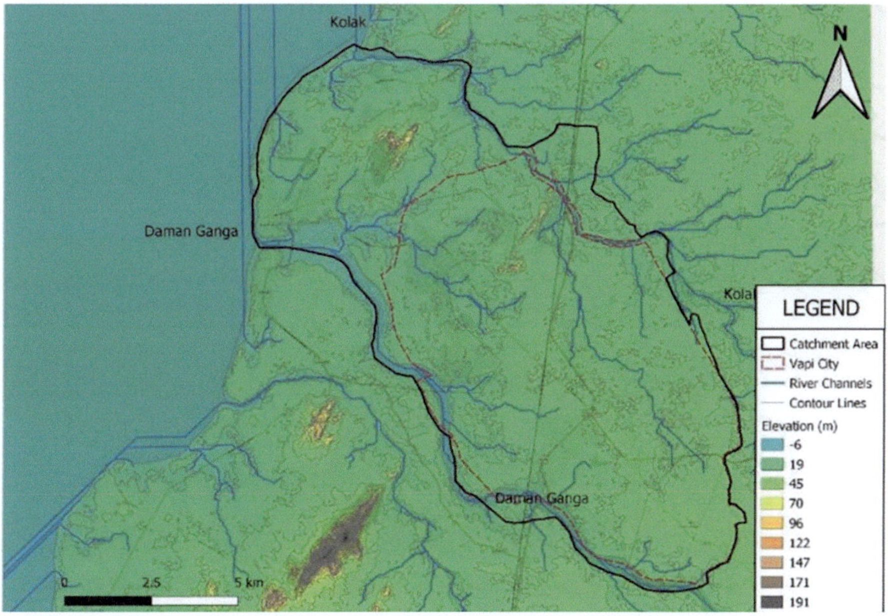

Fig. 15 Topographic map of Vapi

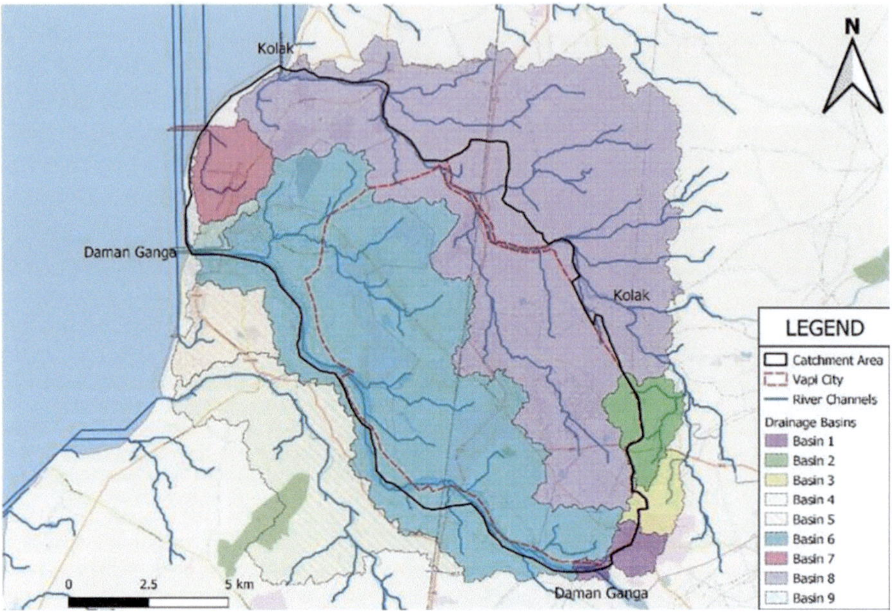

Fig. 16 Drainage basin map of Vapi

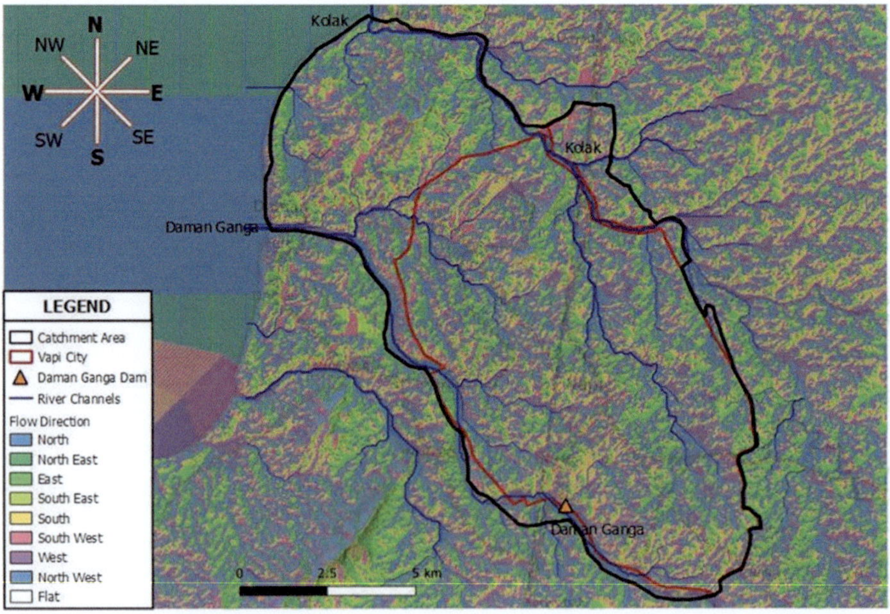

Fig. 17 Flow direction map for Vapi

4 Mapping Major Formal and Informal Waste Repositories in the Catchment

Existing systems of waste recovery in India feature an array of actors linked to each other by an internal logic dictated by economic, social, and cultural relationships (Gill 2010; Doron and Jeffrey 2018). The processes involved with picking, trading, and recovering waste constitute what is referred to as the "waste economy" (Harriss-White 2020). The relationships underlying the Indian waste economy are oftentimes informal; structured around unregulated activities which are not formally registered nor recognized by official waste management (WM) structures (Deepthi et al., 2018). The informal waste sector (IWS) is estimated to be responsible for recycling between 50% and 80% of India's waste (Annepu 2012; Bhaskar and Chikarmane 2012) but figures can vary widely due to limitations of data availability and methodologies. The pivotal role of informal structures in providing essential WM services across Indian cities has been emphasized in numerous studies and country-specific reports (Deepthi et al. 2018; ESCAP 2019; UNEP et al. 2019). However, national plastic WM strategies since the 1990s have so far not sufficiently tapped into the sector's potential (SBM Urban 2019). This illustrates the state's paradoxical reliance on, but yet distance from the IWS economy (Harriss-White 2020). However, this aspect contradicts the fact that at times, such waste management related services by ragpickers in India tend to be paid as waged or performance-based labor like segregation of value-added plastics and other materials.

Waste pickers gather various plastics from the bottom tier of the recycling hierarchy and are often called "scavengers" or "rag pickers," with terminology varying based on local language and the types of materials they collect. (Gill 2010; Doron and Jeffrey 2018). IWS work in developing countries has conventionally been portrayed as small-scale, unskilled labor providing a source of livelihood for poor individuals (Haan et al. 1998; Moreno-Sanchez and Moldanodo 2006; Wilson et al. 2006). However, an increasing body of literature challenges the narrative of the marginalized and unorganized view of IWS pickers by highlighting the interconnected socioeconomic relations and value networks surrounding the trade and distribution of PW (Gill 2010; Harriss-White 2020). These narratives allude the need to rethink categorical views of informal and formal labor.

Structural challenges that affect IWS workers include the absence of labor regulations, lack of access to social welfare and government schemes, and minimal protection against health and unemployment shocks. The COVID-19 pandemic illuminates these vulnerabilities, as informal workers are not reached by government relief efforts (ICTD 2020). It is thus relevant to engage with argument, that IWS workers are particularly exposed to emerging risks and economic precariousness, as a result of being invisible to, or discriminated by, development policies under the official state system. This highlights that IWS workers' social networks and political voice are often unheard and de-legitimized within legal and political structures, from which the need to empower IWS workers' voice, visibility and validity has become a reoccurring theme (Pillay 2017; Goelnitz 2018).

Beyond the recognition of delegitimized and devalued IWS labor, informality has stereotypically been criminalized as a homogenous entity (Haan et al. 1998; Wilson et al. 2006). However, the informal economy as a whole is more heterogeneous and complex than these narratives suggest (Chen 2012). Bhan et al. (2018) criticize literature across the board of conventional disciplines for failing to distinguish between illicit informal activities and legal but unrecorded informality. Accordingly, informal labourers can be considered entrepreneurs and micro-business operators who pay local informal fees and are highly skilled in their occupations. This is important, because at the same time as exposing the misinterpretations of informal labor as being illegitimate, it highlights the skills and knowledge that can be utilized by finding more inclusive ways to optimize local WM systems. In the context of informal value networks of PW in Delhi, Gill (2010) argues that entrepreneurial acumen is an essential requisite to survive within the competitive capitalist structures of the IWS economy. There is a growing body of literature that builds up under the entrepreneurial view of IWS workers (Gunsilius 2010; ESCAP 2019; Harriss-White 2020). While these narratives highlight opportunities for developing entrepreneurial skillsets, other studies have underlined the persistent barriers to utilizing the skills within informal structures, including challenges related to socioeconomic insecurity and vulnerability (Toxic Link 2016; Gidwani and Maringanti 2016). Nevertheless, as a cohort, these studies challenge the stereotypical view of marginalized and vulnerable IWS workers, thus supporting the need to reconceptualize what is meant by informality. However, as Gill (2010) and Chen (2012) stated that the optimistic entrepreneurial domain can perhaps be better understood as a result of survival within a system of structural constraints and exclusion.

4.1 Plastics and the Informal Sector

This is especially important in the context of economic and environmental losses and damages that stem from poor waste management. An average Indian uses 11 kg of plastics every year (FICCI, 2017). It is estimated (Ellen MacArthur Foundation, 2016–2017) that 80–120 billion USD is lost annually to the economy from the material value of plastic packaging because of the limited (one-time) use cycle. In India, a significant portion of the informal workforce has long participated in various stages of the recycling process, including the collection, dismantling, sorting, shredding, cleaning, compounding, pelletizing, and manufacturing of new products. This informal recycling sector addresses the gaps often left by municipal systems, which frequently lack the necessary infrastructure, resources, and capacity for effective waste collection and treatment. Additionally, it plays a vital role in minimizing waste and costs while recovering valuable materials that would otherwise be discarded. The informal sector is estimated to divert approximately 4.7 million tons of plastic from public waste collection systems each year (WBCSD, 2017).

Nevertheless, plastic waste that circulates within the informal sector faces difficulties, with 15–25% deemed inappropriate for subsequent recycling efforts

(Nizzetto & Sinha, 2020). This includes composite plastics (with different polymers that cannot be segregated), and plastics that have deteriorated in quality over time (after much recycling). These materials are not bought by the pelleting industries because they are considered valueless and therefore a loss for the shredding and sorting units. In 2019, the estimated annual fluxes into the environment were from recyclable plastics [shredding and sorting (3.3–4.8 Mt), grinding and pelleting (2.4–3.6 Mt), remanufacturing (2.4–3.6 Mt)], and from nonrecyclable plastics including single-use plastics (3.3 to 4.8 Mt., due to littering, landfilling or incineration), end-of-life recyclables (0.8 to 1.2 Mt), and losses during production and transportation (0.2 to 0.4 Mt) (Nizzetto & Sinha, 2020).

Delhi, the capital of India, stands out as a significant hub for recycling activities in the country. Although it accounts for only 5% of the nation's plastic waste, it contributes approximately 25% of India's recycled plastic output. Tikri Kalan, a key trading and sorting area, sees about 1900 tons of materials exchanged daily. This region features three distinct segments: one organized PVC market alongside two informal satellite markets known as Badi Tikri and Choti Tikri, which together support the primary market's operations. In Surat, Gujarat, the Surat Municipal Corporation (SMC) implements a segregation process for plastic waste prior to its disposal in landfills. The SMC utilizes rag pickers at each transportation hub to sort valuable plastics and other materials. Initial plastic waste sorting occurs at collection sites, with further segregation taking place at transportation locations (CPCB, 2015).

The Indian government has taken significant steps to integrate the informal sector with formal waste management services. As previously noted, the Swachhta Kendras have engaged with 5500 Safai Sathis (workers). With a commitment to enhancing social security for these workers, Self-Help Groups have been established. Swachhta Kendras regularly host workshops aimed at helping them open bank accounts, obtain Aadhar cards, access health check-ups, and receive personal training (UNDP, 2018–2024). Expanding such initiatives nationwide would benefit both the individuals involved and the environment.

It is essential to gain a deeper insight into the material aspects of the informal sector's contribution to waste management, as well as the ways in which socioeconomic effects related to informal waste management in India are perceived and felt by those at the grassroots level of the recycling process. In line with national and international goals and regulations to reduce plastic pollution at its source and adopt more circular economy initiatives, it is key to consider how existing IWS structures can be best included in the development of future WM strategies. However, as evidence has shown, top-down policies often overlook the interests and voices of IWS pickers at a local level and fail to account for socioeconomic and structural differences across scales. As a consequence, vulnerabilities faced by livelihoods in the informal PW sector are exacerbated, which potentially reproduces adverse socioeconomic and environmental impacts.

5 Domestic Plastic Waste Generation Data

India generates 62 million tonnes of municipal solid waste (MSW), out of which 19 million tonnes are not collected, and 31 million tonnes of the collected waste is dumped in landfills (Mohanty, 2018). Of this dumped waste, some plastics take hundreds of years to break down; the estimated half-lives for HDPE range from 58 years (bottles) to 1200 years (pipes). (Chamas et al., 2020).

Approximately 9.4 Million tonnes of plastic waste is generated in India per annum, which amounts to 26,000 tonnes of waste per day (MoHUA, 2019). Out of this approximately 5.6 Million tonnes per annum of plastic waste is recycled (15,600 tonnes of waste per day) most of which is in the informal sector and 3.8 Million tonnes per of annum plastic waste is left uncollected or littered (9400 tonnes of waste per day) (MoHUA, 2019). Although the average per capita consumption of plastic (11 kg/annum) in India in 2011 was much lower compared to the world's average (28 kg/person) and other western countries, the impact of plastic waste generation will be high in India given the high population and improper management of plastic waste (Sheelanere et al., 2019). It makes India the 15th biggest plastic waste generator in the world (Sharma, 2019).

The plastic industry in India has seen steady growth since its establishment in the 1960s and has grown at a Compound Annual Growth Rate (CAGR) of 10% in terms of volume from 8.3 Mega Metric Tonnes Per Annum (MMTPA) in 2010 to 13.4 MMTPA in 2015 and was expected to reach 22 MMTPA by 2020 (with a growth rate of CAGR of 10.5%) (FICCI, 2017). The India packaging (predominantly plastics) market was valued at USD 50.5 billion in 2019, and is expected to reach USD 204.81 billion by 2025, with a CAGR of 26.7% during 2020–2025 (Anon 2020a). Comparatively, in 2018, the global production of plastics had increased to nearly 360 million metric tons, and is expected to triple and account for a fifth of global oil consumption, by 2050 (Tiseo, 2021). The global plastic market size was valued at USD 568.9 billion in 2019 and is expected to grow at a CAGR of 3.2% from 2020 to 2027 (Anon 2020b). The plastic industry engulfs the entire spectrum of daily use items and covers almost every sector of life such as clothing, housing, construction, furniture, automobiles, household items, agriculture, horticulture, irrigation, packaging, medical appliances, electronics, and electrical items. The packaging industry in India has seen a strong penetration of plastics as compared to global standards (FICCI, 2017). It is one of the fastest growing industries, and has an influence on all other industries, either directly or indirectly.

India observes a significant regional diversity in consumption of plastics in the four key geographical regions, with the Western region accounting for 47%, the Northern region for 23%, the Eastern region for 9% and the Southern region for 21% (FICCI, 2012). The bulk of the consumption in Northern India is from end-use industries of auto, packaging (including bulk packaging), plasticulture applications, and electronic appliances, which are concentrated mostly in UP and Delhi-NCR. However, plastic processing in other parts like Rajasthan, Punjab, Haryana, Uttarakhand, J&K and Himachal Pradesh is expected to grow based on increased availability of feedstock and higher focus on the manufacturing sector (FICCI, 2012).

As per the Un-Plastic Collective (UPC) Report of the United Nations Environment Programme (UNEP), Confederation of Indian industries (CII), and WWF India, among the total plastic waste generated, 40% remains uncollected and 43% of the collected part contains single-use plastics such as carry bags, straws, packaging, bottle caps, garbage bags, PET bottles, and snacks and confectionery packets (Anon 2019). Figure 18 shows the contribution of single-use plastics to plastic waste generation in India. A 2015 report of CPCB on 60 Indian cities showed that a daily average of 3501 tonnes of plastic waste is generated, out of which 94% plastic waste is composed of thermoplastics or recyclable materials such as polyethylene terephthalate (PET), and polyvinyl chloride (PVC) (CPCB, 2015). Although the majority of the plastic waste are recyclable, they are either left littered and uncollected in the open environment or a significant part of the collected waste is dumped in landfills. Therefore, most plastic waste are not recycled or reused or disposed of properly, which points towards the inefficiency of the waste management system. These effortlessly dumped waste are either landfilled or end up in polluting streams or groundwater sources.

Various survey reports have assessed the annual plastic waste generated by different states in India (see Fig. 19). Gujarat, recognized as a significant industrial and commercial center, is home to 1002 registered plastic manufacturing and recycling facilities, four compostable units, and ten multilayered plastic units (CPCB Annual Report, 2018–2019). In the fiscal year 2018–2019, the state ranked second in plastic waste generation, contributing approximately 356,873 tonnes annually (CPCB Annual Report, 2018–2019). In addition to industrial contributions, single-use plastic (SUP) waste and littering during religious events further exacerbate the plastic waste problem in the state (UNEP, 2018). Common domestic items such as empty water bottles, garbage bags, carry bags, and snack packaging significantly contribute to the accumulation of plastic waste in Gujarat.

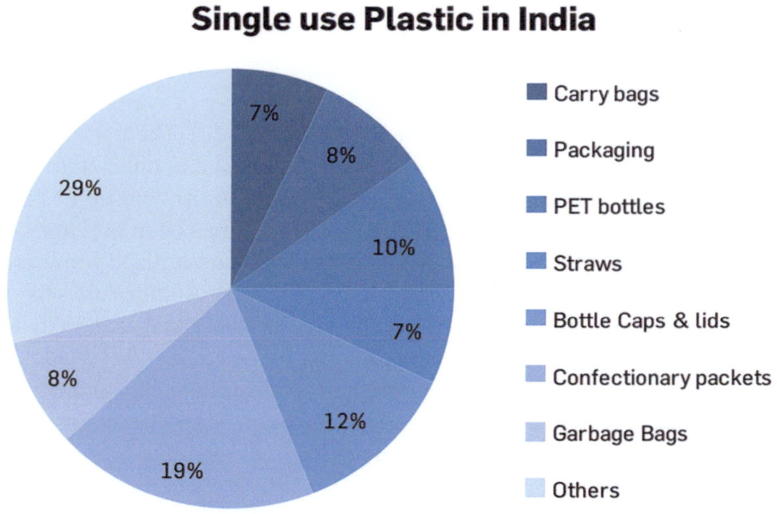

Fig. 18 Contribution of single-use plastics in plastic waste generation in India (Reddy n.d.)

Fig. 19 Major plastic waste producing Indian States (TPA) (Swachh Bharat Mission (Urban), 2019)

6 Other Sources of Plastic Waste Generation Data

As of November 2017, Gujarat produced 10,145 metric tons of solid waste daily (Anon 2018a). The municipal corporations, operating under state government regulations, are responsible for the collection, transportation, segregation, processing, and disposal of municipal solid waste. Among the eight major municipal corporations, Ahmedabad and Surat generated the largest volumes of waste, followed by Vadodara and Rajkot (CPCB, 2018).

As stated earlier, the estimated plastic waste generation in Gujarat was 3,56,873 T/annum during 2018–2019 (CPCB Annual Report, 2018–2019). The cities of Gujarat generated an average of 345 gm/capita/day of waste as against the national average of 500 gm/capita/day in 2012 (Anon 2012). There are 1002 registered plastic manufacturing/recycling units, four compostable units, and 10 multilayered plastic units (CPCB Annual Report, 2018–2019). According to provision “5(b)” of the PWM Rules, 2016 (amended in 2018), it has been previously mentioned that around 86,553.62 MT of plastic waste has been utilized for coprocessing in cement plants (CPCB Annual Report, 2018–2019). Additionally, the use of plastic carry bags is entirely prohibited in the city of Gandhinagar (CPCB Annual Report, 2018–2019).

Industries such as chemicals, petrochemicals, pharmaceuticals, and textiles are primarily located in coastal districts like Surat and Vapi. These sectors play a major role in water pollution, exerting pressure on marine and coastal ecosystems (Gujarat ENVIS, n.d.).

As evident in Table 9, the predominant form of plastic waste produced by the industrial clusters in Surat and Vapi is packaging waste. Key industries located in

Table 9 Some of the plastic waste generating industries in Surat and Vapi

S.No.	Industry type	Few industries	Raw materials	Waste generated	Possible plastic waste generated
1.	Agro based	AnupamRasayan India ltd., Mahavir synthesis pvt. ltd., Aether industries ltd., Centre point industries	Jute, sugarcane, vegetable oil, wheat, rice, oil seeds, barley	Fruits waste peels, coffee pulp, husks, wheat straw residues, rice straws, field residues like (leaves, stalks, seed pods, stems), process residues (husks, seeds, roots, bagasse, molasses) Industrial residues (potato peel, orange peel, cassava peel, groundnut oil cake, coconut oil cake, and soybean oil Cake)	Polyethylene films Plastic canisters Packaging plastic waste Polymers used in Plasticulture applications (Srinidhi et al., 2018)
2.	Textile	Toray Kusumgar Advanced textile Pvt. ltd., Vactech composites Pvt. ltd., reliance industries ltd, colosperse dyes &intermediates, M/S Shree Durga Syntex private limited, M/S spectrum dyes and chemicals private limited, Centre point Industries	Cotton, wool, silk, jute, manmade fiber	Wastewater is generated (Carboxymethyl cellulose, polyvinyl alcohol, mineral acid, caustic Soda, soda ash, detergent, sodium hypochlorite, dyes, starches, synthetic waxes	Microplastic pollution from plastic microfibers and nanofibers (Henry et al., 2019) Plastic packaging waste (Mukherjee 2017)
3.	Paper, publishing, printing	Aryan paper Mills Pvt. ltd., best paper Mills Pvt. ltd., Devang paper mills Pvt. ltd., Mehali papers Pvt. ltd., Shree Ajit pulp & paper ltd., Shree Gajanan paper & boards Pvt. ltd., Mahavir synthesis pvt. Ltd.	Pulp (Wood pulp, non-wood pulp, recycled paper pulp), sizing elements, filler agents, pigments and coloring agents	Solid waste such as lime mud, lime slaker grits, green liquor dregs, boiler and furnace ash, scrubber sludges, wood processing residuals and wastewater treatment sludges are generated from Different mills	Plastic cartridges plastic containers Packaging plastic waste (Hayta et al., 2019)

(continued)

Table 9 (continued)

S.No.	Industry type	Few industries	Raw materials	Waste generated	Possible plastic waste generated
4.	Chemical & chemical based	Bayer Pvt. ltd., Dakle reinforced plastics Pvt. ltd., Huber group India Pvt. ltd., Sarna chemicals Pvt. ltd., Supreet chemicals Pvt. ltd., Vertellus Speciality materials India Pvt. ltd., Colourtex industries ltd., reliance industries ltd, Pravin laboratories, Acetochem Pvt. ltd, Kanaiya dye chem industries, Harish chemicals, spectrum dye And chemicals pvt ltd.,	Benzene, toluene, xylene and naphthalene altogether called BTXN And many more chemicals are being used	Harmful wastewater, plastic packaging waste	Plastic containers Plastic packaging waste
		Nova dyestuff industries pvt. ltd., Hindusthan chemicals co. ltd., Macson products, Mahavir synthesis Pvt. ltd., Aether industries ltd., colosperse dyes & intermediates, M/S spectrum dyes and chemicals private Limited, M/S Mangalmurti bio chem Pvt. ltd, Bilag industries ltd, Aarti industries ltd, Bayer Vapi Pvt. ltd., Chemisynth Vapi, C.S. Specialty chemicals, Ganesh polychem limited, Hemanichemiorganic pvt. ltd., hemani industries limited, K.V. Arochem Pvt. ltd., Manglam drugs & Organics ltd			

5.	Construction infrastructure	M/S Mata Sitadevi infrastructure and research private limited, M/S Sanghi industries limited	Concrete, bricks, ceramics, Soil and sand, limestone, llay, marl or shale, sand, fly ash/ pulverized Fuel ash (PFA), or ironstone, concrete, asphalt, wood, metals, gypsum, plastics and salvaged building Components	Sand, gravel, concrete, stone, bricks, wood, metal, glass, plastic, paper, metal waste like pipes, conduits, and light sheet material	Plastic waste generated from electric cable, water pipes, building materials (Kwesiga, 2018)
6.	Logistics, supply chain, packaging, Transport	Eltete TPM ltd., daman ganga packaging solutions LLP, Maheshwari Pvt. ltd.,			Plastic packaging Waste
7.	Machinery and instruments	Gaurniflon India pvt. ltd., ITEC measures Pvt.ltd., Nesstech Instruments Pvt.ltd.,			Plastic packaging waste
8.	Pharmaceuticals	Virdev intermediates Pvt. ltd., Mahavir synthesis Pvt. ltd., Ami organics Pvt. ltd., Aether industries ltd., Manglam drugs ltd, Amoli organics p ltd, Centre point industries, Heranba industries ltd	Plastic & polymers, glass, paper, aluminum foil and paper boards	Expired drugs, patients' discarded personal medications, syringes, IV bags, tubing, vials, open Containers of drugs that cannot be used, containers that held acute hazardous waste drugs, drugs that are discarded, contaminated garments, absorbents and spill Cleanup material,	Plastic waste from Pharmaceutical bottles, disposable cups, disposable syringes, packaging materials, laboratory ware, and certain electronic use (Banerjee et al., 2014)

(continued)

Gujarat's coastal areas, including dyestuff, manufacturing and engineering, textiles, chemicals and petrochemicals, pharmaceuticals, and diamond processing, play a significant role in generating plastic waste. Given the geographical positioning of Surat and Vapi, it is crucial to address waste leakage at the source to prevent it from entering waterways and causing detrimental impacts across the source-to-sea system.

7 Plastic Waste Collection Rate

7.1 *Registered Plastic Waste Recyclers in Gujarat*

There are 4773 registered units (including plastic manufacturing, compostable, MLP, and recycling units) in 30 States/UTs in India (CPCB Annual Report, 2018–2019) (Fig. 20). Out of these, as mentioned earlier, Gujarat has 1002 plastic manufacturing, 4 compostable, and 10 multilayer plastic registered units. There are 18 registered recycling units in Vapi and three in Surat, with others distributed in various other districts of the state (Table 10). In addition to the above, there are many unregistered plastic manufacturing/recycling units/waste processing units in India/Gujarat. Such units operate illegally within residential or unapproved areas, and are not legally registered as per the requirement of Rule 13 of the Plastic Waste Management Rules, 2018.

7.2 *Plastic Waste Collection and Recycling*

The Surat Municipal Corporation has initiated a strategy to collect and segregate plastic waste at disposal locations to prevent its accumulation in landfills. In collaboration with Eco Vision Environmental Resources LLP, a recycling facility has

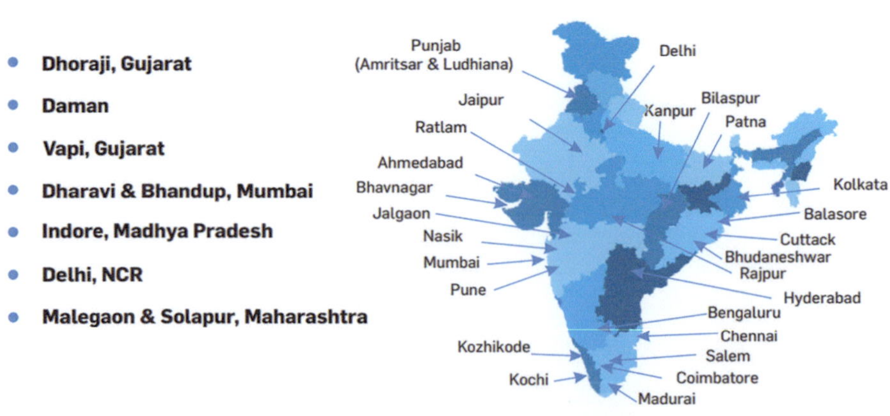

Fig. 20 Map showing the location of various plastic recycling units in India (Anon 2018b)

Table 10 List of registered plastic recycling units in Surat and Vapi as per the GPCB (Department of Drinking Water and Sanitation, Ministry of Jal Shakti, Government of India, n.d.; Central Pollution Control Board, n.d.)

Sl. No.	Name and Addresses of recyclers	Name of product produced by recycling	Quantity of plastic product produced (MT/M)
Surat			
1	Ecovision environmental resources Llp. (O-46995) plot no. 10,11,12,53, Gamtal. Bhatar, Surat,	Extrusion and shredding of plastic waste/extrusion and shredding of plastic waste from industrial sources (decontaminated containers, industrial plastic Scrap/waste etc.)	3000
2	J. B. Ecotexllp, (O-45785) PLOT NO: Block no. 195,198,199, NH-8, Vill. Dhamrod, Dhamrod—394,125 DIST: Surat, TAL: Mangrol	Recycled polyester staple fiber by using waste PET bottles as raw material through recycling	5046
3	Alliance Fibres private limited (O-31399) plot No-1,2,7&8, 1/A,2/A,3/ A,4/B,&9/D, survey no-269, block No-203 &208 MotaBorasara,, Mangrol -395,002 Dist: Surat, Tal: Mangrol	1. Recycle polyester staple fiber	1. 1500
		2. Recycle polyester yarn from pet bottle scrape	2. 500
Vapi			
1	M/s. Mahavir plastic—plot noJ/1901/15, fourth phase, GIDC, Vapi	Reprocessed plastic granules	50
2	Ashtavinayak Plastopack Pvt. ltd.—1002/1/D, GIDC, phase IV, Vapi—396,195	Plastic bags/rolls/sheets and packing materials	250
3	Parvati poly Plast—plot-c/4, Echachha industrial estate, Abrama	Plastic pipe plastic reprocess granules	10 70
4	Uma TechoPlast—Echchha industries estate, plot no. 8B, ValiaFalia, Abrama—396,001	Engineering work plastic pipe Plastic reprocess granules	200 kg/month 80
5	Navrang plastic—C-1/B, 2012, IIIrd phase, GIDC-Vapi	LD /HDP El PP /plastic granules	150
6	Savita polymers—plot No-11, Sunil Ind. Estate, Chikhla Road, village-Lilapore	Reprocessed plastic granules	120
7	Vimal plastic industries—AtulParnera road, Parnera -396,020	1. Plastic carboys 2. Reprocessed plastic granules	1. 20 2. 50

(continued)

Table 10 (continued)

Sl. No.	Name and Addresses of recyclers	Name of product produced by recycling	Quantity of plastic product produced (MT/M)
8	JagnathPlastics(old NameMahadev Chemicals) -plot no5103/5, phase-IV, GIDC Vapi	Reprocessed plastic granules	50
9	Krishna poly Plast—plot no. 8/B, Ichchha Industrial Park, Abrama—Valsad, ValiyaFaliya, Valsad	Reprocessed plastic granules	200
10	Khushi polymers—plot no. 6, Ichcha estate, ValiyaFaliya, Abrama, Valsad	Reprocessed plastic granules	200
11	Kiran trading—SR. NO. 3392 (old Sr. no.425/1/2/3), Plot no. 146, Damanganga Industrial Park—II, Vapi	1.Reprocessed plastic granules 2. Grinding all type of recyclable plastic waste /scrap & insulation cables	1. 200 2. 300
12	Hasti Enterprise—plot no. 9/A, part: B, survey no. 198/2, Paki-1, Karvad, Degam road, Vapi	Reprocessed plastic granules	200
13	Nova Industries—C-1/80, plot no527, 100 shed area, GIDC, Vapi	Plastic granules	150
14	Avani Industries—block no. 53, Vill. Sarodhi, Ashwin Estate, Chikhla	Reprocessed plastic granules	15
15	Indian power cable trading—Sr. no. 278/Pk.1/7, plot no. 7, N, A. Order no. 88, Gowdown No. 3, Karvad	Reprocessed plastic granules	3500
16	A. K. Barrel suppliers—shed no. L-715/9, 40 shed area, G.I.D.C., Vapi—396,195	Plastic granules	0
17	Shree Ganesh Acrylite industries—plot no-C-1-1812/2, phase-III, GIDC, Vapi—396,195	Acrylic plastic sheets	15
18	Dalmia dye Chem industries limited, P NO: 780/3,40 shed Area, GIDC, VAPI	Pet flakes (washing)	250

been set up to process 5000 kilograms of plastic waste, with future plans to expand this capacity to 100,000 kilograms. Additionally, efforts are underway to convert shredded plastic waste into diesel (Karelia, 2017).

Given the significant calorific value of plastic waste relative to coal, the Gujarat Pollution Control Board has implemented measures for co-processing by using plastic waste as alternative fuels or raw materials to recover energy and resources. The state of Gujarat is home to numerous cement plants that can convert plastic waste into fuel. However, several challenges hinder this process. For instance, approximately 70% of plastic waste has high moisture content, which limits its viability as fuel. In addition to practical issues like transportation and storage, obstacles include a lack of incentives for the cement industry to adopt plastic waste as fuel and resistance to embracing unconventional methods (Shah, 2018). Nevertheless, co-processing plastic waste offers several benefits, such as decreasing reliance on traditional fuels or raw materials, lowering greenhouse gas emissions, and promoting waste utilization. To address the challenges of waste generation and recycling, it is crucial to educate industrialists about the benefits of recycling and co-processing. Equally important is raising public awareness regarding the detrimental effects of littering and improper waste disposal, encouraging them to reduce consumption to mitigate environmental pollution.

8 Plastic Waste Recycling Industries

See Table 10. We have provided a list of major plastic waste recycling industries in Surat and Vapi. The list includes those engaged in the shredding of plastic waste to granule making and reprocessing into new products. These industries are involved in recycling a reasonable quantity of plastic waste every month and produce recycled plastic products.

References

Aminur Rahman, M., Mojumdar, S., Rahman, S. A., & Marimuthu, K. (2023). Plastic pollutions in the ocean: Their sources, causes, effects and control measures. *Journal of Biological Studies, 6*(1), 37–52. https://onlinejbs.com/index.php/jbs/article/view/7755

Anon. (2012). Status of waste management in Urban & rural areas. In *State of environment report Gujarat 2012, Urban, rural & built environment*. Gujarat ENVIS. Retrieved March 3, 2021, from http://gujenvis.nic.in/PDF/waste.pdf

Anon. (2013). *Saida Garbage Mountain, Lebanon. Environmental justice atlas, 2017-03-31*. Retrieved March 2, 2021, from https://ejatlas.org/conflict/garbage-mountainsaida

Anon. (2018a). *Generation of solid waste*. MNRE India. Retrieved March 3, 2021, from http://www.indiaenvironmentportal.org.in/files/file/GENERATION%20OF%20SOLID%20WASTE.pdf

Anon. (2018b). *Report on the Indian plastic industry*. 10th International Plastics Exhibition, Plast India, 2018. Retrieved March 1, 2021, from https://plastindia.org/pdf/Indian-Plastics-Industry-Report-2018-2.pdf

Anon. (2019). *The Un-PLASTIC collective brochure*. UNEP, CII, WWF. 2019. Retrieved March 1, 2021, from https://d2391rlyg4hwoh.cloudfront.net/downloads/unplastic_brochure.pdf

Anon. (2020a). *Global newswire news release*. Retrieved March 2, from https://www.globenewswire.com/news-release/2020/06/05/2044182/0/en/The-India-Packaging-Market-was-valued-at-USD50-5-billion-in-2019-and-it-is-expected-to-reachUSD-204-81-billion-by-2025-registering-a-CAGRof-26-7-during-the-period-of-2020-2025.html

Anon. (2020b). *Plastic market size, share & trends analysis report by product (PE, PP, PU, PVC, PET, polystyrene, ABS, PBT, PPO, epoxy polymers, LCP, PC, polyamide), by application, by region, and segment forecasts, 2020–2027*. Grandview Research. Retrieved March 2, 2021, from https://www.grandviewresearch.com/industryanalysis/global-plastics-market

Banerjee, T., Srivastava, R., & Hung, Y.-T. (2014). *Plastics waste management in India: An integrated solid waste management approach*. World Scientific Publishing Co.

Bhadiyadra, K. (2015). *Storm water drainage problem of Surat city & its solution due to flood in river Tapi*. Surat, Gujarat, India. https://shorturl.at/Ej8ZZ

Bhan, et al. (2018). Identifying the User in an Informal Trade Ecosystem, https://dl.designresearchsociety.org/drs-conference-papers/drs2018/researchpapers/65/

Bhaskar, A., & Chikarmane, P. (2012). The story of waste and its reclaimers: Organising waste collectors for better lives and livelihoods. *Indian Journal of Labour Economics, 55*, 595–620.

Bouissou, J. (2014, September 15). *Indian city of Surat anticipates worst effects of climate change*. Retrieved November 12, 2020, from the Guardian web site: https://www.theguardian.com/cities/2014/sep/15/indian-cities-climate-changesurat

Business World. (2017, June 28). *Richest Cities of India*. Retrieved from Business World Web site: http://www.businessworld.in/article/Richest-CitiesOf-India/28-06-2017-121011/

Census of India. (2011a). *District census handbook Surat*. Census of India.

Census of India. (2011b). *District census handbook Valsad*. Gujarat. https://gujecostat.gujarat.gov.in/uploads/mediafiles/2424-PART-B-DCHB-VALSAD.pdf

Census of India, Central Ground Water Board. (2013). *Groundwater brochure—Valsad District*. https://www.cgwb.gov.in/old_website/District_Profile/Gujarat/Valsad.pdf

Central Ground Water Board. (2015). *Valsad District*. Ministry of Water Resources. Retrieved from https://www.cgwb.gov.in/old_website/Annual-Reports/Annual%20Report%202015-16.pdf

Central Pollution Control Board. (n.d.). *Notice: PWP Registration* [PDF]. Retrieved from https://eprplastic.cpcb.gov.in/plastic/downloads/notice_pwp_registration.pdf

Central Pollution Control Board. (2015). *Assessment and characterization of plastic waste generation in 60 major cities* [PDF]. Government of India. https://cpcb.nic.in/displaypdf.php?id=cGxhc3RpY3dhc3RlL1BXXzYwX2NpdGllc19yZXBvcnQtSmFuLTIwMTUucGRm

Central Pollution Control Board (CPCB) Annual Report. (2018–2019*) On the Implementation of Plastic Waste Management Rules (as per Rule '17(4)' of PWM Rules*, 2018). Retrieved March 1, 2021, from https://cpcb.nic.in/uploads/plasticwaste/Annual_Report_2018–19_PWM.pdf

Central Pollution Control Board (CPCB) & Central Institute of Plastics Engineering & Technology (CIPET). (2015, January). *Assessment & quantification of plastics waste generation in major cities*. Ministry of Environment & Forests, Government of India. https://optoce.no/wp-content/uploads/2019/03/CPCBCIPET-Assessment-and-Quantification-of-Plastics-Waste_2015.pdf

Central Water Commission. (2014). *Tapi Basin*. Ministry of Water Resources. Retrieved November 20, 2020, from https://indiawris.gov.in/downloads/Tapi%20Basin.pdf

Central Water Commission. (2016). *Water year book 2015–2016: Tapi Basin*. CWC.

Chakraborty P., Sakthivel S., Masafumi N., Balasubramanian P., Shunkei K., & Bommanna L. (2016) Contamination Status. Electronic Waste and associated hazardous substances in South-Asian countries of Asia Pacific region: ACS Symposium Series-Persistent Organic Chemicals in the Environment: Status and Trends in the Pacific Basin Countries I. ISBN13: 9780841231979; eISBN: 9780841231962; https://doi.org/10.1021/bk-2016-1243.

Chakraborty, P., Harish G., Balasubramanian P., Moitraiyee M., Nirmala K.S., Masafumi N., & Scott S.N. (2021). Passive Air Sampling of PCDD/Fs, PCBs, PAEs, DEHA, and PAHs from Informal Electronic Waste Recycling and Allied Sectors in Indian Megacities. 55, 14, 9469–9478, Environmental Science & Technology; ISSN: 1520–5851.

Chamas, A., Moon, H., Zheng, J., Yang, Q., Tabassum, T., Jang, J. H., Abu-Omar, M., Scott, S., & Suh, S. (2020). Degradation rates of plastics in the environment. *ACS Sustainable Chemistry & Engineering, 8*(9), 3494–3511. https://doi.org/10.1021/acssuschemeng.9b06635

Chen, M. A. (2012). *The informal economy: Definitions, theories and policies* (WIEGO Working Paper No. 1). WIEGO. https://www.wiego.org/wp-content/uploads/2019/09/Chen_WIEGO_WP1.pdf

Christian, H. (2016). Fuzzy ranking for landfill site selection in Indian context. *International Journal of Geomate, 11*. https://doi.org/10.21660/2016.26.5382

CPCB. (2015). *Assessment & quantification of plastics waste generation in major cities*. Ministry of Environment & Forests. https://optoce.no/wp-content/uploads/2019/03/CPCBCIPET-Assessment-and-Quantification-of-Plastics-Waste_2015.pdf

CPCB. (2018). *Trend of solid waste generation in 46 cities*. Retrieved from CPCB: https://cpcb.nic.in/uploads/MSW/trend_46_cities_list.pdf

CPCB. (n.d.). *Waste generation and composition*. Retrieved from CPCB web site: https://cpcb.nic.in/waste-management/

CPCB ENVIS. (n.d.). *Municipal solid waste*. Retrieved from CPCB ENVIS web site: http://www.cpcbenvis.nic.in/cpcb_Newsletter/SOLID%20WASTE.pdf

Dean, S. 2017. *A widely reported study on the dangers of microplastics in fish is about to be retracted*. Retrieved March 1, 2021, from Https://www.sciencealert.com/a-widely-reported-study-on-theeffects-of-microplastics-in-fish-is-about-to-beretracted

Deepthi, M. V., Rrn, S., Ameen, K. M., Roy, P., & Chandrasekhar, K. (2018). *TERI analysis on challenges and opportunities—Plastic waste management in India*. The Energy and Resources Institute.

Department of Drinking Water and Sanitation, Ministry of Jal Shakti, Government of India. (n.d.). *List of State-wise Plastic Waste Processors* [PDF]. Swachh Bharat Mission. Retrieved from https://swachhbharatmission.ddws.gov.in/sites/default/files/What's-New-file/List-of-State-wise-Plastic-Waste-Processors.pdf.

Deshpande S. 2019. *Former UN environment chief joins Tapi clean-up*. The times of India. Retrieved March 1, 2021, from https://timesofindia.indiatimes.com/city/surat/former-un-environment-chief-joins-tapiclean-up/articleshow/69949890.cms

Doron, A., & Jeffrey, R. (2018). *Waste of a nation: Garbage and growth in India*. Harvard University Press.

Dubey, M., & Ujjania, N. (2013). Water quality and pollution status of Tapi River, Gujarat, India. *International Journal of Pure and Applied Zoology, 1*, 2320–9577.

DW. (n.d.). *Saving Mumbai's natural mangrove buffer against rising tides*. Retrieved November 23, 2020, from DW https://www.dw.com/en/savingmumbais-natural-mangrove-buffer-against-risingtides/a-47992557

Erickson-Davis, M. (2018, May 2). *New study finds mangroves may store way more carbon than we thought*. Retrieved November 16, 2020, from Mongabay web site: https://news.mongabay.com/2018/05/new-study-finds-mangroves-maystore-way-more-carbon-than-we-thought/

Ferraro, G., & Failler, P. (2020). Governing plastic pollution in the oceans: Institutional challenges and areas for action. *Environmental Science & Policy, 112*, 453–460. https://linkinghub.elsevier.com/retrieve/pii/S146290112030513X

FICCI. (2009). *Survey on the current status of municipal solid waste Management in Indian Cities and the potential of landfill gas to energy projects in India*. Federation of Indian Chambers of Commerce and Industry. https://www.nswai.org/docs/292_2_ficci_survey_09.pdf

FICCI, Plastic Infrastructure. (2017). Retrieved October 30, 2020, from http://ficci.in/spdocument/20872/report-Plasticinfrastructure-2017-ficci.pdf

GCPC ENVIS. (2018). *Cleaner production in wastepaper recycling industrial sector*. Retrieved March 2, 2021, from http://gcpcenvis.nic.in/Monographs/cpwastepapersector.pdf

Gervais, Christopher J. (2018). *Plastic pollution facts*. Retrieved March 1, 2021, from: christopherjgervais.wordpress.com/2018/03/06/plastic-pollution-facts/

Gidwani, V., & Maringanti, A. (2016). The waste-value dialectic: Lumpen urbanization in contemporary India. *Comparative Studies of South Asia, Africa and the Middle East, 36*(1), 112–133. https://doi.org/10.1215/1089201x-3482159

Gill, K. (2010). *Of poverty and plastic: Scavenging and scrap trading entrepreneurs in India's urban informal economy*. Oxford University Press.

Goswami, M., and Patel, B. (2015). *CPR Namati environment justice program in Gujarat*. Retrieved March 2, 2021, from https://namati.org/news-stories/can-indias-most-pollutedenvironment-be-saved/

Government of Gujarat. (n.d.). *Economy*. Retrieved November 15, 2020, from Valsad District: https://valsad.nic.in/economy/

GPCB. (2012). *Annual report 2011-2012*. Gujarat Pollution Control Board. Retrieved November 29, 2020, from Http://Gujenvis.nic.In/Pdf/Gpcb%20ar%202011_12.Pdf

Gregory, M. R. (2009). Environmental implications of plastic debris in marine settings--entanglement, ingestion, smothering, hangers-on, hitch-hiking and alien invasions. *Philosophical Transactions of the Royal Society of London. Series B, Biological Sciences, 364*(1526), 2013–2025. https://doi.org/10.1098/rstb.2008.0265

Grotal. (n.d.). Asias largest plant Vapi. Retrieved November 15, 2020, from Grotal: https://www.grotal.com/I/Vapi/Asias-Largest-Plant

Central Pollution Control Board (CPCB). (2017). *Guidelines for Co-processing of plastic waste in Cement Kilns (As per Rule '5(b)' of plastic waste management rule*s, 2016. Retrieved March 3, 2021, from https://cpcb.nic.in/uploads/plasticwaste/Co-processing_Guidelines_Final_23.05.17.pdf

Gujarat Pollution Control Board (2008) *Guidelines for Environmentally Sound Management of e-Waste*. Retrieved March 2, 2021, from https://gpcb.gujarat.gov.in/uploads/e_waste_rules_2016_guidlines.pdf

Gujarat ENVIS. (n.d.). *Status of Industrial Environment in Gujarat*. Retrieved March 1, 2021, from http://www.gujenvis.nic.in/PDF/soe-industrial.pdf last

Gujarat Pollution Control Board. (2019). *River action plan of Kolak*. GPCB. https://shorturl.at/V4s9a

Gundaliya, P. J. (2016). Floodplain delineation using HECRAS model—A case study of Surat City. *Open Journal of Modern Hydrology, 6*, 34–42. https://doi.org/10.4236/ojmh.2016.61004

Haan, H. C., Coad, A., & Lardinois, I. (1998). *Municipal waste management: Involving micro-and-small enterprises. Guidelines for municipal managers*. International Training Centre of the ILO, SKAT, WASTE. http://www.skat-foundation.org/publications/waste.Htm

Harriss-White, B. (2020). Waste, social order, and physical disorder in small-town India. *The Journal of Development Studies, 56*(2), 239–258. https://doi.org/10.1080/00220388.2019.1577386

Haward, M. (2018). Plastic pollution of the world's seas and oceans as a contemporary challenge in ocean governance. *Nature Communications, 9*(1), 667. https://doi.org/10.1038/s41467-018-03104-3

Hayta, P., & Oktav, M. (2019). The importance of waste and environment management in printing industry. *Journal EJENS—European Journal of Engineering and Natural Sciences; Istanbul, 3*(2), 18–26.

Henry, B., Laitala, K., & Klepp, I. (2019). Microfibres from apparel and home textiles: Prospects for including microplastics in environmental sustainability assessment. *Science of the Total Environment, 652*, 483–494.

Jambeck, J. R., Geyer, R., Wilcox, C., Siegler, T. R., Perryman, M., Andrady, A., & Law, K. L. (2015a, February 13). Plastic waste inputs from land into the ocean. *Science, 347*(6223), 768–771.

Jambeck, J. R., Geyer, R., Wilcox, C., Siegler, T. R., Perryman, M., Andrady, A., Narayan, R., & Law, K. L. (2015b Feb 13). Marine pollution. Plastic waste inputs from land into the ocean. *Science., 347*(6223), 768–771. https://doi.org/10.1126/science.1260352.35

Jamwal, N. (2015, July 04). C*itizens' reports indict Gujarat government for Surat floods*. Retrieved November 19, 2020, from DownToEarth: https://www.downtoearth.org.in/reviews/citizens-reports-indictgujarat-government-for-surat-floods-6606

Jay Chemicals. (n.d.). *Science & Sustainability*. Retrieved November 23, 2020, from Jay Chemicals: https://jaychemicals.co.in/sustainability.html

Jordan C. (2013) "Laysan Albatrosses' Plastic Problem," *Smithsonian Ocean*. [Online]. Available: https://ocean.si.edu/ocean-life/seabirds/laysan-albatrosses-plastic-problem. Date Accessed 4th August 2025.

Kamal, P., Hemangi, D., & Hitesh, D. (2014). Impact of industrialization and urbanization on water quality of Tapi Surat, Gujarat, India. *Journal of Environmental Research and Development, 9*, 306–317.

Karelia G. (2017). *Road to a plastic free City: Surat aims to produce diesel out of plastic waste*. Retrieved March 3, 2021, from https://swachhindia.ndtv.com/roadplastic-free-city-surat-aims-produce-diesel-plasticwaste-9665/

Kaza, Silpa; Yao, Lisa C.; Bhada-Tata, Perinaz; Van Woerden, Frank. 2018. What a Waste 2.0: A Global Snapshot of Solid Waste Management to 2050. Urban Development. © World Bank. Retrieved from http://hdl.handle.net/10986/30317 License: CC BY 3.0 IGO.

Keep America Beautiful. (2021). *End littering*. Keep America Beautiful (KAB). 2021. Retrieved March 2, 2021, from https://kab.org/goals/end-littering/

Kurniawan et al. (2017). *Current issues in applied biology*. February 2017. Retrieved March 1, 2021, from https://www.slideshare.net/edy0kurniawan/plastic-wastemanagement-72541520

Kwesiga, D. (2018). Strategies for managing plastic waste from construction and manufacturing projects. *Rwanda Journal of Engineering, Science, Technology and Environment, 1*, 1. https://doi.org/10.4314/rjeste.v1i1.9S

Li, W. C., Tse, H. F., & Fok, L. (2016). Plastic waste in the marine environment: A review of sources, occurrence and effects. *Science of the Total Environment, 566*, 333–349.

Luo, W., Su, L., Craig, N., Du, F., Wu, C., & Shi, H. (2018). Comparison of microplastic pollution in different water bodies from urban creeks to coastal waters. *Environmental Pollution, 246*, 174. https://doi.org/10.1016/j.envpol.2018.11.081

mangrove.org. (n.d.). *Ecological importance*. Retrieved November 11, 2020, from mangrove.org: http://mangrove.org/video/mangroves.html

Martin, C. (2019, May 13). Mangrove forests trap floating litter. Retrieved November 13, 2020, from discovery kaust: https://discovery.kaust.edu.sa/en/article/811/mangrove-forests-trap-floating-litter

Mathews, R. E., & Stretz, J. (2019). *Source-to-Sea framework for marine litter prevention: Preventing plastic leakage in river basins*. SIWI. https://siwi.org/publications/source-sea-framework-marine-litter-prevention-preventing-plastic-leakage-river-basins/

Mehta Y.B. (2019), *Activists to flag off 'save Tapi' campaign today*. Retrieved March 1, 2021 from https://timesofindia.indiatimes.com/city/surat/activists-to-flag-off-save-tapi-campaign-today/articleshow/69076572.cms

Ministry of Water Resources. Central Ground Water Board. (2015). Valsad District. Ministry of Water Resources. Retrieved from https://www.cgwb.gov.in/old_website/Annual-Reports/Annual%20Report%202015-16.pdf

MoEFCC. (2016). *Solid waste management rules*. Ministry of Environment, Forest and Climate Change. https://cpcb.nic.in/uploads/MSW/SWM_2016.pdf

Mohanty, S. (2018). *Recycling of plastics in Indian perspective*. UNIDO. Retrieved March 1, 2021, from https://www.unido.org/sites/default/files/files/2018-11/Plenary%202%20-%20Plastics%20-%20Mohanty.pdf

Moreno-Sánchez, R., & Maldonado, J. (2006). Surviving from garbage: The role of informal waste-pickers in a dynamic model of solid-waste management in developing countries. *Environment and Development Economics, 11*, 371–391. https://doi.org/10.1017/S1355770X06002853

MSME. (2013). *Brief industrial profile of Valsad District*. https://dcmsme.gov.in/dips/BIP%20VALSAD%20SEPT2912.pdf

MSME-Development Institute. MSME Development Institute. (n.d.). *Brief industrial profile of Daman District*. Ministry of MSME. https://dcmsme.gov.in/dips/DIPS%20DAMAN.pdf

MSME-DI. (n.d.). *Brief Industrial Profile of Valsad District*.

Mukherjee, A. (2017). *Clothes from plastic bottles*. NISCAIR-CSIR. Retrieved March 1, 2021, from http://nopr.nniscair.rres.iin/handle/123456789/42128

National Oceanic and Atmospheric Administration. (2019, July 24). *What is blue carbon?* Retrieved November 16, 2020, from National Ocean Service web site: https://oceanservice.noaa.gov/facts/bluecarbon.html

National Productivity Council and UN Environment Programme. (2019). *Plastic leakage pathways report*. https://www.npcindia.gov.in/NPC/Uploads/file%20upload/Plastic%20Leakage%20Pathways.pdf

National Water Development Agency. (2014). *Damanganga–Pinjal Link Project*. Government of India. Retrieved August 20, 2025, from https://nwda.gov.in/content/innerpage/damanganga-pinjal-link-project.php

Norwegian Institute for Water Research (NIVA), Mu Gamma Consultants (MGC), & Toxics Link. (2020). *Pilot study for plastic and biomedical waste management during the COVID-19 pandemic in Delhi and Surat* [PDF]. India-Norway Cooperation Project on Capacity Building for Reducing Plastic and Chemical Pollution in India (INOPOL). https://www.mugamma-consultants.com/wpcontent/uploads/2024/06/INOPOL_2020_Pilot_Study_For_Plastic_and_Biomedical_Waste_Management_during_Covid_19.pdf

Nizzetto, L., & Sinha, S. (2020). Top priority to curb plastic pollution: Empowering those at the bottom. *One Earth, 2*, 11–15. https://doi.org/10.1016/j.oneear.2020.01.005. https://www.cell.com/one-earth/pdf/S2590-3322(20)30005-1.pdf

OECD. (2018). Improving plastics management: Trends, policy responses, and the role of international co-operation and trade: Policy perspectives. In *OECD Environment Policy Paper No.12*. OECD. Retrieved March 3, 2021, from https://www.oecd.org/environment/waste/policyhighlights-improving-plastics-management.pdf

Pillay, K. (2017) *Power to the waste pickers*. 27th April 2017. Independent Online. Retrieved March 12, 2021, from https://www.iol.co.za/mercury/environment/powerto-the-waste-pickers-8838618

Plastics Europe. (2019). *Plastics—the facts*. Retrieved March 2, 2021, from https://www.plasticseurope.org/application/files/9715/7129/9584/FINAL_web_ version_Plastics_the_facts2019_14102019.pdf

Prabhune, T. (2018). *Gujarat tops in putting plastic to cleaner use*. The Times of India. Retrieved March 2, 2021, from https://timesofindia.indiatimes.com/city/ahmedabad/gujarat-tops-in-putting-plasticto-cleaner-use/articleshow/64954235.cms

Prasad, G., et al. (2023). A scientific approach to the occurrence, isolation, and characterization of existing microplastic pollution in the marine environment—A review. *Water, Air, & Soil Pollution, 234*(7), 480. https://link.springer.com/10.1007/s11270-023-06494-3

Reddy, K. (n.d.). *Tips to reduce single-use plastic*. Retrieved March 1, 2021, from https://www.kupendrareddy.com/tips-to-reduce-single-use-plastic/

Reddy, S. M., Basha, S., Subbarayappa, A., & Ramachandraiah, G. (2006). Description of the small plastics fragments in marine sediments along the Alang-Sosiya shipbreaking yard, India. *Estuarine, Coastal and Shelf Science, 68*, 656–660. https://doi.org/10.1016/j.ecss.2006.03.018

Resilient Cities Network; Ocean Conservancy; The Circulate Initiative; Centre for Environment and Education; in collaboration with Surat Municipal Corporation. (2024, June). *City waste management profile: Surat, India*. The Circulate Initiative. https://www.thecirculateinitiative.org/wp-content/uploads/City-Waste-Management-Profile-Surat-India.pdf

Shah, S. (2018) *How Gujarat succeeded in coprocessing plastic waste in cement kilns and promoting circular economy; down to earth*. 03 September 2018. Retrieved March 1, 2021, from Https://www.downtoearth.org.in/blog/environment/how-Gujarat-succeeded-in-co-processing-plasticwaste-in-cement-kilns-and-promoting-circulareconomy-61518

Sharma, N. (2019). *India's plastic waste crisis is too big, even for Modi*. Retrieved March 1, 2021 from https://qz.com/india/1693117/indias-plastic-waste-crisis-is-too-bigeven-for-modi/

Sheelanere, P., Kumar, R., Kumar, S., Julian, J., Latheef, A., Raghunathan, N., Vivekanandan, V., & Chakraborty, S. (2019). *Reducing ocean plastics-formative research leading to intervention planning in India. Research Report* (p. 70). Natural

Resources Institute, University of Greenwich. https://ikhapp.org/all-publications/reducing-ocean-plastics-formative-research-leading-to-intervention-planning-in-india/
Spalding, M. (2013, October 11). *Science: Mangrove forests as incredible carbon stores*. Retrieved November 16, 2020, from the nature conservancy web site: https://blog.nature.org/science/2013/10/11/new-science-mangroveforests-carbon-store-map/
Smith, M., Love, D. C., Rochman, C. M., & Neff, R. A. (2018). Microplastics in seafood and the implications for human health. *Current Environmental Health Reports, 5*(3), 375–386. http://link.springer.com/10.1007/s40572-018-0206-z
Sreenivas Janyala. (2006, October 28). Gujarat brings out the scrubbers. Retrieved November 23, 2020, from the Indian express: Http://archive.indianexpress.com/news/gujarat-brings-out-thescrubbers/15509/
Srinidhi, A., & Nazareth, D. (2018). *Beating plastic pollution in agriculture—World environment day special*. Retrieved March 1, 2021, from Https://thewotrblog.wordpress.com/2018/06/05/beating-plasticpollution-in-agriculture-world-environment-dayspecial/
Surat & Tapi District Profiles. (2010). Retrieved March 1, 2021, from https://www.slideshare.net/vibrantgujaratsummit/surat-tapi?next_slideshow=1
Surat District Profile. (January 2011). Retrieved March 1, 2021, from https://www.slideshare.net/ourvibrantgujarat/surat-district-profile
Surat Municipal Corporation. (2014). *Annual report*. SMC. https://www.suratmunicipal.gov.in/Content/Documents/Departments/VectorBorneDiseasesControl/VBDC_Annual_Report.pdf
Surat Municipal Corporation. (2015). *Public health bye-Laws 2015 for the Surat municipal Corporation of Gujarat State*. Surat Municipal Corporation. Retrieved November 25, 2020, from http://suratmunicipalcorporation.org:8020/News/3926.pdf
Surat Municipal Corporation. (2018). *Annual report*. Surat Municipal Corporation. Retrieved November 25, 2020, from https://www.suratmunicipal.gov.in/Content/Documents/Departments/VectorBorneDiseasesControl/VBDC_Annual_Report.pdf
Surat Smart City. (n.d.). *Surat—Smart city mission*. Retrieved November 25, 2020, from Surat Smart City Web site: https://www.suratsmartcity.com/SmartCityMission/SuratSmartCityMission
Surat Urban Development Authority. (2017). *Development Plan 2035*. Surat Urban Development Authority. https://www.sudaonline.org/wp-content/uploads/2017/03/SUDA-DP-2035-report-final-22.02.2017.pdf
Swachh Bharat Mission (Urban). (2019). *Plastic waste management: Issues, solutions and case studies, ministry of housing and urban affairs (MoHUA)*. Retrieved March 1, 2021, from http://164.100.228.143:8080/sbm/content/writereaddata/SBM%20Plastic%20Waste%20Book.pdf
The Energy and Resources Institute (TERI). (2018). *Fact sheet on plastic waste in India*. Retrieved March 1, 2021, from https://www.teriin.org/sites/default/files/files/factsheet.pdf
Times of India. (2020, August 15). *Overflowing creeks flood Surat with sewage*. Retrieved November 23, 2020, from Times of India: https://timesofindia.indiatimes.com/city/surat/overflowing-creeks-flood-surat-with-sewage/articleshow/77559252.cms
Tiseo, I. (2021). *Global plastics industry—Statistics & facts*. January 27, 2021. Retrieved March 3, 2021, from https://www.statista.com/topics/5266/plasticsindustry/#:~:text=As%20of%202018%2C%20the%20global,uses%20and%20its%20 environmental%20impact
U.T. Administration of Daman & Diu. (2019). *Action plan for Damanganga River*. Union Territory Administration of Daman & Diu. Retrieved November 20, 2020, from https://daman.nic.in/websites/Pollution-ControlCommittee/2019/Action-Plan-on-DamangangaRiver-03-04-2019.pdf.
UN Environment. (2018) *Our planet is drowning in plastic pollution*. https://www.unenvironment.org/interactive/beat-plasticpollution/
UNEP (2018) *How the Indian State of Gujarat is taking on plastic pollution*. Retrieved March 1, 2021, from https://www.unenvironment.org/news-and-stories/story/howindian-state-gujarat-taking-plastic-pollution
Valsad District. (n.d.). *Economy*. https://valsad.nic.in/economy/

Valsad District Committee. (2019). *District environmental action plan*. Valsad District. https://shorturl.at/Vlk63

Waghwala, R. K., & Agnihotri, P. G. (2019). Flood risk assessment and resilience strategies for flood risk management: Case study of Surat city. *International Journal of Disaster Risk Reduction, 40*, 101155.

Weyler, R. (2017). *The ocean plastic crisis*. October 2017. Greenpeace. Retrieved March 1, 2021, from https://www.greenpeace.org/international/story/11871/theocean-plastic-crisis/

Wilson, D. C., Velis, C., & Cheeseman, C. (2006). Informal sector recycling in waste management in developing countries. *Habitat International, 30*, 797–808.

Wood, J. (2019, February 15). *5 reasons to protect mangrove forests for the future*. Retrieved November 20, 2020, from World Economic Forum Web site: https://www.weforum.org/stories/2019/02/5-reasons-to-protect-mangrove-forests-for-the-future/

World Bank. (2018). *What a waste 2.0*. The World Bank.

World Economic Forum and PWC. (2017). *Migration and its impact on cities*. World Economic Forum and PWC.

World Economic Forum, in collaboration with PwC. (2017, October). Migration and its impact on cities. Geneva: World Economic Forum. Retrieved from https://www3.weforum.org/docs/Migration_Impact_Cities_report_2017_low.pdf

Chemicals in Plastic and Plastic Wastes

Sarath Chandra and Paromita Chakraborty

1 Introduction

Plastics are a broad class of materials renowned for their adaptability and extensive use. Based on their chemical makeup and characteristics, they may be roughly divided into numerous groups. Within the world of polymers, thermosetting plastics and thermoplastics are two separate groups. Both have distinct qualities, but their heat-related behaviours considerably set them apart. Thermosetting polymers maintain their shape and integrity even when heated to high degrees after being manufactured and cured. The reason for this is that during the curing process, their crosslinked molecular structure hardens permanently. In contrast, because they don't have a crosslinked structure, thermoplastics soften when heated and may be moulded several times. They become solid upon cooling, making them incredibly adaptable for a wide range of applications. Thermoplastics outperform thermosetting plastics in high-temperature situations (Bîrcă et al., 2019; Harper, 2006).

There are seven major types of plastics of which the commonly used ones worldwide, are polyethylene (PE) of both high and low density, polypropylene (PP), polystyrene (PS), polyvinyl chloride (PVC), polyethylene terephthalate (PET) (Young & Lovell, 2011). Plastic additives are crucial parts of the production process for plastics because they improve the functionality, aesthetics, and safety of plastic goods. These additives include plasticizers that increase flexibility and elasticity,

S. Chandra
Department of Civil Engineering, College of Engineering and Technology, SRM Institute of Science and Technology Kattankulathur, Chengalpattu, Tamil Nadu, India

P. Chakraborty (✉)
Environmental Science and Technology Laboratory, Centre for Research in Environment, Sustainability Advocacy and Climate Change, Directorate of Research, SRM Institute of Science and Technology Kattankulathur, Chengalpattu, Tamil Nadu, India
e-mail: paromitc@srmist.edu.in

P. Chakraborty et al. (eds.), *Efficient and Enhanced Management of Plastic Waste in India*, https://doi.org/10.1007/978-3-031-89398-8_2

fillers and reinforcements that increase strength and durability, colorants and pigments for aesthetic appeal, flame-retardants to lessen flammability, and antistatic agents to dissipate static charges. Stabilizers shield plastics from deterioration brought on by heat and UV radiation. Other additions, such lubricants, antioxidants, and UV absorbers, also improve the general caliber and durability of plastic materials. It is impossible to emphasize how useful these additives are for enabling plastics to suit the various requirements of sectors including construction, automotive, healthcare, and packaging (Barrick et al., 2021; Coleman, 2017; Pritchard, 2012). During several phases of the polymer processing, including the polymer synthesis process inside the reactor and the processing stage of the completed polymer pellets by mixing or blending, additives can be included into the polymer matrix. Additionally, additives may be sprayed directly onto the completed product's surface. A lot of the management of solid additives has been examined. For instance, high-performance additives (such as FRs) are often introduced at the compounding step, whereas stabilizing additives are typically introduced during the production stage of polymers. To chemically bond reactive additives to the polymeric backbone for specific applications, reactive compounding processes are used (Coleman, 2017). Studies estimate that the average mass composition of nonfiber plastics is 93% polymer resin and 7% additives. The amount of nonfiber plastics (hence defined as resins plus additives) produced since 1950 rose to 7300 million metric tons (Mt). when additives were taken into account in the computation. Again 1000 Mt. more are added by polyphthalamide (PP&A) fibers. About three-quarters of all additives are plasticizers, fillers, and flame-retardants. The three classes with the biggest output of nonfiber plastics are PE (36%), PP (21%) and PVC (12%), followed by PET, PUR, and PS (10%). Seventy percent of all PP&A fiber output is polyester, the majority of which is PET. These seven categories account for 92% of all plastics ever produced. Packaging, which is the major plastic consuming sector, has utilized around 42% of all nonfiber plastics (Geyer et al., 2017).

The variety of formulas and formulations is essentially limitless and starts with a relatively small number of fundamental plastic kinds. From food and cosmetics to the environment, these compounds have the potential to migrate from plastic packaging into a variety of items. Several chemicals used in the manufacture of plastics, notably those used in packaging, pose serious risks. Because of this, these chemicals may be ingested by people and only a small number of them have a serious negative impact on health. This movement happens during the use, disposal, and recycling phases, increasing the likelihood of significant chemical exposure in both the environment and human populations. Recycling practises may exacerbate this issue by causing a buildup of hazardous chemicals in secondary materials. Thus the objectives of this study are to elucidate:

1. The types of chemicals used in the plastic products.
2. The pathway for these chemicals to release in the riverine and marine environment.
3. The human health effects of exposure to these chemical additives.

2 Types of Chemicals Used in Plastic Products

2.1 Polymers

Polymer is a substance composed of molecules that have long sequences of one or more species of atoms or groups of atoms linked to each other by covalent bonds (Young & Lovell, 2011). Polymers' adaptability comes from their capacity to be customized for particular purposes by a variety of elements including monomer makeup, molecular weight, and processing methods. This enables a wide range of qualities, including elasticity, heat resistance, flexibility, and strength. Industries have been transformed by polymers, which have an influence on everything from packaging materials to electronics and medical equipment. Their pervasiveness in contemporary life emphasizes their crucial contribution to the development of the tools and technology we use on a daily basis. Additionally, breakthroughs are being sparked by advances in polymer science, which will eventually result in the creation of high-performance and more environmentally friendly materials. The 20 majorly used polymers along with their structures, glass transition, and polymerization temperatures are given in Table 1. Table 2 lists the most commonly used additives in each of these polymers.

2.2 Stabilizers

Stabilizers are essential additives used in the production of plastics because they protect polymers against deterioration brought on by exposure to heat, light, or oxygen. Stabilizers considerably increase the lifespan of the material by keeping the mechanical and aesthetic characteristics of the plastic throughout time. Antioxidants, UV stabilizers, and heat stabilizers are the three main categories of stabilizers that are frequently utilized. Antioxidants are crucial for battling oxidation's negative consequences. They perform the role of scavengers, removing free radicals that can cause polymer degradation. Manufacturers make sure that the plastic is durable and useful even under adverse environmental conditions by adding antioxidants. They either scatter or absorb the UV rays, avoiding photo-induced deterioration and preserving the structural integrity of the material. Applications exposed to direct sunlight, such as outdoor items and building materials, require UV stabilizers in particular. In order to protect polymers from sunlight's harmful effects, photochemical reactions in polymers and plastics must be stopped or delayed. Reduce the rate of photoinitiation or the length of the kinetic chain during the propagation stage of the photooxidation process to achieve this retardation. Visible and infrared light are rather unharmful; therefore, photostabilization of polymer items involves the protection against destructive reactions mainly caused by high-energy UV radiation (Yousif & Haddad, 2013). Thermal deterioration is combated with heat stabilizers. Heat stabilizers interact with these intermediates when the molecular structure of

Table 1 Majorly used 20 polymers along with their structures, glass transition, and polymerization temperatures

S. no.	Short form	Polymer name	Monomer	Temperature (°C)		Processing method	Application
				Glass transition	Polymerization		
1	PE	Polyethylene	$-\!\left[CH_2\right]_n\!-$	−128 to −20	–	Thin wall injection molding, blown film extrusion, extrusion, cast film extrusion, molding, rotational molding, injection molding	Medical and orthopaedic implants film and packaging are the major applications
2	HDPE	High-density polyethylene	$-\!\left[CH_2CH_2\right]_n\!-$	−133 to −118	80–100	Cast film extrusion, blow molding, extrusion, extrusion coating, blown film extrusion, rotational molding, injection molding	Distribution pipes for natural gas and oil, industrial pipes, slurry and leachate pipes, portable water pipes, wires and cable, freezer bags
3	LDPE	Low density polyethylene	$-\!\left[CH_2CH_2\right]_n\!-$	−133 to −103	132–332	Cast film extrusion, blown film extrusion, extrusion, coating, injection molding, rotational molding, molding, lamination, coextrusion	Single-use plastic bags, car covers, cling wrap, moisture barriers and liners, squeeze bottles
4	LLDPE	Linear low-density polyethylene	$CH_2{=}CH_2$; $CH_3CH_2CH{=}CH_2$; $CH_3(CH_2)_3CH{=}CH_2$	−110	180–337	Blow molding, blown film, coating, slot cast extrusion, injection molding, extrusion	Films, hoses, lids, automotive parts, bags, bottles, diaper backsheet, closures, liners, industrial container parts, display cabinets, playground equipment, toys, potable water tanks, tubing, trash cans

5	PET	Poly(ethylene terephthalate)	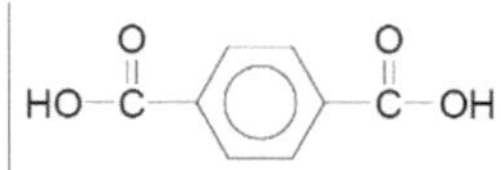	60–85	260–300	Extrusion, injection blow molding, monofilament extrusion, injection molding, blow molding	Fiber, film, medical implants, automotive parts, housewares and textiles, sporting goods, bottles, brush bristles, lighting and electrical housings, packaging, plumbing, motor parts, power tools, support brackets
6	PP	Polypropylene	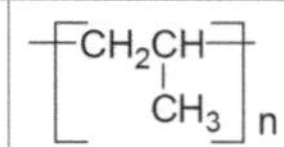	–51 to –3.2	70	Injection-stretch blow molding, blow molding, extrusion, thermoforming, injection molding	Furniture, automotive, packaging, electrical components, fibers, tapes, mechanical lungs, sutures, orthopaedic bandages
7	PS	Polystyrene	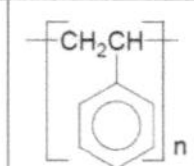	85–102	100–180	Thermoforming, electrospinning, foaming, extrusion, injection molding, blow molding	Polymer blends, household items, electrotechnical components, packaging, insulating films and foams, toys, and other applications

(continued)

Table 1 (continued)

S. no.	Short form	Polymer name	Monomer	Temperature (°C)		Processing method	Application
				Glass transition	Polymerization		
8	PC	Polycarbonate	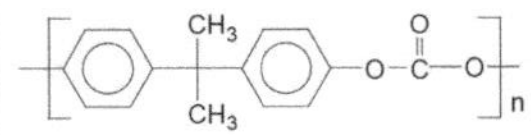	137–154	30–40 (polycondensation) 160–240 (transesterification)	Injection molding, electrospinning, extrusion, solution casting, gas-assisted injection molding, calendering, thermoforming, blow molding	Blood bags, corrective eyeglasses, dental applications, disposable syringes, medical tubing, optical lenses, pacemaker components, bearings, camera components, computer printers, xerox machine components, data storage devices, speedometer needles, head lamp covers, gears, helmets, windscreens, drinking cups, dinnerware, goggles, glazing, golf tees, guide pins, microfibers, projection screens, instrument panels, rollers, safety glasses, tool boxes, toys, water bottles, laminated walls, solar modules, roofing, skylights

9	PVC	Poly(vinyl chloride)	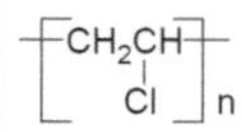	−52 to 87 (depending on plasticizer)	55–73	Thermoforming, extrusion, plastisol coating, injection molding, rotational molding, calendering, blow molding	Protective clothing, footwear, gloves, coated fabrics, bottles, cables, domestic appliances, film and sheet, fittings, drain pipes, furniture trim, rigid sheet, flooring, tubing, foam backings of carpets, gutters, pipes, windows, wallpaper, pipe and fitting, flooring, roofing, wire and cable, office equipment, packaging, toys, siding, flexible film, bottles, coating
10	ABS	Poly(acrylonitrile-co-butadiene-co-styrene)	$H_2C{=}CHC{\equiv}N$; $H_2C{=}CHCH{=}CH_2$	102–107 (acrylonitrile-styrene mesophase) and −58 (butadiene component)	62–75	Injection molding, casting, electroplating, extrusion, rotational molding, calendering, vacuum forming, thermoforming, vacuum metallization, film lamination	Lunch and tool boxes, medical applications, military use, luggage, appliance components, vacuum cleaners, power tools, automotive parts, office equipment parts and machines, hot tubs, packaging, lawn and garden equipment, pipes, furniture

(continued)

Table 1 (continued)

S. no.	Short form	Polymer name	Monomer	Temperature (°C) Glass transition	Temperature (°C) Polymerization	Processing method	Application
11	EP	Epoxy resin	Product of $C_{15}H_{16}O_2$ (bisphenol A); C_3H_5ClO (epichlorohydrin); oxirane groups can be generated from peroxidation of C=C bonds	37–127 (thermoplastic) 130–246 (adhesives) 54.5–62 (commercial)	120–160 (baking) 5–150 (adhesives)	Transfer molding, coatings, spraying, dipping, infusion molding, in situ polymerization, compounding, casting, lamination, pultrusion, sheet molding	Aircraft parts, wear resistant tools adhesives, fuel cells, biosensors, coatings, composite consumer products, electronic printed circuit boards, surface protective coatings, semiconductor packaging, flooring, tooling, and casting
12	PA-6	Polyamide-6	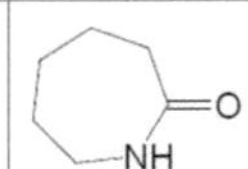	−22 to 75	190–275	Injection molding, extrusion	Weed trimmer parts, gears, washers, engine and motor parts, safety helmet parts, covers, receptacles, fasteners, clips, flanges, flexible tubing, fans, key housing, film, hand-held power tools
13	PES	Poly(ether sulfone)	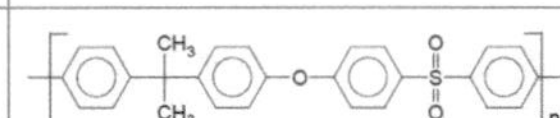	220–246	340–390	Compression molding, blow molding, film extrusion, machining, casting, profile extrusion, wire and cable extrusion spinning, injection molding, thermoforming	Dip switches, integrated sockets for circuits, printed circuit boards, transformer wire coatings, aircraft interiors, hollow fiber, automotive fuses, coatings, coil bobbins, fiber optics connectors, multipin connectors, membranes, medical applications, microwave cookware, sight glasses

14	PMA	Poly(methyl acrylate)	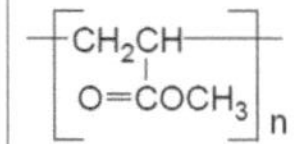	3.5–60.7	–	–	Leather finishing and textiles
15	PMMA	Poly(methyl methacrylate)	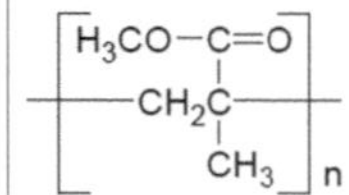	82–122	210–260	Extrusion, injection molding, compression molding, casting	Injection-molded bath sinks, artificial stones for kitchen worktops, automotive lenses for headlights and rear lights, dials, Fresnel lenses, solar collector lenses, optical components, light guides, optical fibers, composites, household items, prosthetics, displays, rods
16	PPT	Poly(propylene terephthalate)	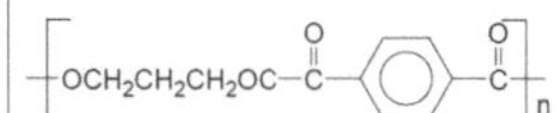	67–69	225–250	Spinning, injection molding	Fibers
17	PTFE	Polytetrafluoroethylene		40–90	400	Paste extrusion, ram extrusion, compression molding, casting, fiber spinning, coating, dip coating, flow coating, film coating, solid phase forming, isostatic molding, sintering, spraying	Gaskets, nonstick coatings, paints and coatings, geo membranes, filtration membranes, friction and wear reduction applications such as oils and greases as well as dry lubricants, electric insulation applications, aircraft insulated wires, bearings, piston rings, seals, tank lining, valve and pump parts, pipes, tubes, film, sutures, vascular grafts, laboratory equipment, printing inks, tapes

(continued)

Table 1 (continued)

S. no.	Short form	Polymer name	Monomer	Temperature (°C) Glass transition	Temperature (°C) Polymerization	Processing method	Application
18	PU	Polyurethane	Diols, triol, tetrafunctional polyols; 1-3 functional isocyanates	−60.3 to −19	150–200	Compounding, chemical or moisture cure, coatings, blow molding, electrospinning, mixing; coating, calendaring, extrusion, solution coating of fabrics, injection molding	Automotive parts, film and sheet, hose and tubes, wire and cable, coatings, sealants and adhesives, foams, mortars, primers, footwear, and other products
19	PVAc	Poly(vinyl acetate)	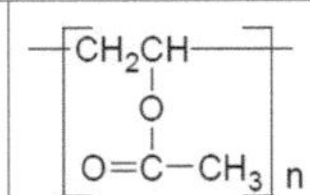	24–40	–	Mixing/ compounding	Production of poly(vinyl alcohol), adhesives, paper, paints
20	PVDC	Poly(vinylidene chloride)	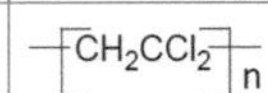	−20 to −17	60	Coating, extrusion, fiber spinning, molding	Film layer, fiber, coatings, filters, membranes, tape, shower curtains, screens, stuffed animals, monofilaments

Table 2 Commonly used additives in majorly used polymers along with their blending suitability with other polymers

S. No.	Abbreviation	Polymer Name	Stabilizers	Additives	Suitable Polymers
1	PE	Polyethylene	UVA: Benzotriazole; screener: Titanium dioxide; zinc oxide; carbon black; acid scavenger: Hydrotalcite; fiber: Carbon nanotube; HAS: Triazine and sebacates; phenolic antioxidant; Phosphite; Thiosynergist; quencher; optical brightener; vitamin E in medical applications	Sand, red mud, kaolin, talc, mica, marble, diatomaceous earth, soot, natural and synthetic silica, hollow and zirconium silicates, nickel fibers, glass fiber, glass spheres, starch, sunflower oil, mineral oil, paraffin oil, carbon black, ground tire rubber, aluminum, calcium carbonate, barium sulfate, ferromagnetic powder, calcium sulfate whiskers, ground tire rubber, hydrotalcite, lignin, magnesium hydroxide, wollastonite, superconductor, EPDM, wood flour, dioctyl phthalate, EVA, glycerin, polyethylene glycol, ethoxylated amines, copper complex of polyacrylic acid, fatty diethanol amines, ionomer, glycerol monostearate, glyceryl tribenzoate, graphite, lauric diethanolamide, quaternary ammonium compound, oleamide, ethylene bis-oleamide, trineoalkoxy zirconate, synthetic zeolite, stearyl erucamide, erucamide, siloxane spheres	–

(continued)

Table 2 (continued)

S. No.	Abbreviation	Polymer Name	Stabilizers	Additives	Suitable Polymers
2	HDPE	High-density polyethylene	UVA: Benzotriazole screener: Titanium dioxide; zinc oxide; carbon black; acid scavenger: Hydrotalcite; fiber: Carbon nanotube; HAS: Triazines and sebacates; phenolic antioxidants, Phosphite, Thiosynergist, quencher, optical brightener	Sand, red mud, kaolin, talc, mica, marble, diatomaceous earth, soot, natural and synthetic silica, hollow and zirconium silicates, nickel fibers, glass fiber, glass spheres, starch, sunflower oil, mineral oil, paraffin oil, carbon black, ground tire rubber, aluminum, calcium carbonate, barium sulfate, ferromagnetic powder, calcium sulfate whiskers, ground tire rubber, hydrotalcite, lignin, magnesium hydroxide, wollastonite, superconductor, EPDM, wood flour, dioctyl phthalate, EVA, glycerin, polyethylene glycol, ethoxylated amines, copper complex of polyacrylic acid, fatty diethanol amines, ionomer, glycerol monostearate, glyceryl tribenzoate, graphite, lauric diethanolamide, quaternary ammonium compound, oleamide, ethylene bis-oleamide, trineoalkoxy zirconate, synthetic zeolite, stearyl erucamide, erucamide, siloxane spheres	PMMA, PP, PS, PVC, SEBS, TPU, UHMWPE PA12, PET, SBS, LDPE, LLDPE, NR, EPDM, EVA, EVOH, PA6

3	LDPE	Low-density polyethylene	UVA: Benzatriazole screener: Titanium dioxide; zinc oxide; carbon black; acid scavenger: Hydrotalcite; fiber: Carbon nanotube; HAS: Triazines and sebacates; phenolic antioxidants, Phosphite, Thiosynergist, quencher, optical brightener	Sand, red mud, kaolin, talc, mica, marble, diatomaceous earth, soot, natural and synthetic silica, hollow and zirconium silicates, nickel fibers, glass fiber, glass spheres, starch, sunflower oil, mineral oil, paraffin oil, carbon black, ground tire rubber, aluminum, calcium carbonate, barium sulfate, ferromagnetic powder, calcium sulfate whiskers, ground tire rubber, hydrotalcite, lignin, magnesium hydroxide, wollastonite, superconductor, EPDM, wood flour, dioctyl phthalate, EVA, glycerin, polyethylene glycol, ethoxylated amines, copper complex of polyacrylic acid, fatty diethanol amines, ionomer, glycerol monostearate, glyceryl tribenzoate, graphite, lauric diethanolamide, quaternary ammonium compound, oleamide, ethylene bis-oleamide, trineoalkoxy zirconate, synthetic zeolite, stearyl erucamide, erucamide, siloxane spheres	EEA, EPDM, NBR, NR, PA6, PET, starch, chitosan, EVA, HDPE, PP, PS, PVDF, SBR, HIPS, LLDPE, iPP
4	LLDPE	Linear low-density polyethylene	UVA: Benzatriazole screener: Titanium dioxide; zinc oxide; carbon black; acid scavenger: Hydrotalcite; fiber: Carbon nanotube; HAS: Triazines and sebacates; phenolic antioxidants, Phosphite, Thiosynergist, quencher, optical brightener	Carbon black, lauric diethanolamide, ethoxylated amines, fatty diethanol amines, quaternary ammonium compound, glycerol monostearate, copper complex of polyacrylic acid, graphite, ionomer, polyethylene glycol, trineoalkoxy zirconate, zeolite, natural and synthetic silica, diatomaceous earth, talc, erucamides, siloxane spheres, oleamides and stabilizers	EPDM, PP, SBS, PA6, EVA, LDPE, NR, PC, PLA, PMMA, PVC, EVOH, starch

(continued)

Table 2 (continued)

S. No.	Abbreviation	Polymer Name	Stabilizers	Additives	Suitable Polymers
5	PET	Poly(ethylene terephthalate)	UVA: Benzatriazole Screener: Titanium dioxide; zinc oxide; carbon black; acid scavenger: Hydrotalcite; fiber: Carbon nanotube; HAS: Triazines and sebacates; phenolic antioxidants, Phosphite, Thiosynergist, quencher, optical brightener	Polyisonaphthalene, polyaniline, antimony doped tin oxide, multiwalled carbon nanotubes, carbon fiber, carbon black, fly ash, clays, glass spheres, glass fiber, aramid, wollastonite, natural and synthetic silica, mica, montmorillonite, talc, calcium stearate, calcium carbonate, titanium dioxide, diatomaceous earth, silicone fluid, oil and resin, pentaerythritol ester, fluorine compounds, glycerol bistearate, zinc stearate	–
6	PP	Polypropylene	UVA: Benzatriazole screener: Titanium dioxide; zinc oxide; carbon black; acid scavenger: Hydrotalcite; fiber: Carbon nanotube; HAS: Triazines and sebacates; phenolic antioxidants, Phosphite, Thiosynergist, quencher, optical brightener	Clay, diatomaceous earth, hydromagnesite-huntite, wollastonite, aluminum flakes, glass flakes, glass beads, glass fiber, stainless steel fiber, poly(alkylene terephthalate) fiber, antimony trioxide, barium sulfate, calcium sulfate, hydrotalcite, nano-calcium carbonate, calcium carbonate, bismuth carbonate, magnesium hydroxide, carbon nanotube, carbon black, fly ash, montmorillonite, metal powders (aluminum, iron, silver, nickel), mica, phenolic microspheres, potassium-magnesium aluminosilicate, red phosphorus, sepiolite, silicium carbide, talc, silica flour, wood fiber and flour, zinc borate, isooctyl tallate, erucamide, behenamide, oleamides, silicone oil, paraffinic, naphthenic, and aromatic processing oils, glycerin, polybutenes, dioctyl sebacate, alkyl-bis(2-hydroxyethyl)amine, lauric diethanol amide, crosslinked silicone spheres, N,N-bis(2-hydroxyethyl) alkoxypropylbetaine, polypyrrole, synthetic and natural silica, stearyl diethanolamine, calcium and glyceryl stearates	SEBS, PCL, PHB, EOC, EPDM, PA6, PPy, PE, PS, PANI

7	PS	Polystyrene	UVA: Benzatriazole screener: Titanium dioxide; zinc oxide; carbon black; Acid scavenger: Hydrotalcite; fiber: Carbon nanotube; HAS: Triazines and sebacates; phenolic antioxidants, Phosphite, Thiosynergist, quencher, optical brightener	Talc, kaolin, montmorillonite, mica, zinc oxide whisker, glass fibers, glass beads, barium sulfate, barium titanate, nano calcium carbonate, carbon nanotubes, copper, magnesium hydroxide, zinc stearate, glycerol monostearate, aluminum distearate, polyether esteramide, ethylene bis stearamide, PTFE, red phosphorus, silica, titanium dioxide, zeolites; zinc borate, adipates, glutarates, phthalates, di-and tri-isopropylbiphenyls, dioctyl sebacate, liquid paraffins, mineral oil, polybutenes, tricresyl phosphate, calcium montanate, butylene glycol montanate, cocobis(2-hydroxyethyl)amine, conductive carbon black, N,N-bis(2-hydroxyethyl)-N-(3-dodecyloxy-2-hydroxypropyl) methylammonium methosulfate, quaternary ammonium compounds, polyaniline, blowing agents, PDMS	PEDOT, PA6, LDPE, POSS, PANI, PDMS, PMMA, PP, PVC, PVME, PE
8	PC	Polycarbonate	UVA: Benzatriazole screener: Titanium dioxide; zinc oxide; carbon black; acid scavenger: Hydrotalcite; fiber: Carbon nanotube; HAS: Triazines and sebacates; phenolic antioxidants, Phosphite, Thiosynergist, quencher, optical brightener	Nanosilica, single-walled carbon nanotubes, glass beads, carbon black, graphite, glass and steel fibers, nickel-coated carbon fiber, siloxane particles, amorphous silica, tri-(2-ethylhexyl) phosphate, tritolyl phosphate, molybdenum sulfide, indium tin oxide, boric oxide, copper oxide, titanium dioxide, calcium carbonate, pentaerythritol tetraborate, wollastonite, zinc stearate, pentaerythritol tetrastearate, glycerol monostearate, glycerol mono-iso-stearate, phthalates, mineral oil, trimellitic acid tridecyloctyl ester, tetraethylene glycol dimethyl ether, polyetheresteramide, polyoxyethylene fatty acid ester, dimethylsiloxane grafting, siloxane, impact modifiers, release agents, UV stabilizers, PTFE	PET, PBT, PVDF, PHEMA, ABS, PLA

(continued)

Table 2 (continued)

S. No.	Abbreviation	Polymer Name	Stabilizers	Additives	Suitable Polymers
9	PVC	Poly(vinyl chloride)	UVA: Benzatriazole screener: Titanium dioxide; zinc oxide; carbon black; acid scavenger: Hydrotalcite; fiber: Carbon nanotube; HAS: Triazines and sebacates	Clay, montmorillonite, sand, silica, talc, graphite, carbon fiber, wood fiber, aluminum fiber, copper powder, carbon black, titanium dioxide, antimony trioxide, aluminum hydroxide, magnesium hydroxide, calcium carbonate, benzoates, azelates, citrates, phosphates, adipates, sebacates, phthalates, zinc or calcium stearate, propanesultone, epoxidized soybean oil, ester wax, aluminosilicate, synthetic and natural silica, polyester-type polymeric plasticizers, ethylene interpolymers, glyceryl monostearate, ethylene N,N′-bisoleamide, stearamide, ethylene N,N′-bisstearamide, chlorinated polyethylene, polyethylene glycol monolaurate, ethoxylated fatty dimethyl ethylammoniumethosulfate,	PVDF, PMMA, ENR, epoxy, EVA, NR, PANI, PUR, PS, PVA, PVB, SAN, SBR, NBR
10	ABS	Poly(acrylonitrile-co-butadiene-co-styrene)	UVA: Benzatriazole screener: Titanium dioxide; zinc oxide; carbon black; acid scavenger: Hydrotalcite; fiber: Carbon nanotube; HAS: Triazines and sebacates; phenolic antioxidants, Phosphite,	Carbon black, fluorocarbon, glass beads, talc, paraffin wax, bis-stearamide wax, magnesium hydroxide, antimony oxide, sodium alkyl sufonate, nickel or copper coated carbon fibers, phosphate esters, hydrocarbon processing oil, long chain fatty acid esters, aromatic sulfonamide, ethanol,2,2′-iminobis-, Ncoco alkyl derivatives, glycerol monostearate, polyaniline, polyesteramide, cetyl palmitate, methyl behenate,	PC, PVC, PANI-EB, chitosan, ground rubber, EPDM, SAN, PA6, PLA, PTT

11	EP	Epoxy resin	UVA: 2,4-dihydroxybenzophenone; 2-(2H-benzotriazol-2-yl)-p-cresol; 2-benzotriazol-2-yl-4,6-di tert-butylphenol; screener: Nano-ZnO; nano-silica-titania	Condensation products of adipic acid and glycols, isodecyl pelargonate, cyclohexyl pyrrolidone, epoxidized oils, low molecular polyamides, polysulfides dibutyl phthalate, flame-retardants, modifiers, diluents (glycide ether), rheological additives, alkyl dipolyoxyethylene ethyl ammonium ethyl sulfate, tin oxide, silver-coated basalt, carbon black, graphite, carbon monofiber, quaternary ammonium compound, calcium carbonate, ethylene bis stearoformamide, carnauba wax, montan wax, ceramic microspheres, silicone oil, sorbitan tristearate, carbon fiber, PTFE	PSU, PEO, PMMA, PA, PCL, PP, PVP, PBT, PC
12	PA-6	Polyamide-6	–	–	PVDF, PC, NR, ABS, chitosan, HDPE, PA66, PBT, PP, PPO, PS, LDPE
13	PES	Poly(ether sulfone)	–	Carbon fiber, graphite, glass fiber	PEO, PBI, PC, PEEK, epoxy
14	PMA	Poly(methyl acrylate)	–	–	PS, PVF, PEO, PLA, PMMA
15	PMMA	Polymethylmethacrylate	UVA: Benzatriazole HAS: Triazines and sebacates screener: Zinc oxide phenolic antioxidant optical brightener	Silica, aluminum, nickel, mica, glass fiber, titanium fiber, barium sulfate, titanium dioxide, aluminum hydroxide, montmorillonite, Ni-BaTiO3, di-(2-ethylhexyl) phthalate, 4-cyanophenyl 4 heptylbenzoate, 2-hydroxyethyl methacrylate, polyaniline, polypyrrole, copper dimethacrylate, indium tin oxide, glycerol monolaureate, lauramide diethanolamide, crosslinked siloxane particles, stearic acid, magnesium stearate, silicon nitride, methylpolysilsiquioxane, erucamide	PVDF, PVC, PEO, PP, PVAc, PPy

(continued)

Table 2 (continued)

S. No.	Abbreviation	Polymer Name	Stabilizers	Additives	Suitable Polymers
16	PTFE	Polytetrafluoroethylene	–	Alumina, silica, graphite, attapulgite diamond, boron nitride, titanium dioxide, bronze powder, carbon black, carbon fiber, glass fiber, carbon nanofiber, copper powder, molybdenum sulfide, Ni-Zn ferrite	PEEK, PI, PPS, FEP, PA6, PA66, POM
17	PU	Polyurethane	UVA: Benzatriazole HAS: Triazines and sebacates screener: Carbon black Phosphite, Thiosynergist, optical brightener	Aluminum hydroxide, aluminum nitride, calcium carbonate, calcium sulfate, fluoromica, bentonites, graphite, kaolin, mica, sand, sepiolite, silica, montmorillonite, nanosilica, organic fibers, rubber particles, benzoates, adipates, azelates, sebacates, citrates, epoxidized soybean oil, phosphates, phthalates, graphite, carbon black, carbon nanotubes, vanadium pentoxide, polyaniline, polyethylene glycol, polypyrrole, glycerol monostearate, silver-coated basalt particles, sulfonated polyol, tetraorganoboron, diatomaceous earth, natural silica, starch, glass or ceramic spheres, sodium myristate, sodium oleate, silicone fluid, zinc stearate, crosslinked silicone, lecithin, fluorocarbon, isononylphenyl isocyanate, metal soap, montan ester wax, silicone oil, ethylene bisstearamide	SBS, CA, PEG, PCL, PDMS, CHI
18	PVAc	Poly(vinyl acetate)	–	Mica, talc, clay, aluminosilicate, calcium carbonate, castor oil, epoxidized soybean oil, acetyl triethyl citrate, tributyl citrate, triethyl citrate, benzyl butyl phthalate, dibutyl phthalate, dipropylene glycol ethylene and propylene glycols, dibenzoate, glycerin, triacetin, triethanolamine, quaternary ammonium salt, zinc stearate	PVOH, PVF, NR, PCL, PBM, PVC, PE, PEO, PHB, PLA, PMMA, PPy, polyacrylate

19	PVDC	Poly(vinylidene chloride)	–	Polymeric plasticizer of adipic acid and propylene glycol, polymeric condensation product of azelic acid and 1,3-butanediol, acetyl tri-n-butyl citrate, epoxidized soybean oil, imidazoline/metal salt, ionic polymer	PBA

ABS acrylonitrile butadiene styrene, *CA* cellulose acetate, *CHI* chitosan, *EEA* ethylene-ethyl acrylate, *ENR* epoxidized natural rubber, *EPDM* ethylene propylene diene monomer, *EOC* ethylene-octene copolymer, *EVA* ethylene-vinyl acetate, *EVOH* ethylene-vinyl alcohol, *FEP* fluorinated ethylene propylene, *HDPE* high-density polyethylene, *HIPS* high impact polystyrene, *iPP* isotactic polypropylene, *LDPE* low-density polyethylene, *LLDPE* linear low-density polyethylene, *NBR* nitrile butadiene rubber, *NR* natural rubber, *PA* polyamide, *PA6* polyamide 6 (nylon 6), *PA12* polyamide 12 (nylon 12), *PA66* polyamide 66 (nylon 66), *PANI* polyaniline, *PANI-EB* polyaniline-emeraldine base, *PBM* poly(butylene malonate), *PBT* polybutylene terephthalate, *PC* polycarbonate, *PCL* polycaprolactone, *PEEK* polyether ether ketone, *PEG* polyethylene glycol, *PE* polyethylene, *PEDOT* poly(3,4-ethylenedioxythiophene), *PEO* polyethylene oxide, *PFA* perfluoroalkoxy alkane, *PHA* polyhydroxyalkanoate, *PHB* polyhydroxybutyrate, *PHEMA* poly(2-hydroxyethyl methacrylate), *PI* polyimide, *PLA* polylactic acid, *PMMA* polymethyl methacrylate, *POM* polyoxymethylene (acetal), *POSS* polyhedral oligomeric silsesquioxane, *PPO* polyphenylene oxide, *PP* polypropylene, *PPy* polypyrrole, *PS* polystyrene, *PSU* polysulfone, *PTFE* polytetrafluoroethylene, *PTT* polytrimethylene terephthalate, *PVA* polyvinyl alcohol, *PVAc* polyvinyl acetate, *PVB* polyvinyl butyral, *PVC* polyvinyl chloride, *PVDF* polyvinylidene fluoride, *PVF* polyvinyl fluoride, *PVME* polyvinyl methyl ether, *PVOH* poly(vinyl alcohol), *PVP* polyvinylpyrrolidone, *SAN* styrene acrylonitrile, *SBS* styrene–butadiene–styrene, *SEBS* styrene–ethylene–butylene–styrene, *SBR* styrene–butadiene rubber, *PBA* poly(butyl acrylate) the polymer becomes thermally unstable, stopping further breakdown. Manufacturers improve the plastic's resilience to high temperatures by adding heat stabilizers, making it ideal for applications where heat exposure is a problem. Plastics may be specially formulated by manufacturers to address certain environmental issues and increase its service life in a variety of applications by adding antioxidants, UV stabilizers, and heat stabilizers.

2.3 *Plasticizers*

Plasticizers are essential elements in the production of plastic, improving flexibility and processability. These additions make the polymer softer and more malleable by lowering its glass transition temperature (Tg). Plasticizers such asphthalates, adipates, and citrates are some of the most popular ones used in the production of plastic. A plasticizer, as defined by the Council of the International Union of Pure and Applied Chemistry (IUPAC), refers to a chemical or substance incorporated into a material, frequently a plastic or elastomer, with the purpose of enhancing its flexibility, ease of manipulation, or expandability. In line with this explanation, a plasticizer can potentially reduce the material's elastic modulus, the temperature at which a second-order transition takes place, or the viscosity of the molten state (Marturano et al., 2017).

The classification of plasticizers into primary and secondary categories, as per gel theory, is contingent upon the quality of the interaction between the polymer and plasticizer, specifically concerning the leaching or migration of plasticizer molecules from the polymer matrix. Primary plasticizers exhibit solubility characteristics closely resembling those of the polymer, enabling them to readily form a gel with the polymer within the processing temperature range. They seldom ever "bleed" from the plasticized substance in this situation. However, secondary plasticizers are less compatible with the polymer and have a reduced gelation capability (Marturano et al., 2017).

Plasticizers are primarily categorized based on their chemical composition into two classes: phthalates and non-phthalates. Among the commonly employed phthalate plasticizers for PVC, two notable examples are diethyl hexyl phthalate (DEHP) and diisononyl phthalate (DINP).

The main distinction between the two groups of plasticizers—phthalates and non-phthalates—is made based on their chemical makeup. Diethyl hexyl phthalate (DEHP) and diisononyl phthalate are two of the most popular phthalate plasticizers used for PVC.

Because they work well in increasing flexibility and lowering brittleness, phthalates are frequently employed as organic plasticizers. Aside from that, phthalates increase the plastic's resilience to heat and chemicals, making it appropriate for a variety of uses, such as flexible PVC goods, coatings, and adhesives.

Adipates are another type that helps to make plastic flexible and less brittle. Similar to phthalates, adipates improve the plastic's resilience to heat and chemicals, making them ideal for applications that call for increased durability. Citrates are highly valued as organic plasticizers due to their ability to enhance flexibility. Consequently, plasticizers play a pivotal role in molding the physical characteristics of polymers, rendering them more adaptable and simplifying their processing. The careful selection of plasticizers allows manufacturers to tailor the plastic to specific requirements, making it an indispensable component in various industries.

2.4 Fillers

Fillers are essential additives in plastic production, exerting a significant impact on the mechanical properties and manufacturing expenses of polymers. They are introduced into the polymer matrix in varying proportions to enhance specific attributes. Talc, calcium carbonate, and glass fiber are a few popular fillers. Natural mineral talc is frequently used as a filler to increase stiffness and minimize shrinking during moulding. Additionally, its integration increases the plastic's resistance to chemicals and heat, making it appropriate for uses where dimensional stability and thermal resistance are essential. A common inorganic filler used in the production of plastic is calcium carbonate. The polymer's rigidity, impact resistance, and heat resistance are improved by adding calcium carbonate. In addition to these advantages, calcium carbonate gives producers a financial advantage since it lowers production costs. Another prominent class of inorganic fillers used in the production of plastic is glass fibers. Their incorporation into the polymer matrix significantly enhances the material's stiffness, strength, and impact resistance. Glass fibers are a common material option in the automotive and aerospace sectors because they help create lighter weight yet mechanically strong materials. A few examples of fillers that are crucial for creating high-performing polymers and satisfying specific application criteria are talc, calcium carbonate, and glass fibers.

2.5 Pigments

Pigments are essential additives used in the production of plastics. They provide polymers color, opacity, and improved weatherability. The pigments titanium dioxide, iron oxide, and carbon black are frequently employed. A popular inorganic pigment called titanium dioxide gives polymers outstanding whiteness and opacity. Titanium dioxide improves the plastic's weatherability by diffusing and reflecting light, shielding it from UV rays, and delaying early deterioration. Due of this quality, it is frequently used in outdoor applications and items that are exposed to sunlight. Iron oxide pigments provide plastics a variety of hues, such as red, yellow, brown, and black. Due to their remarkable resistance to heat and light, these additives find utility across various applications, such as construction materials, automotive components, and consumer goods. A frequently employed organic pigment is carbon black, which plays a vital role in enhancing the visual appeal and durability of plastic products. Titanium dioxide, iron oxide, and carbon black are just a few examples of pigments that offer manufacturers a palette of colors and improved UV resistance to cater to diverse market demands.

2.6 Flame-Retardants

The incorporation of flame-retardants (FRs) is crucial in plastic manufacturing as it reduces flammability and enhances fire safety. Halogenated chemicals, phosphorus-based compounds, and inorganic materials are all frequently used flame-retardants. The market now offers about 175 distinct FRs, which may be divided into three main chemical categories: halogenated, organophosphorous, and inorganic halogenated chemicals. They are frequently employed as organic FRs, and when exposed to fire, they release halogenated gases that thin the combustible gases around the polymer. Additionally, they produce a shielding layer of char that serves as a barrier to contain the fire.

Brominated flame-retardants (BFRs) are the most widely utilized FRs due to their superior efficiency, cost-effectiveness, and versatility across a wide array of applications. BFRs can exhibit diverse chemical structures, spanning from fully halogenated aliphatic compounds to aromatic carbon substrates with all hydrogen atoms replaced by bromine atoms. Nonetheless, from an environmental perspective, BFRs display exceptional resistance to chemical, biological, and photolytic degradation. They may also migrate over vast distances and bioaccumulate in living tissues (De Wit, 2002; Marturano et al., 2017). Another crucial group of products are FRs made of phosphorus. Phosphorous flame-retardants (PFRs) are a subclass of FRs that contain phosphorus in their structure and operate by generating phosphoric acid or phosphorous gases when exposed to fire. PFRs can have different phosphorus oxidation states (0, +3, +5). These substances efficiently prevent the ignition and spread of fires by forming a protective char layer. PFRs, unlike BFRs, can also function in the condensed phase by promoting the development of char, producing intumescence (foaming up), or creating inorganic glasses (Le Bras et al., 1998; Schartel, 2010). To make polymers flame resistant, inorganic flame-retardants such as metal oxides are used. In order to cool the polymer and dilute flammable gases, they function by releasing water or other gases. This essentially stops the combustion process. Inorganic flame retardants typically include aluminum hydroxide ($Al(OH)_3$) and magnesium hydroxide ($Mg_2(OH)_4$). The mechanisms through which these chemicals function differ significantly from those of organic FRs. In reality, the heat from combustion prevents IFRs from evaporating; nonetheless, endothermic processes can cause them to break down into nonflammable gases, mostly water (Horn, 2000; Marturano et al., 2017). Hence, FRs are critical additives for ensuring the fire safety of plastic products. Halogenated compounds, phosphorus-based compounds, and metal oxides offer a range of flame-resistant properties, allowing manufacturers to create materials suitable for applications that demand high fire safety standards.

2.7 *Impact Modifiers*

Impact modifiers frequently possess a rubbery or elastic nature, exhibiting a lower modulus compared to the polymer matrix, along with a low glass transition temperature (Tg). Furthermore, the rubbery phase must exhibit a small particle size distribution and strong compatibility with the host polymer.

Butadiene-based graft copolymers: On their own, butadiene-based graft copolymers make up 45% of the worldwide market for impact modifiers, with their remarkable success primarily attributed to their exceptionally low glass transition temperature (Tg) of 80 °C.

Acrylonitrile-butadiene-styrene terpolymers, (ABS) is the most widely favored graft terpolymer, gains its unique characteristics from the interplay of its three components: butadiene imparts a soft rubbery quality, while acrylonitrile and styrene supply the terpolymer with the necessary polarity for compatibility with the polymer matrix. As mentioned earlier, protective measures are necessary to shield the butadiene chain from UV degradation, while acrylonitrile contributes to stiffness and chemical resistance. ABS demonstrates excellent reliability within the temperature range of 20 to 80 °C. Moreover, it is exceptionally lightweight, easily amenable to extrusion and injection molding, and amenable to 3D printing processes (Rutkowski & Levin, 1986).

Methacrylate-butadiene-styrene (MBS) impact modifiers are manufactured through two distinct processes: one involves copolymerizing styrene and methyl methacrylate in the presence of polybutadiene, while the other entails polymerizing styrene and butadiene rubber to form impact modifiers. However, it's worth noting that the presence of butadiene makes these materials susceptible to UV degradation, limiting their use to indoor applications. Methacrylate, which replaces acrylonitrile in the blend, imparts a special transparency to the polymer but also reduces its chemical resistance. MBS modifiers find their primary application in PVC for both transparent and opaque packaging purposes. These applications encompass electrical covers, packaging films, and impact-resistant bottles (Tseng & Lee, 2000).

2.8 *Acrylic Modifiers*

For PVC formulations intended for outdoor use, special consideration is essential to ensure long-lasting structural and aesthetic qualities. Acrylic modifiers, in this context, emerge as the preferred choice for impact modification, as they effectively address the challenges related to the limited weather resistance often associated with ABS and MBS (Marturano et al., 2017).

2.9 Elastomers

To enhance the impact strength of polyolefins in homopolymer formulations is achieved by incorporating various elastomers, such as ethylene-propylene copolymer (EPM) or ethylene-propylene diene terpolymer (EPDM) (Karger-Kocsis et al., 1979).

2.10 Antioxidants

By using the right stabilizing chemicals, known as antioxidants (AOs), in the polymer formulation, oxidative breakdown in polymers can be controlled by AOs. Chain-breaking antioxidants are another name for primary antioxidants. These antioxidant compounds employ a process called the "chain-breaking electron donor mechanism" to neutralize free radicals. In general, antioxidants can be categorized into two main groups: radical scavengers (chain-breaking acceptors) and hydrogen donors (chain-branching donors) (Rabek, 2012). To prevent the degradation of the polymer over prolonged period or throughout the lifespan of the plastic product, a "primary antioxidant" is incorporated. Sterically hindered phenols represent the earliest and arguably the most vital category of these primary AOs. Their primary function involves transferring hydrogen from the phenolic OH-group to peroxy radicals, often generated by heat, leading to the formation of phenoxyl radicals. This process results in the production of hydroperoxides. Secondary AOs, also referred to as hydroperoxide decomposers, possess the capability to break down hydroperoxides (ROOH) into harmless secondary substances (Rabek, 2012).

2.11 Antimicrobials

Antimicrobial (AM) compounds find their primary applications in the medical field and food packaging industry. Consequently, the key requirement for AM additives is to ensure minimal toxicity to both humans and the environment during production and use. Moreover, ideal AM additives should seamlessly integrate with processing aids and other additives employed in plastic formulations, without affecting the properties or visual aesthetics of the final product (Nichols, 2004).

3 Pathways for Release of Plastic Chemicals and Additives from Plastic Products in the Environment

3.1 Factors Driving the Release of Plastic Chemicals into the Environment

The release of plastic chemicals and additives from plastics in the environment can occur through various pathways, primarily influenced by the surrounding conditions and the type of plastic used. Some of the main factors driving the release of plastic chemicals are as follows:

3.1.1 Direct Contact with Water and Food

Plastics that come into direct contact with water, such as bottles, containers, and fishing equipment, can undergo a phenomenon known as "aqueous leaching," which may result in the release of chemicals and additives into the water. Water bodies may get contaminated as a result of these compounds dissolving in water and moving away from the plastic surface. When in close contact with food, plastics used for food preparation, storage, or packaging can leach chemicals and additives into the meal. Migratory leaching is the mechanism that causes these chemicals to be ingested by both people and animals. Water distribution plastic pipes experience polymer chain rupture and reduced antioxidant contents often caused by long-term exposure to disinfectants in the supply water (He et al., 2023). Their presence leads to vinyl, hydroxyl, and/or carbonyl group formation as well as increase in plastic structure crystallinity (Khan et al., 2022). pH, temperature, free chlorine exposure time, pressure, and plastic constituent properties control morphological and mechanical and surface characteristics of plastic pipes (Yu et al., 2011). This inevitably leads to the formation of micro-and nanoplastics which also get delivered along with the supply water (Hammodat et al., 2023).

3.1.2 Weathering and Photodegradation

Plastics may degrade physically and chemically when they are exposed to sunlight and environmental elements including heat, oxygen, and moisture. The "weathering" process has the potential to leak additives and breakdown products into the environment. The highest temperatures to which a plastic material may be exposed outdoors (170 °F) do not supply sufficient thermal energy (70–90 kcal/mole) to cause bond cleavage in commercial polymers. Thus, although temperature is important as a rate controlling parameter in many of the degradative processes (e.g. oxidation and hydrolysis), polymer degradation by pure thermal energy is not a critical factor in weatherability studies. Plastic systems containing ingredients or impurities which absorb light in this wavelength range are susceptible to photodegradation by sunlight (Kämpf et al., 1991). Carbonyl groups are particularly vulnerable since they absorb lights at a wavelength of 280 nm. Chain scission (dissociation) and crosslinking (photodimerization) are two of the important photodegradation processes. Direct photolysis occurs when a compound absorbs sunlight and the absorption process triggers transformation by, for instance, ionization, bond breaking, and excited-state reactivity (Fabbri et al., 2023). Chain cleavage during the early stages of oxidation permits the smaller chains to crystallize, usually accompanied by a rapid increase in density. In the second stage, density rises less rapidly as the chemical reaction levels off in regions still accessible to oxygen. Photo-oxidation rates depend to a significant level on ambient temperature. The rate of oxidation is controlled by the rate of diffusion of oxygen into the polymer and any compromise to the plastic surface in terms of scratches and cracks can lead to acceleration of oxidative degradation. The crystalline structure caused by such degradation processes is the reason for plastic to become brittle and lose its strength, leading to microplastic and nanoplastic production. Hydrolytic degradation usually leads to the deterioration of tensile properties by chain scission. While chain scission is often seen to lead to a decrease in density of the plastic material, it was found to increase with the degree of hydrolysis, indicating that other factors such as solubility and concentration of water in the polymer phase also play a role.

3.1.3 Types of Leaching Processes

It is significant to note that the kind of plastic, the temperature, the pH, and the length of exposure all affect how much leaching occurs. There are several regulatory measures in effect to limit the utilization of certain chemicals recognized as harmful to human health and the environment. Certain polymers and additives exhibit a higher tendency to leach than others. To mitigate the release of plastic chemicals and additives into the environment, it is crucial to employ proper waste management practices, promote recycling, and explore alternative materials. In most cases, additives are not chemically bonded to the polymer matrix, allowing them to potentially seep out of the material. The leaching process, when in contact with aqueous substances, can be elucidated through three primary mechanisms: diffusion within the polymer matrix, partitioning at the solid–liquid interface, and dispersion into the bulk medium (Lau and Wong, 2000). Diffusion within the polymer is typically governed by Fick's rules of diffusion, which depend on the concentration gradient and the diffusion coefficient specific to the additive-polymer system. The molecular weight, size, and interaction of organic additives with the polymer matrix collectively influence this diffusion coefficient. Furthermore, the solubility of the additive in the surrounding medium determines its partitioning behavior from the solid phase to the liquid phase. Solubility and dilution are often used to regulate how much dispersion enters the bulk medium (Arvanitoyannis & Kotsanopoulos, 2014; Reynier et al., 2001). In practical leaching tests, it is common to observe a two-phase release pattern. Initially, there is a relatively rapid release of the labile surface fraction, which is followed by an extended period characterized by slower, diffusion-controlled release (Bandow et al., 2017; Beldi et al., 2012; Rani et al., 2017; Suhrhoff & Scholz-Böttcher, 2016; Sun et al., 2019; Turner & Lau, 2016). Leaching is typically undesirable and can impact the functionality of a material. Plasticizers from water bottles and plastic water pouches might migrate due to differences in packaging material quality and storage conditions (Mukhopadhyay et al., 2022). Furthermore, when it comes to materials utilized in food packaging, pharmaceuticals, and medical devices, such migration can lead to the contamination of consumer products (Cuadros-Rodríguez et al., 2020; Giuliani et al., 2020). Regarding their safety for public health, a variety of additives have come under examination in several fields (Groh et al., 2019; Ong et al., 2020).

Leaching experiments have been carried out on a range of polymers with diverse physicochemical properties to gain deeper insights into potential exposure to leachables stemming from plastic pollution. Among the frequently studied thermoplastics are PE, PVC, and PS, aligning with their prevalence in the environment, high production volumes, and perceived impact (Andrady, 2017; Lithner et al., 2009). Nevertheless, there is limited research available on the leaching of additives from thermoset polymers. Yet, various other physicochemical attributes of the plastic, including molecular weight, crystallinity, and glass transition temperature, can significantly influence the extraction and leaching processes (Poças et al., 2008). It has been demonstrated that the rate of additive release increases as the size of plastic particles decreases (Chen et al., 2019; Liu et al., 2020; Sun et al., 2019; Ye et al., 2020) Phthalates are gradually emitted from products into the air or other mediums since they are not chemically bound to the polymer matrix. This characteristic makes phthalates highly prevalent and one of the most commonly found semivolatile organic compounds (SVOCs) in indoor environments (Clausen et al., 2012). In a passive flux sampler, Fujii and colleagues investigated the temperature-dependent emissions of various phthalate esters. They observed a remarkable increase in the highest emission rate, with a 100-fold rise when the temperature increased from 20 to 80 °C (Fujii et al., 2003).

Ekelund and coauthors conducted a study to examine the aging process of PVC plasticized with DEHP by measuring weight loss over time at temperatures spanning from 60 to 155 °C. Their findings indicated that emissions below 100 °C were primarily influenced by evaporation, specifically diffusion within the boundary layer. Notably, the activation energy for DEHP evaporation below 100 °C was measured at 98 kJ/mol, which matched the activation energy for the evaporation of pure liquid DEHP. It's worth mentioning that the trials conducted in vented ovens did not control the air exchange rate, a known factor in the release of DEHP from PVC (Clausen et al., 2010; Ekelund et al., 2010) Desorption of the molecules from the polymer surface can cause chemical additives included in plastics to migrate to people (Li et al., 2018a).

3.2 *Wastewater Discharge and Effluents*

Wastewater contamination can be released into the environment by businesses that employ plastic additives in their operations. Plastic additives and chemicals may also be found in small amounts in wastewater treatment plant effluents. Plasticizers and plastic monomers are ubiquitously found in all environmental waters and wastewater ranging in concentration from ng/L to µg/L (Kumar, 2023). These chemicals also undergo biological or nonbiological transformations, and these transformed products are also frequently detected in wastewater. Municipal wastewater treatment plants (WWTPs) are important point sources of NPs in the environment (Carr et al., 2016). A significant portion of these particles in surface runoff and industrial and domestic effluents are removed at WWTPs. These particles end up accumulating in sewage sludge and in its processed by-products called biosolids (Crossman et al., 2020). In developed countries, efficiencies of removal of microplastics and PBTs of $\geq$90% are commonly quoted for effluents, but wastewaters of developing countries only receive primary or sometimes little to no treatment (Hale et al., 2022).

3.3 *Open Burning, Informal recycling of waste plastics-Associated Atmospheric Emissions*

Globally, plastic production was estimated to be 380 million tons in 2018. Since 1950 to 2018, plastics of about 6.3 billion tonnes have been produced worldwide, 9% and 12% of which have been recycled and incinerated, respectively (Alabi et al., 2019). In low and middle income countries and developing economies, open burning of dumped plastic waste is a frequently practiced for reducing its volume (Chakraborty et al., 2022). Burning plastics can cause toxic chemicals, such as additives, to be released into the environment as particulate matter and volatile and semi-volatile organic compounds (VOCs & SVOCs). These pollutants may be harmful to human health and increase air pollution. Both primary and secondary emissions. In the context of biomedical waste treatment, incineration is a widely employed method for disposal, especially for plastics and other potentially hazardous materials. Plastic additives play a crucial role in this process. Various types of plastics, including PVC, PP, and other thermoplastics, are laden with additives like stabilizers, plasticizers, and flame-retardants. Persistent organic pollutants (POPs) are also frequently found to occur interlinked with waste plastics (Chakraborty et al., 2022). Plastic wastes laden with various additives, undergo incomplete combustion either via open burning or improper incineration and it can lead to the release of this range of potentially harmful chemicals into the atmosphere. These emissions encompass a diverse set of compounds, some of which can have adverse effects on both the environment and human health.

Plastic additives like plasticizers and flame-retardants can be released during open burning of dumped waste along with volatile organic compounds (VOCs).

VOCs are precursors to ground-level ozone and can contribute to smog formation. Prolonged exposure to VOCs can lead to respiratory problems, eye irritation, and, in some cases, adverse neurological effects. Many plastic additives, such as lead stabilizers and cadmium-based pigments, can contain heavy metals. During incineration, these metals can be released into the atmosphere in the form of fine particulate matter. Inhalation or ingestion of these particles can lead to a range of health problems, including neurological disorders, kidney damage, and developmental issues in children.

Fine particulate matter (PM) is generated during the combustion of plastics, including both primary particles and secondary aerosols formed through chemical reactions in the atmosphere. Inhalation of PM can lead to respiratory and cardiovascular problems, and fine particles can also carry absorbed chemical pollutants, increasing their potential to cause harm. While plastics themselves do not emit significant amounts of green house gases (GHGs) during incineration, the process does release carbon dioxide (CO_2) and potentially other GHGs if the incineration is not conducted under controlled conditions. This contributes to overall greenhouse gas emissions and climate change. Addressing the atmospheric emissions associated with plastic incineration necessitates the implementation of advanced incineration technologies, strict emission control measures, and effective waste management practices. Additionally, efforts to reduce plastic consumption, promote recycling, and develop more environmentally friendly materials are crucial steps towards mitigating the environmental impact of plastic-associated chemicals.

3.3.1 Open Burning of Waste Plastic

In plastics, FRs and plasticizers are not generally chemically bonded to the polymers but occupy the space between. They inhibit combustion and therefore when the host polymer is burned, they are released as gas, as airborne particulates, and in the residual ash. Several studies have investigated plasticizers such as phthalates transmission from waste incinerator plants, finding that they have the potential to be emitted intact from facilities without adequate air pollution control and management. Their low molecular weight and tendency for noncovalent bonding to polymers means that some formulations are very sensitive to changes in temperature and pH and readily escape from their host products into the environment, where they have potential for long-range transport. Several examples also exist to indicate toxic compounds in plastic through unintentional contamination, such as during the reprocessing of e-waste or end-of-life vehicles. Highly toxic compounds are formed during the incomplete combustion of plastics, especially those containing chlorine (Zhang et al., 2011). Black carbon (BC) is a particularly serious air pollutant emitted from the uncontrolled burning of waste in open fires because it has a global warming potential up to 5000 times greater than CO_2 and is also linked to detrimental health impacts (Reyna-Bensusan et al., 2019). Plastics in particular when they are burnt in an uncontrolled manner openly, can result in dioxin-like compounds, polycyclic aromatic hydrocarbons and plastic additives (Chakraborty et al., 2018).

Dioxins and furans are persistent organic pollutants (POPs) known for their harmful effects on humans, including carcinogenicity, reproductive issues, and immune system disruption (Zhang et al., 2017). They can also accumulate in the food chain, posing risks to wildlife and human populations.

During incomplete combustion, there is break down of the polymers and release these additives into the atmosphere. Stabilizers, which protect plastics from heat and oxidation, can contribute to the release of potentially harmful chemicals. Plasticizers, responsible for enhancing flexibility, may also vaporize and become part of the emissions. Additionally, FRs, designed to reduce flammability, can be released in the form of fine particles and gases. These emissions can pose environmental and health risks, especially if not properly managed (Verma et al., 2016). Efforts have been made to mitigate the impact of incineration on atmospheric emissions. Advanced incineration technologies, such as controlled air incinerators and modern filtration systems, can capture a significant portion of pollutants. Moreover, stringent regulations and guidelines are in place to limit emissions and ensure proper waste incineration practices. Additionally, there is an ongoing push to develop alternative waste treatment methods that reduce reliance on incineration and prioritize more sustainable approaches, such as recycling and controlled biological treatment. In conclusion, it is crucial to recognize the role of plastic additives in the process of open burning and incineration and the potential impact on atmospheric emissions.

3.3.2 Informal Recycling

Waste plastics are majorly recycled by the informal sector. Toxic chemical compounds are released into the atmosphere via gaseous phase or through carbonaceous PM mainly driven by the partial and incomplete burning of plastic waste by the informal recycling sector (Tue et al., 2016). This smoke, soot, and ash emitted from such processes reaches the atmosphere and eventually finds its way into water and soil, or is deposited as ash (Li et al., 2018b). Health risk via inhalation from these toxic chemicals is a major concern for workers in the informal recycling sector due to their proximity and prolonged exposure to such incomplete combustion activities. Due to inexpensive manpower in developing countries, e-waste is imported for recycling purposes mostly by the informal sector and therefore possess serious public health concerns (Rex et al., 2019). Plastics associated with E-waste recycling are often a major problem in developing economies, often contaminating the surrounding environment (Chakraborty et al., 2016a, b, c, 2018, 2019, 2021). E-waste associated regions in India have been evidenced as pollution hotspots, spreading these chemicals via direct or secondary emissions affecting the air and water, during the COVID-19 pandemic (Chandra and Chakraborty, 2023; Rex and Chakraborty, 2022). Proper waste management practices, advanced recycling technologies, and sustainable alternatives are all essential components of a comprehensive approach to handling plastic waste recycling in an environmentally responsible manner.

3.4 Landfills and Municipal Waste Disposal

Plastics discarded in landfills may encounter a range of environmental conditions, potentially leading to the leaching of chemicals and additives into the groundwater and soil. Precipitation can get into landfills while they're open and accepting waste, and even after the cell is sealed, the plastic caps develop holes over time, letting in more rain and snow. The water that gets into landfill cells picks up contaminants from the waste and becomes "leachate" which acts as a conveyor of toxic contaminants into the environment. The leakage of chemical additives and other toxic heavy metals from discarded plastics including industrial plastics, synthetic textiles, road markings, and tire wear is also a serious concern. This problem can be made worse by improper plastic garbage disposal and littering.

3.5 Degradation into Microplastics and Nanoplastics

As time passes, plastics can undergo degradation, breaking down into minuscule fragments referred to as microplastics and, in some cases, nanoplastics. Various polymer products have been found to produce multiple sizes of micro-and nanoplastics as a result of weathering processes including polystyrene (Song et al., 2020), polyethylene (Kalogerakis et al., 2017), and polypropylene (Hu et al., 2022). In the environment, these microscopic plastic particles are easily dispersed via the air, water, and soil and can contain harmful chemicals and additives on their surfaces, possibly spreading pollution to far-off sites.

4 Health Effects

Extrusion-based 3D printers release ABS, which has been associated with systemic and localized adverse effects such as oxidative stress and inflammatory reactions, as evidenced by both in vitro and animal studies. Research on the impact of 3D printer use on human health is limited. A study involving employees of businesses utilizing 3D printers found a significant increase in respiratory problems, particularly among those working over 40 hours per week with 3D printers. Some specific cases have demonstrated the risk of developing asthma due to exposure to ABS and contact dermatitis resulting from working with epoxy resins. In a chamber study, Almstrand et al. observed a slight increase in exhaled nitric oxide levels following exposure to ABS, as opposed to PLA (Almstrand et al., 2023).

Certain chemical substances added to plastics possess a partition coefficient exceeding 5, indicating a significant affinity for tissues. Consequently, they have the capacity to accumulate and potentially magnify within the food web. Examples of such substances include high molecular weight phthalates, stabilizers, antioxidants,

UV stabilizers, and specific biocides like tributyltin and triclosan (Sendra et al., 2021).

Additives like bisphenol A (BPA) and bisphenol S (BPS), phthalates, polybrominated diphenyl ethers (PBDEs), and tetrabromobisphenol A (TBBPA) have the potential to adversely affect the mammalian endocrine system (Godswill & Godspel, 2019). Within these effects, both BPA and BPS can influence the catabolism and metabolism of sex hormones. Additionally, BPA can affect gene expression associated with the thyroid hormone axis, which in turn has repercussions on biological processes like development and metabolism (Rubin, 2011).

Phthalates have been associated with a range of chronic illnesses, including diabetes, insulin resistance, obesity, breast cancer, immune system dysfunction, endocrine disruption, metabolic disorders, and disruption of the endocannabinoid system. They can also function as endocrine disruptors, affecting hormone levels, especially estrogen and thyroid hormones. PBDEs have been shown to reduce fertility at low doses by altering hormone levels, leading to hyperactivity and hyperthyroidism. On the other hand, TBBPA is swiftly eliminated by animals and does not have adverse effects related to endocrine system disruptions (Sendra et al., 2021).

Certain plastic additives, including BPA, phthalates, and nonylphenols, directly activate estrogen receptors (Watson et al., 2005). Multiple studies indicate that exposure to both low and high molecular weight phthalates, as well as BPA, is associated with compromised semen quality, including reduced sperm concentration, motility, and morphology. Additionally, exposure to BPA and phthalates has been linked to detrimental effects on female reproductive health, including conditions such as polycystic ovarian syndrome, endometriosis, uterine fibroids, and reproductive system malignancies (Sendra et al., 2021).

Inorganic pigments like iron oxide, biocides such as triclosan and arsenic, curing agents like 4,4′-diaminodiphenylmethane and 4,4′-methylenedianiline, flame-retardants including tetrabromobisphenol A, plasticizers like diethylhexyl phthalate, and slip agents have all been linked to cases of spontaneous abortion, with oleylamide also being associated (Sendra et al., 2021).

Phthalates, have been associated with various mental health conditions, including Alzheimer's disease, bipolar disorder, autism spectrum disorders, learning difficulties, dementia, depressive disorders, attention deficit hyperactivity disorder (ADHD), and intellectual disabilities (Chopra et al., 2014; Ejaredar et al., 2015; Gascon et al., 2015; Testa et al., 2012). Considering the associations between phthalates and conditions like preterm birth and reduced anogenital distance in male offspring, it's possible that these illnesses are influenced, either directly or indirectly, by phthalate exposure (Kahn et al., 2020). A study conducted in South Korea unveiled a higher incidence of attention-deficit disorders in children whose mothers had been exposed to elevated levels of flame-retardants (Kim et al., 2018). Several studies have provided evidence of a connection between prenatal exposure to BPA and childhood obesity, as well as prenatal exposure to phthalates and conditions such as gestational diabetes, impaired glucose tolerance, and increased child adiposity (Kahn et al., 2020).

As per a recent study, exposure to BPA and phthalates leads to changes in reproductive hormones, characterized by an increase in gonadotropins and estradiol levels, along with a decrease in testosterone levels. This exposure also impacts spermatogenesis and elevates oxidative stress (Pallotti et al., 2020). Among the plastic additives examined in this study, several have been associated with various diseases. Biocides such as antimony, arsenic, tributyltin, and triclosan, curing agents like 4,4-diaminodiphenylmethane, 4,4-methylenedianiline, hydrazine, and triazine, fillers including calcium carbonate, FRs such as alkanes, antimony, boric acid, decabromodiphenyl ethane, tetrabromobisphenol A, and tris(chloroethyl) phosphate, heat stabilizers like nonylphenol, barium, cadmium, and lead, inorganic pigments such as iron oxide and titanium dioxide, plasticizers including adipates, butylbenzyl phthalate, dibutyl phthalate, dicyclohexyl phthalate, diethyl phthalate, diethylhexyl phthalate, diheptyl phthalate, diisobutyl phthalate, di-n-pentyl phthalate, and tris(chloroethyl) phosphate, and stabilizers like BPA, cadmium, Irganox 1010, lead, nonylphenol, and octylphenol, have been linked to diseases including endometriosis, male and female infertility, follicular cysts, ovarian and testicular diseases, ovarian cysts, ovarian neoplasms, anovulation, and polycystic ovarian syndrome. These diseases are among the top 50 diseases with the highest associations for each chemical compound (Sendra et al., 2021).

Category	Name	Structure	Health effects
Plasticizers	Phthalates	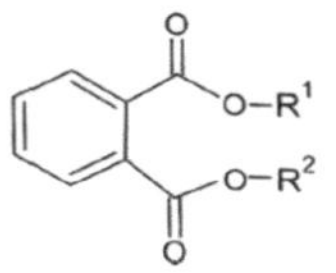	Endocrine disruption Reproductive effects-reduced sperm quality, lower testosterone levels, and genital abnormalities Developmental effects (prenatal exposure)-altered neurodevelopment, behavioral problems, cognitive deficits Obesity and metabolic effects-affect adipose tissue function and disrupt metabolic pathways Asthma and allergies
	Non-phthalate (Adipates)	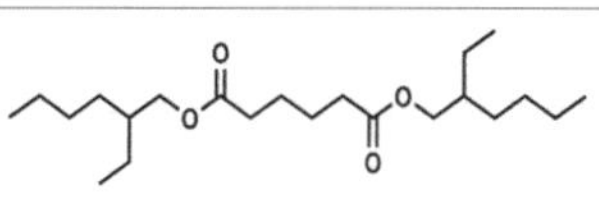	Skin sensitization and irritation Oral toxicity (on high levels of exposure through ingestion)

Category	Name	Structure	Health effects
Flame-retardants (halogenated FRs)	PBDEs	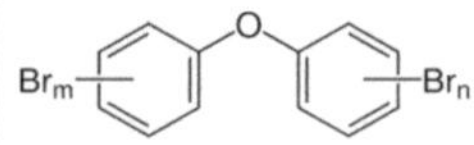	Endocrine disruption Developmental effects-cognitive deficits, reduced IQ, attention deficits, and behavioral problems in children. Reproductive effects—altered reproductive hormone levels, decreased fertility, impaired sperm quality Carcinogenicity—liver tumors, thyroid cancer Neurological effects Thyroid disruption
	HBCD	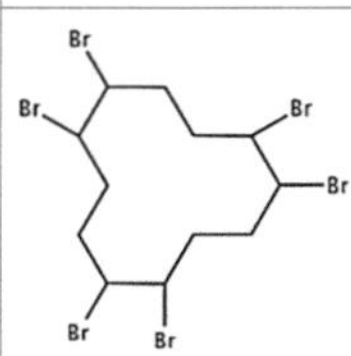	Endocrine disruption Neurological and behavioral effects Reproductive and developmental effects—effects on fertility, prenatal development, and postnatal growth Carcinogenicity (group 2B)
Impact modifiers	Butadiene-based modifiers (acrylonitrile-butadienestyrene)	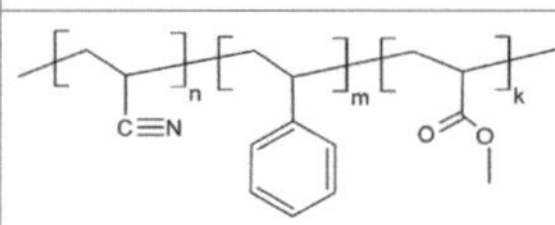	Inhalation exposure—respiratory irritation and adverse respiratory effects (high conc.) Occupational exposure—Causes leukemia and lymphoma Carcinogenicity (group 1)
	Acrylic modifiers (acrylonitrile-styreneacrylate)	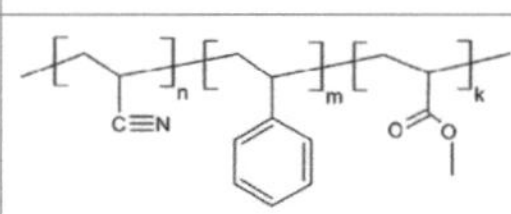	Respiratory irritation Allergic eeactions Eye irritation
	Elastomers (ethylene-propylene Copolymer)	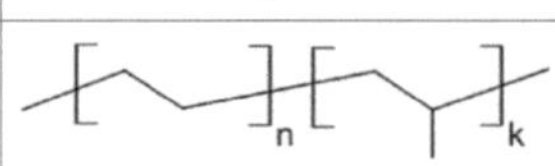	Occupational exposure—allergies and irritation
Antioxidants	Phenolic-based antioxidants (curcumin)	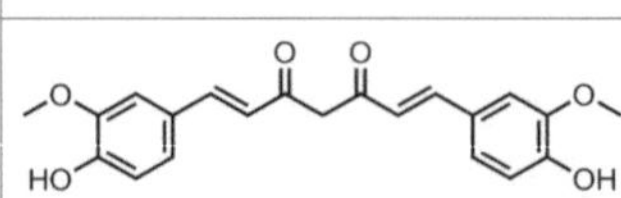	–

Category	Name	Structure	Health effects
UV stabilizers	UV absorbers Phenolic benzotriazole (Bisoctrizole)	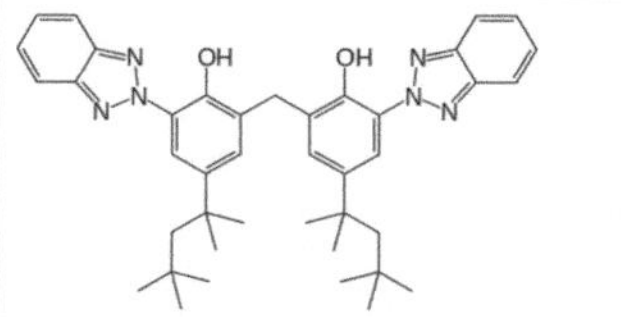	Skin irritation and allergic reactions for sensitive individual Respiratory irritation Possible endocrine disruption
Antimicrobials	Organic antimicrobial agents (Triclosan)	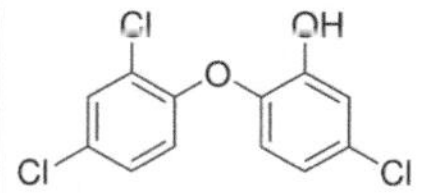	Skin irritation, itching, redness, or other allergic symptoms Skin sensitization Respiratory irritation, coughing Endocrine disruption (eg: Triclosan)
	Inorganic antimicrobial agents	Zeolites loaded with silver ions	–

(Anttinen-Klemetti et al., 2006; Hazardous Substances Fact Sheet 2000; Khare et al., 2023; Linares et al., 2015; Marques & Cairrao, 2023; Wang & Qian, 2021; Weatherly & Gosse, 2017)

Acknowledgments This research was supported by SRMIST's institutional fund given to the Centre for Research in Environment, Sustainability Advocacy and Climate Change (REACH), Directorate of Research, SRMIST and by the Royal Norwegian Embassy in New Delhi through the INOPOL Project (Project No. IND-19/0005 and Project No: IND-22/0007).

References

Alabi, O. A., Ologbonjaye, K. I., Awosolu, O., & Alalade, O. E. (2019). Public and environmental health effects of plastic wastes disposal: A review. *Journal Toxicological Risk Assessment, 5*(021), 1–13.

Almstrand, A. C., Bredberg, A., Runstrom Eden, G., Karlsson, H., Assenhoj, M., Koca, H., Olin, A. C., & Tinnerberg, H. (2023). An explorative study on respiratory health among operators working in polymer additive manufacturing. *Frontiers in Public Health, 11*, 1148974. https://doi.org/10.3389/fpubh.2023.1148974

Andrady, A. L. (2017). The plastic in microplastics: A review. *Marine Pollution Bulletin, 119*(1), 12–22.

Anttinen-Klemetti, T., Vaaranrinta, R., Mutanen, P., & Peltonen, K. (2006). Inhalation exposure to 1, 3-butadiene and styrene in styrene–butadiene copolymer production. *International Journal of Hygiene and Environmental Health, 209*(2), 151–158.

Arvanitoyannis, I. S., & Kotsanopoulos, K. V. (2014). Migration phenomenon in food packaging. Food–package interactions, mechanisms, types of migrants, testing and relative legislation—A review. *Food and Bioprocess Technology, 7*, 21–36.

Bandow, N., Will, V., Wachtendorf, V., & Simon, F.-G. (2017). Contaminant release from aged microplastic. *Environmental Chemistry, 14*(6), 394–405.

Barrick, A., Champeau, O., Chatel, A., Manier, N., Northcott, G., & Tremblay, L. A. (2021). Plastic additives: Challenges in ecotox hazard assessment. *PeerJ, 9*, e11300.

Beldi, G., Pastorelli, S., Franchini, F., & Simoneau, C. (2012). Time-and temperature-dependent migration studies of Irganox 1076 from plastics into foods and food simulants. *Food Additives & Contaminants: Part A, 29*(5), 836–845.

Bîrcă, A., Gherasim, O., Grumezescu, V., & Grumezescu, A. M. (2019). Introduction in thermoplastic and thermosetting polymers. In *Materials for biomedical engineering* (pp. 1–28). Elsevier.

Carr, S. A., Liu, J., & Tesoro, A. G. (2016). Transport and fate of microplastic particles in wastewater treatment plants. *Water Research, 91*, 174–182.

Chakraborty, P., Selvaraj, S., Nakamura, M., Prithiviraj, B., Ko, S., & Loganathan, B. (2016a). E-waste and associated environmental contamination in the Asia/Pacific region (part 2): A case study of dioxins and furans in E-waste recycling/dump sites in India. In *ACS symposium series* (Vol. 1243, pp. 139–154). American Chemical Society. https://doi.org/10.1021/bk-2016-1243.ch007

Chakraborty, P., Selvaraj, S., Nakamura, M., Prithiviraj, B., Ko, S., & Loganathan, B. G. (2016b). E-waste and associated environmental contamination in the Asia/Pacific region (part 1): An overview. In *ACS symposium series* (Vol. 1243, pp. 127–138). American Chemical Society. https://doi.org/10.1021/bk-2016-1243.ch006

Chakraborty, P., Prithiviraj, B., Selvaraj, S., & Kumar, B. (2016c). Polychlorinated biphenyls in settled dust from informal electronic waste recycling workshops and nearby highways in urban centers and suburban industrial roadsides of Chennai city, India: Levels, congener profiles and exposure assessment. *Science of the Total Environment, 573*, 1413–1421. https://doi.org/10.1016/j.scitotenv.2016.07.129

Chakraborty, P., Selvaraj, S., Nakamura, M., Prithiviraj, B., Cincinelli, A., & Bang, J. J. (2018). PCBs and PCDD/fs in soil from informal e-waste recycling sites and open dumpsites in India: Levels, congener profiles and health risk assessment. *Science of the Total Environment, 621*, 930–938. https://pubmed.ncbi.nlm.nih.gov/29223911/

Chakraborty, P., Sampath, S., Mukhopadhyay, M., Selvaraj, S., Bharat, G. K., & Nizzetto, L. (2019). Baseline investigation on plasticizers, bisphenol a, polycyclic aromatic hydrocarbons and heavy metals in the surface soil of the informal electronic waste recycling workshops and nearby open dumpsites in Indian metropolitan cities. *Environmental Pollution, 248*, 1036–1045. https://doi.org/10.1016/j.envpol.2018.11.010

Chakraborty, P., Gadhavi, H., Prithiviraj, B., Mukhopadhyay, M., Khuman, S. N., Nakamura, M., & Spak, S. N. (2021). Passive air sampling of PCDD/fs, PCBs, PAEs, DEHA, and PAHs from informal electronic waste recycling and allied sectors in Indian megacities. *Environmental Science & Technology, 55*(14), 9469–9478. https://doi.org/10.1021/acs.est.1c01460

Chakraborty, P., Chandra, S., Dimmen, M. V., Hurley, R., Mohanty, S., Bharat, G. K., Steindal, E. H., Olsen, M., & Nizzetto, L. (2022). Interlinkage between persistent organic pollutants and plastic in the waste management system of India: An overview. *Bulletin of Environmental Contamination and Toxicology, 109*(6), 927–936.

Chandra, S., & Chakraborty, P. (2023). Air-water exchange and risk assessment of phthalic acid esters during the early phase of COVID-19 pandemic in tropical riverine catchments of India. *Chemosphere, 341*, 140013.

Chen, Q., Allgeier, A., Yin, D., & Hollert, H. (2019). Leaching of endocrine disrupting chemicals from marine microplastics and mesoplastics under common life stress conditions. *Environment International, 130*, 104938.

Chopra, M., Kaur, P., Bernela, M., & Thakur, R. (2014). Surfactant assisted nisin loaded chitosan-carageenan nanocapsule synthesis for controlling food pathogens. *Food Control, 37*, 158–164.

Clausen, P. A., Liu, Z., Xu, Y., Kofoed-Sørensen, V., & Little, J. C. (2010). Influence of air flow rate on emission of DEHP from vinyl flooring in the emission cell FLEC: Measurements and CFD simulation. *Atmospheric Environment, 44*(23), 2760–2766.

Clausen, P. A., Liu, Z., Kofoed-Sorensen, V., Little, J., & Wolkoff, P. (2012). Influence of temperature on the emission of di-(2-ethylhexyl)phthalate (DEHP) from PVC flooring in the emission cell FLEC. *Environmental Science & Technology, 46*(2), 909–915. https://doi.org/10.1021/es2035625

Coleman, E. A. (2017). Plastics additives. In *Applied plastics engineering handbook* (pp. 489–500). Elsevier.

Crossman, J., Hurley, R. R., Futter, M., & Nizzetto, L. (2020). Transfer and transport of microplastics from biosolids to agricultural soils and the wider environment. *Science of the Total Environment, 724*, 138334.

Cuadros-Rodríguez, L., Lazúen-Muros, M., Ruiz-Samblás, C., & Navas-Iglesias, N. (2020). Leachables from plastic materials in contact with drugs. State of the art and review of current analytical approaches. *International Journal of Pharmaceutics, 583*, 119332.

De Wit, C. A. (2002). An overview of brominated flame retardants in the environment. *Chemosphere, 46*(5), 583–624.

Ejaredar, M., Nyanza, E. C., Ten Eycke, K., & Dewey, D. (2015). Phthalate exposure and childrens neurodevelopment: A systematic review. *Environmental Research, 142*, 51–60.

Ekelund, M., Azhdar, B., & Gedde, U. W. (2010). Evaporative loss kinetics of di (2-ethylhexyl) phthalate (DEHP) from pristine DEHP and plasticized PVC. *Polymer Degradation and Stability, 95*(9), 1789–1793.

Fabbri, D., Carena, L., Bertone, D., Brigante, M., Passananti, M., & Vione, D. (2023). Assessing the photodegradation potential of compounds derived from the photoinduced weathering of polystyrene in water. *Science of the Total Environment, 876*, 162729. https://doi.org/10.1016/j.scitotenv.2023.162729

Fujii, M., Shinohara, N., Lim, A., Otake, T., Kumagai, K., & Yanagisawa, Y. (2003). A study on emission of phthalate esters from plastic materials using a passive flux sampler. *Atmospheric Environment, 37*(39–40), 5495–5504.

Gascon, M., Casas, M., Morales, E., Valvi, D., Ballesteros-Gómez, A., Luque, N., Rubio, S., Monfort, N., Ventura, R., & Martínez, D. (2015). Prenatal exposure to bisphenol a and phthalates and childhood respiratory tract infections and allergy. *Journal of Allergy and Clinical Immunology, 135*(2), 370–378. e377.

Geyer, R., Jambeck, J. R., & Law, K. L. (2017). Production, use, and fate of all plastics ever made. *Science Advances, 3*(7), e1700782.

Giuliani, D., Dickson, M. M., Espa, G., & Santi, F. (2020). Modelling and predicting the spatio-temporal spread of COVID-19 in Italy. *BMC Infectious Diseases, 20*(1), 1–10.

Godswill, A. C., & Godspel, A. C. (2019). Physiological effects of plastic wastes on the endocrine system (bisphenol a, phthalates, bisphenol S, PBDEs, TBBPA). *International Journal of Bioinformatics and Computational Biology, 4*, 11–29.

Groh, K. J., Backhaus, T., Carney-Almroth, B., Geueke, B., Inostroza, P. A., Lennquist, A., Leslie, H. A., Maffini, M., Slunge, D., & Trasande, L. (2019). Overview of known plastic packaging-associated chemicals and their hazards. *Science of the Total Environment, 651*, 3253–3268.

Hale, R. C., King, A. E., Ramirez, J. M., La Guardia, M., & Nidel, C. (2022). Durable plastic goods: A source of microplastics and chemical additives in the built and natural environments. *Environmental Science & Technology Letters, 9*(10), 798–807. https://doi.org/10.1021/acs.estlett.2c00417

Hammodat, A. R., Nassar, S., Mortula, M. M., & Shamsuzzaman, M. (2023). Factors affecting the leaching of micro and nanoplastics in the water distribution system. *Journal of Environmental Management, 345*, 118779. https://doi.org/10.1016/j.jenvman.2023.118779

Harper, C. A. (2006). *Handbook of plastics technologies: The complete guide to properties and performance*. McGraw-Hill Education.

Hazardous Substances Fact Sheet (2000)

He, H., Li, F., Liu, K., Zhan, J., Wang, X., Lai, C., Yang, X., Huang, B., & Pan, X. (2023). The disinfectant residues promote the leaching of water contaminants from plastic pipe particles. *Environmental Pollution, 327*, 121577. https://doi.org/10.1016/j.envpol.2023.121577

Horn, W. (2000). Inorganic hydroxides and hydroxycarbonates: Their function and use as flame-retardant additives. *Fire retardancy of polymeric materials, 2000*, 285–352.

Hu, J., Xu, X., Song, Y., Liu, W., Zhu, J., Jin, H., & Meng, Z. (2022). Microplastics in widely used polypropylene-made food containers. *Toxics, 10*(12), 762.

Kahn, L. G., Philippat, C., Nakayama, S. F., Slama, R., & Trasande, L. (2020). Endocrine-disrupting chemicals: Implications for human health. *The Lancet Diabetes & Endocrinology, 8*(8), 703–718.

Kalogerakis, N., Karkanorachaki, K., Kalogerakis, G. C., Triantafyllidi, E. I., Gotsis, A. D., Partsinevelos, P., & Fava, F. (2017). Microplastics generation: Onset of fragmentation of polyethylene films in marine environment Mesocosms [original research]. *Frontiers in Marine Science, 4*, 84. https://doi.org/10.3389/fmars.2017.00084

Kämpf, G., Sommer, K., & Zirngiebl, E. (1991). Studies in accelerated weathering: Part I. Determination of the activation spectrum of photodegradation in polymers. *Progress in Organic Coatings, 19*(1), 69–77. https://doi.org/10.1016/0033-0655(91)80011-7

Karger-Kocsis, J., Kallo, A., Szafner, A., Bodor, G., & Senyei, Z. (1979). Morphological study on the effect of elastomeric impact modifiers in polypropylene systems. *Polymer, 20*(1), 37–43.

Khan, I. A., Lee, K. H., Lee, Y.-S., & Kim, J.-O. (2022). Degradation analysis of polymeric pipe materials used for water supply systems under various disinfectant conditions. *Chemosphere, 291*, 132669. https://doi.org/10.1016/j.chemosphere.2021.132669

Khare, A., Jadhao, P., Vaidya, A. N., & Kumar, A. R. (2023). Benzotriazole UV stabilizers (BUVs) as an emerging contaminant of concern: A review. *Environmental Science and Pollution Research, 30*(58), 121370–121392. https://doi.org/10.1007/s11356-023-30567-9

Kim, S., Eom, S., Kim, H.-J., Lee, J. J., Choi, G., Choi, S., Kim, S., Kim, S. Y., Cho, G., & Kim, Y. D. (2018). Association between maternal exposure to major phthalates, heavy metals, and persistent organic pollutants, and the neurodevelopmental performances of their children at 1 to 2 years of age-CHECK cohort study. *Science of the Total Environment, 624*, 377–384.

Kumar, R. (2023). Plastic chemical constituents in wastewater, surface water, and drinking water. In A. Sinha, S. P. Singh, & A. B. Gupta (Eds.), *Persistent pollutants in water and advanced treatment technology* (pp. 71–95). Springer Nature. https://doi.org/10.1007/978-981-99-2062-4_4

Le Bras, M., Bourbigot, S., Camino, G., & Delobel, R. (1998). *Fire retardancy of polymers: The use of intumescence* (Vol. 224). Elsevier.

Li, J., Liu, H., & Chen, J. P. (2018a). Microplastics in freshwater systems: A review on occurrence, environmental effects, and methods for microplastics detection. *Water Research, 137*, 362–374.

Li, T.-Y., Zhou, J.-F., Wu, C.-C., Bao, L.-J., Shi, L., & Zeng, E. Y. (2018b). Characteristics of Polybrominated diphenyl ethers released from thermal treatment and open burning of E-waste. *Environmental Science & Technology, 52*(8), 4650–4657. https://doi.org/10.1021/acs.est.8b00780

Linares, V., Bellés, M., & Domingo, J. L. (2015). Human exposure to PBDE and critical evaluation of health hazards. *Archives of Toxicology, 89*(3), 335–356.

Lithner, D., Damberg, J., Dave, G., & Larsson, Å. (2009). Leachates from plastic consumer products–screening for toxicity with Daphnia magna. *Chemosphere, 74*(9), 1195–1200.

Liu, W., Zhao, Y., Shi, Z., Li, Z., & Liang, X. (2020). Ecotoxicoproteomic assessment of microplastics and plastic additives in aquatic organisms: A review. *Comparative Biochemistry and Physiology Part D: Genomics and Proteomics, 36*, 100713.

Marques, M. L., & Cairrao, E. (2023). Occurrence and health effects of Hexabromocyclododecane: An updated review. *Toxics, 11*(5), 409.

Marturano, V., Cerruti, P., & Ambrogi, V. (2017). Polymer additives. *Physical Sciences Reviews, 2*(6), 20160130.

Mukhopadhyay, M., Jalal, M., Vignesh, G., Ziauddin, M., Sampath, S., Bharat, G. K., Nizzetto, L., & Chakraborty, P. (2022). Migration of plasticizers from polyethylene terephthalate and low-density polyethylene casing into bottled water: A case study from India. *Bulletin of Environmental Contamination and Toxicology, 109*(6), 949–955. https://doi.org/10.1007/s00128-022-03474-x

Nichols, D. (2004). *Biocides in plastics* (Vol. 15). iSmithers Rapra Publishing. ISBN 9781859575123.

Ong, S. W. X., Tan, Y. K., Chia, P. Y., Lee, T. H., Ng, O. T., Wong, M. S. Y., & Marimuthu, K. (2020). Air, surface environmental, and personal protective equipment contamination by severe acute respiratory syndrome coronavirus 2 (SARS-CoV-2) from a symptomatic patient. *JAMA, 323*(16), 1610–1612.

Pallotti, F., Pelloni, M., Gianfrilli, D., Lenzi, A., Lombardo, F., & Paoli, D. (2020). Mechanisms of testicular disruption from exposure to bisphenol a and phtalates. *Journal of Clinical Medicine, 9*(2), 471.

Poças, M. F., Oliveira, J. C., Oliveira, F. A., & Hogg, T. (2008). A critical survey of predictive mathematical models for migration from packaging. *Critical Reviews in Food Science and Nutrition, 48*(10), 913–928.

Pritchard, G. (2012). *Plastics additives: An AZ reference* (Vol. 1). Springer Science & Business Media.

Rabek, J. F. (2012). *Photostabilization of polymers: Priciples and application*. Springer Science & Business Media.

Rani, M., Shanker, U., & Jassal, V. (2017). Recent strategies for removal and degradation of persistent & toxic organochlorine pesticides using nanoparticles: A review. *Journal of Environmental Management, 190*, 208–222.

Rex, K. R., & Chakraborty, P. (2022). Legacy and new chlorinated persistent organic pollutants in the rivers of south India: Occurrences, sources, variations before and after the outbreak of the COVID-19 pandemic. Journal of Hazardous Materials, 437, 129262.

Reyna-Bensusan, N., Wilson, D. C., Davy, P. M., Fuller, G. W., Fowler, G. D., & Smith, S. R. (2019). Experimental measurements of black carbon emission factors to estimate the global impact of uncontrolled burning of waste. *Atmospheric Environment, 213*, 629–639. https://doi.org/10.1016/j.atmosenv.2019.06.047

Reynier, A., Dole, P., Humbel, S., & Feigenbaum, A. (2001). Diffusion coefficients of additives in polymers. I. Correlation with geometric parameters. *Journal of Applied Polymer Science, 82*(10), 2422–2433.

Ronnie Rex, K., Rajakumaran, R., Gaonkar, O., Dsouza, E., & Chakraborty, P. (2019). Recycling processes and plastic in electronic waste is an emerging problem for India: Implications for future Prospect. In M. Z. Hashmi & A. Varma (Eds.), *Electronic waste pollution: Environmental occurrence and treatment technologies* (pp. 171–186). Springer International Publishing. https://doi.org/10.1007/978-3-030-26615-8_12

Rubin, B. S. (2011). Bisphenol a: An endocrine disruptor with widespread exposure and multiple effects. *The Journal of Steroid Biochemistry and Molecular Biology, 127*(1–2), 27–34.

Rutkowski, J. V., & Levin, B. C. (1986). Acrylonitrile–butadiene–styrene copolymers (ABS): Pyrolysis and combustion products and their toxicity—A review of the literature. *Fire and Materials, 10*(3–4), 93–105.

Schartel, B. (2010). Phosphorus-based flame retardancy mechanisms—Old hat or a starting point for future development? *Materials, 3*(10), 4710–4745.

Sendra, M., Pereiro, P., Figueras, A., & Novoa, B. (2021). An integrative toxicogenomic analysis of plastic additives. *Journal of Hazardous Materials, 409*, 124975. https://doi.org/10.1016/j.jhazmat.2020.124975

Song, Y. K., Hong, S. H., Eo, S., Han, G. M., & Shim, W. J. (2020). Rapid production of micro-and Nanoplastics by fragmentation of expanded polystyrene exposed to sunlight. *Environmental Science & Technology, 54*(18), 11191–11200. https://doi.org/10.1021/acs.est.0c02288

Suhrhoff, T. J., & Scholz-Böttcher, B. M. (2016). Qualitative impact of salinity, UV radiation and turbulence on leaching of organic plastic additives from four common plastics—A lab experiment. *Marine Pollution Bulletin, 102*(1), 84–94.
Sun, J., Dai, X., Wang, Q., Van Loosdrecht, M. C., & Ni, B.-J. (2019). Microplastics in wastewater treatment plants: Detection, occurrence and removal. *Water Research, 152*, 21–37.
Testa, C., Nuti, F., Hayek, J., De Felice, C., Chelli, M., Rovero, P., Latini, G., & Papini, A. M. (2012). Di-(2-ethylhexyl) phthalate and autism spectrum disorders. *ASN Neuro, 4*(4), AN20120015.
Tseng, W. T., & Lee, J. S. (2000). Functional MBS impact modifiers for PC/PBT alloy. *Journal of Applied Polymer Science, 76*(8), 1280–1284.
Tue, N. M., Goto, A., Takahashi, S., Itai, T., Asante, K. A., Kunisue, T., & Tanabe, S. (2016). Release of chlorinated, brominated and mixed halogenated dioxin-related compounds to soils from open burning of e-waste in Agbogbloshie (Accra, Ghana). *Journal of Hazardous Materials, 302*, 151–157. https://doi.org/10.1016/j.jhazmat.2015.09.062
Turner, A., & Lau, K. S. (2016). Elemental concentrations and bioaccessibilities in beached plastic foam litter, with particular reference to lead in polyurethane. *Marine Pollution Bulletin, 112*(1–2), 265–270.
Verma, R., Vinoda, K. S., Papireddy, M., & Gowda, A. N. S. (2016). Toxic pollutants from plastic waste-a review. *Procedia Environmental Sciences, 35*, 701–708. https://doi.org/10.1016/j.proenv.2016.07.069
Wang, Y., & Qian, H. (2021). *Phthalates and their impacts on human health*. Healthcare.
Watson, C. S., Bulayeva, N. N., Wozniak, A. L., & Finnerty, C. C. (2005). Signaling from the membrane via membrane estrogen receptor-α: Estrogens, xenoestrogens, and phytoestrogens. *Steroids, 70*(5–7), 364–371.
Weatherly, L. M., & Gosse, J. A. (2017). Triclosan exposure, transformation, and human health effects. *Journal of Toxicology and Environmental Health, Part B, 20*(8), 447–469.
Ye, X., Wang, P., Wu, Y., Zhou, Y., Sheng, Y., & Lao, K. (2020). Microplastic acts as a vector for contaminants: The release behavior of dibutyl phthalate from polyvinyl chloride pipe fragments in water phase. *Environmental Science and Pollution Research, 27*, 42082–42091.
Young, R. J., & Lovell, P. A. (2011). *Introduction to polymers*. CRC Press.
Yousif, E., & Haddad, R. (2013). Photodegradation and photostabilization of polymers, especially polystyrene. *Springerplus, 2*(1), 1–32.
Yu, W., Azhdar, B., Andersson, D., Reitberger, T., Hassinen, J., Hjertberg, T., & Gedde, U. W. (2011). Deterioration of polyethylene pipes exposed to water containing chlorine dioxide. *Polymer Degradation and Stability, 96*(5), 790–797.
Zhang, T., Fiedler, H., Yu, G., Ochoa, G. S., Carroll, W. F., Gullett, B. K., Marklund, S., & Touati, A. (2011). Emissions of unintentional persistent organic pollutants from open burning of municipal solid waste from developing countries. *Chemosphere, 84*(7), 994–1001. https://doi.org/10.1016/j.chemosphere.2011.04.070
Zhang, M., Buekens, A., & Li, X. (2017). Open burning as a source of dioxins. *Critical Reviews in Environmental Science and Technology, 47*(8), 543–620. https://doi.org/10.1080/10643389.2017.1320154

Health, Exposure and Risks: A Review

Moitraiyee Mukhopadhyay, Avanti Roy Basu, Manisha Jain, Twinkle Dev, Mrinal Mathur, and Paromita Chakraborty

1 Introduction

India, despite its significant growth trajectory, grapples with a common challenge: establishing an environmentally friendly model of economic development. This challenge stems from environmental pollution issues, mainly related to the improper disposal of plastic waste and its open burning at landfill sites (Chakraborty et al., 2016, 2018, 2019). Mishandling and mismanagement of plastic waste become prevalent when inadequately planned or under-equipped waste management systems must contend with a complex array of waste materials.

Of particular concern are the plastic additives that can modify plastic properties within terrestrial ecosystems, as they exhibit biological activity that may impact reproduction and development of both plants and animals (Meeker et al., 2009; Oehlmann et al., 2009). There are various types of plastic additives, each serving a specific purpose. It is important to note that the choice of additives depends on the specific requirements of the plastic product and the intended application. Additionally, some additives may raise environmental and health concerns, leading

M. Mukhopadhyay
Environmental Science and Technology Laboratory, Centre for Research in Environment, Sustainability Advocacy and Climate Change, Directorate of Research, SRM Institute of Science and Technology, Kattankulathur, Chengalpattu, Tamil Nadu, India

Nuevo Chakra Pvt. Ltd, Thane, Maharashtra, India

A. R. Basu · M. Jain · T. Dev · M. Mathur
Mu Gamma Consultants, Gurugram, India

P. Chakraborty (✉)
Environmental Science and Technology Laboratory, Centre for Research in Environment, Sustainability Advocacy and Climate Change, Directorate of Research, SRM Institute of Science and Technology, Kattankulathur, Chengalpattu, Tamil Nadu, India
e-mail: paromitc@srmist.edu.in

P. Chakraborty et al. (eds.), *Efficient and Enhanced Management of Plastic Waste in India*, https://doi.org/10.1007/978-3-031-89398-8_3

to increased scrutiny and regulation in certain cases. Since plastic additives are not covalently bonded, they can leach into the environment. They occur in a variety of environmental matrices, and have a high potential to contaminate soil, air and water. However, only 25% of plastic additives have reportedly been assessed for their possible environmental impact. Although global market statistics highlight the significant environmental burden posed by the unregulated manufacturing and use of plastic additives, knowledge on their ecotoxicity is limited. In terms of human health, toxic chemical additives in plastics can disrupt hormone functioning affecting reproduction, growth and cognitive function in the human body.

The list of plastic additives include:

- **Plasticizers (10–70%):** These additives make plastics more flexible and less brittle by increasing their flexibility and reducing their stiffness. Common plasticizers include phthalates, adipates, and citrates.
- **Stabilizers:** Antioxidant Stabilizers (0.1–3%), Light Stabilizers (0.1–3%), UV-absorbing Stabilizers (0.1–3%), Thermal Stabilizers (0.5–3%).
- **Flame Retardants (1–25%):** Flame retardant additives reduce the flammability of plastics and inhibit the spread of flames. They are often used in applications where fire safety is critical, such as in electronics and construction materials. Common retardants include organobromide, organophosphate, and organochlorides.
- **Blowing Agents (1–5%):** Foaming agents introduce gas bubbles into plastics during processing, creating foam materials with reduced density and improved insulation properties.
- **Colorants and Pigments (0.01–10%):** These additives provide color to plastics and are essential for achieving desired aesthetic effects and reduce light permeability. They come in various forms, including soluble dyes, and inorganic and organic pigments.
- **Fillers (up to 50%) and Reinforcements (15–30%):** Fillers, such as calcium carbonate and talc, are added to plastics to reduce cost and improve properties like stiffness. Reinforcements, like glass fibers and carbon fibers, enhance strength and rigidity.
- **Antistatic Agents (0.1–1.0%):** These additives help reduce the buildup of static electricity on the surface of plastic products, which can attract dust and cause other issues.
- **Others:**
  - **Impact Modifiers:** Impact modifiers enhance the toughness and impact resistance of plastics, making them less prone to fracture or shattering.
  - **Biocides:** Biocides are added to plastics to inhibit the growth of microorganisms, such as bacteria and fungi, on the surface of plastic products. This is particularly important in applications like food packaging.
  - **Processing Aids:** These additives improve the processing and manufacturability of plastics. They can aid in extrusion, injection molding, and other fabrication processes.

- **Slip Agents:** Slip agents reduce the friction between plastic surfaces, making them smoother and more resistant to sticking, which is useful in applications like packaging films.
- **Nucleating Agents:** Nucleating agents promote the crystallization of plastics, which can improve their stiffness, transparency, and other properties.
- **Plastic Lubricants:** Lubricants are added to plastics to reduce friction and improve their flow characteristics during processing.
- **Mold Release Agents:** Mold release agents are used to prevent plastic parts from adhering to the mold surface during molding process.
- **Cross-Linking Agents:** Cross-linking agents promote the formation of chemical bonds between polymer chains, increasing the strength and stability of the plastic.
- **Fluorinated Additives:** These additives can provide plastics with enhanced resistance to chemicals, heat, and other harsh conditions. Fluorinated polymers are often used in these applications.
- **Matting Agents:** Matting agents are used to reduce the gloss or shine of plastic surfaces, creating a matte or satin finish.

In addition to plastic-derived chemicals, the combustion of fragmented plastic or polymeric materials during open burning can result in the creation of micro-or nano-sized plastics, raising environmental and public health apprehensions. According to recent research, microplastics and nanoplastics may cause serious human health risks. These tiny particles can be inhaled and travel deep into the lungs, causing noncommunicable diseases like cancer, pulmonary dysfunction, and metabolic issues. Also, the disruption of organismal physiology and behaviour due to micro-and nanoplastics has been extensively documented in marine invertebrates, with some of these effects also observed in larger marine vertebrates like fish (Covello et al., 2024). Research indicates that microplastic toxicity causes intestinal oxidative damage in *Caenorhabditis elegans*, *Artemia parthenogenetica* zooplankton larvae, and *Eisenia fetida* earthworm. Exposure to microparticles is linked to the progression of cellular deformations and enterocyte decomposition. Additional evidence shows the linkage of oral exposure to microplastics in aquatic vertebrates such as juvenile large yellow croaker *Larimichthys crocea*, or *Oryzias melastigma,* juvenile intertidal fish *Girella laevifrons*, and rainbow trout. Polystyrene microbeads and nanoparticles reportedly cause a decrease in digestive enzyme activity, induce goblet cell enlargement and increased mucus secretion, proinflammatory cytokines secretion, leukocyte infiltration, hyperemia, and the loss of villi and crypt cells of the intestines.

More recently, research has suggested potential repercussions of micro-and nanoplastics on mammalian gut microbiota, as well as host cellular and metabolic toxicity, as evidenced in mouse models. Given the extensive range of concerns associated with macro and microplastics, as well as chemicals carried by plastics, this chapter intends to focus on the following essential aspects: (i) interlinkage between plastics and chemicals, (ii) routes of exposure, (iii) plastic-borne chemicals, and (iv) human, societal, and environmental risks associated with plastic and chemical interlinkages.

The potential research gaps include the inadequate understanding about the volume and composition of microplastics entering the environment, the ways by which the properties of microplastics influence toxicity, and the behavior/fate of microplastics under various environmental conditions. It is imperative that future research should develop standardized techniques to allow for data comparability; this would primarily address the technical challenges associated with microplastics research. Identifying these research priorities will assist shape the design of future research, determining both the magnitude and potential ecological implications of microplastic pollution in freshwater and terrestrial settings. (Horton et al., 2017).

2 Materials and Method

The keywords "plastic additives," "microplastic," "human exposure," and "exposure route" were chosen either individually or in combination to conduct searches for pertinent information on Web of Science, Scopus, and Google Scholar.

3 Results and Discussion

3.1 Plastic-Borne Chemicals

Plastic additives can leach out over time, especially when plastics are exposed to heat, UV radiation, or certain environmental conditions. This leaching can result in exposure to these chemicals when they enter the surrounding environment, food, or beverages. When a substance comes in contact with an aqueous environment, the leaching process can be explained through three fundamental mechanisms: diffusion within the polymer matrix, partitioning at the interface between the solid and liquid phases, and dispersion into the surrounding medium (Lau & Wong, 2000). There are several factors that influence the leaching of chemicals, including contact time, temperature, and area and plastic product characteristics, such as chemical properties, thickness, and polymeric structure. As a result, chemical migration and exposure levels of chemicals will vary with the changing condition or scenario (Zimmerman et al., 2021).

Diffusion Within the Polymer: This process is typically governed by Fick's laws of diffusion and is influenced by the concentration gradient and the diffusion coefficient specific to the additive-polymer system. The diffusion coefficient for organic additives relies on factors such as molecular weight, size, and interactions within the polymer matrix.

Partitioning at the Solid–Liquid Interface: Here, it depends on the solubility of the additive in the surrounding medium. The extent to which the additive moves from the solid phase to the liquid phase is influenced by its solubility.

Dispersion into the Bulk Medium: This aspect is generally determined by solubility and dilution.

In practical leaching experiments, a common pattern observed is a biphasic release. Initially, there's a relatively rapid release of the labile surface fraction, followed by an extended period characterized by slower, diffusion-controlled release (Bandow et al., 2017; Beldi et al., 2012; Mukhopadhyay et al., 2022; Rani et al., 2017; Suhrhoff & Scholz-Böttcher, 2016; Sun et al., 2019; Turner & Lau, 2016). Leaching is typically undesirable as it can degrade the performance properties of materials. Moreover, it poses a risk of contaminating consumer products, particularly in the case of materials intended for food contact, pharmaceuticals, and medical devices (Cuadros-Rodríguez et al., 2020; Giuliani et al., 2020). It is worth noting that leaching parameters can vary significantly across studies, encompassing factors like media properties, the ratio of solid to liquid, temperature, duration, and the level of agitation (Bridson et al., 2021).

3.2 Interlinkage Between Plastics and Chemicals

The interlinkage between plastics and plastic-derived chemicals is a complex and multifaceted issue that has significant implications for both the environment and human health. Plastics, which are synthetic polymers made from various organic compounds, have become an integral part of our modern lives due to their versatility and convenience. However, the production, use, and disposal of plastics are closely connected to the release and presence of plastic-derived chemicals, which can have far-reaching consequences. Organic compounds serve as enhancers in polymers, enhancing the characteristics of the end products. The inadvertent discharge of these enhancers into the environment is undesirable for both manufacturers and the ecosystem. This is because the loss of these enhancers reduces the longevity of the polymers. For instance, the depletion of plasticizers diminishes the tensile strength of materials like polyvinyl chloride (PVC) (Boyer, 1951). Moreover, this discharge exposes living organisms to these released enhancers, posing potential risks (Chakraborty et al., 2022). The release may take place during the service life of the plastics or after their disposal. The extent to which an additive can be released from the polymer can be influenced by several factors. These factors encompass the pore size of the polymer, the size of the additives, the physicochemical conditions (such as pH and ionic strength), the biological population, and the type of bond between the chemical and the polymer. Additionally, factors like co-migration and temperature play a role in promoting migration, as do specific physical and chemical properties of the additive and its immediate environment (Teuten et al., 2007).

The sorption and desorption of anthropogenic contaminants from plastics is a crucial aspect of environmental chemistry and pollution management. Anthropogenic contaminants such as polycyclic aromatic hydrocarbons (PAHs), organochlorine pesticides (OCPs), heavy metals, polybrominated diphenyl ethers (PBDEs), fluorinated compounds, and polychlorinated biphenyls (PCBs) are human-made chemicals and pollutants, and plastics can act as both sinks and sources for these substances in the environment (Bakir et al., 2016; Rochman, 2018; Ziccardi et al., 2016). Comprehending the interaction between contaminants and plastics, involving their sorption onto and subsequent desorption from plastic surfaces, is crucial for evaluating environmental risks and crafting effective mitigation plans. While it was once understood that these hydrophobic chemicals tended to impede their mobility within natural organic elements like soil and sediment, recent research has uncovered their greater attraction to a range of plastics, including polyethylene, polypropylene, and PVC (Teuten et al., 2007; Teuten et al., 2009).

PCBs were detected in polypropylene pellets from Japanese coasts (with concentrations ranging from 4 to 117 ng g^{-1}). The study indicated ambient seawater as the source of PCBs, and that adsorption to the pellet surfaces was the enrichment mechanism. Another adsorption experiment in 2002 proved the higher affinity of polyethylene for hydrophobic compounds, as it was found that polyethylene pellets adsorbed four times more PCBs than polypropylene pellets. The results of various scientific studies aligned with similar work done later in 2008. Comparing PCB concentrations in marine plastic resin pellets to those in seawater revealed a high degree of accumulation with an apparent adsorption coefficient of 105–106 (Teuten et al., 2009). Sorbent organic matter in the environment consists of organic polymers with both crystalline and amorphous regions. Sorption of chemicals typically occurs in the amorphous region, which is further classified as either glassy or rubbery based on its internal structure. In this context, sorbent organic matter can be thought of as a blend of glassy and rubbery polymers. The transition between these states is dictated by a polymer's glass transition temperature (T_g). Because of these dual sorption mechanisms, the sorptive behavior of glassy polymers is typically described using the nonlinear Freundlich model, where "q" represents the amount of sorbed compound per unit mass of solid, C_e is the aqueous-phase concentration at equilibrium, K_f is the Freundlich constant indicating sorbent capacity, and "n" is the Freundlich exponent signifying sorbent heterogeneity (Carter, 1995). Absorptive partitioning into an organic matrix is characterized by a linear sorption model ($q = K_pC_e$), where K_p is the partition coefficient. Crystalline polymers possess characteristics that lie between rubbery and glassy polymers due to their unique structure, offering moderate hardness, flexibility, and strength (Treloar, 1974).

3.3 Routes of Exposure (Micro, Macro, Plastic-Borne Chemicals)

Exposure pathway for humans is primarily impacted by five factors viz., potential source regions, transport pathways based on physicochemical properties, exposure routes (inhalation, dermal exposure, and diet) and carcinogenic and noncarcinogenic risk due to exposure (Chakraborty et al., 2023). Exposure to chemicals, including those associated with plastics, can occur through various routes, including the inhalation, ingestion, and dermal contact (Fig. 1):

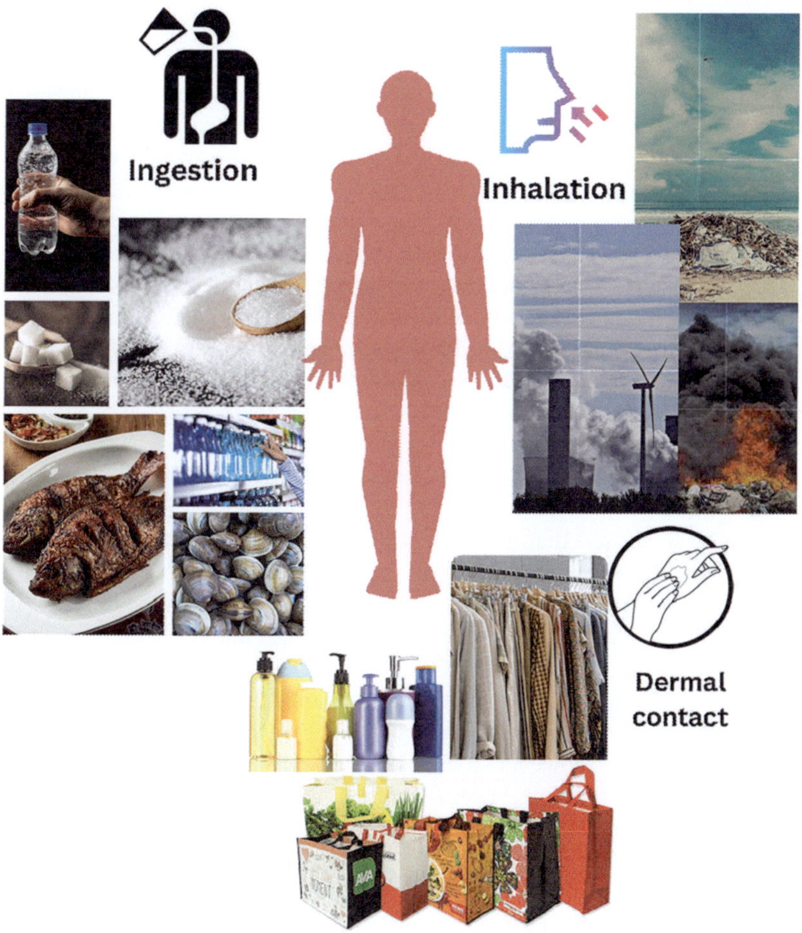

Fig. 1 Routes of exposure (micro, macro, plastic-borne chemicals)

3.3.1 Inhalation

Micro-level exposure can occur when tiny particles or aerosols containing chemicals are inhaled. This is often relevant in situations where microplastics or airborne plastic particles are present in the environment, and they carry or are coated with contaminants. Numerous studies confirm the presence of airborne microplastics in our environment, and humans are exposed to these pollutants daily. Particularly concerning is indoor exposure, as people spend significant hours at home and in offices where microplastic pollution is more concentrated than outdoors.

Previous research investigated the presence of microplastics in urban, suburban, and remote regions but recent research further revealed its occurrence in indoor and outdoor environments in urban settings. These studies allowed the identification of large-quantity alterations based on temporal and spatial patterns. For instance, the average wet and dry deposition of microplastics was 7 particles/m^2/day in Ontario (Canada), 118 particles/m^2/day in Paris (France), 275 particles/m^2/day in Hamburg (Germany), 36 particles/m^2/day in Dongguan (China), and 114 particles/m^2/day in Guangzhou (China). In comparison, the average deposition rate in remote regions such as the French Pyrenees mountains was 365 particles/m^2/day, and the Nam Co Basin in the Tibetan Plateau was 500 kg/year. High microplastic occurrences have been reported in remote regions without a local source, indicating it to be the most probable pathway of long-range transport of particles from urban areas. Urban studies have shown indoor microplastic abundances ranging from 0.4 to 59.4 particles/m^3 in Paris and a deposition rate of 309.40 particles/m^2/day in São Paolo (Brazil). Most microplastics identified in indoor dust samples were synthetic fibers from clothing materials, as well as furniture and surface paint fragments. Indoor quantities were higher than outdoor abundances (0.3–1.5 particles/m^3) in Paris and the deposition rate (123.20 particles/m^2/day) in São Paolo (Peries et al., 2024).

Prata, 2018 suggested that individuals may inhale 26–130 airborne MPs daily, while (Liu et al., 2019) estimated that Shanghai (China) residents inhale around 21 MPs daily from outdoor environments. Children and workers in the polymer industry face higher risks due to the greater abundance of ground-level microplastics and inadequate protection, respectively (Gasperi et al., 2018; Liu et al., 2019). When microplastics enter the human lungs, they encounter clearance mechanisms like sneezing, mucociliary escalators, phagocytosis by macrophages, and lymphatic transport (Prata, 2018). However, some MPs may evade these mechanisms and persist in the lungs (Pauly et al., 1998). Inhaled MPs can trigger inflammation by releasing intracellular messengers, proteases, and reactive oxygen species (ROS) (Gasperi et al., 2018; Wright & Kelly, 2017). They induce oxidative stress and ROS generation in certain species, leading to damage (Barboza et al., 2018; Jeong et al., 2016). Exposure to microplastics increases the level and activity of antioxidant-related enzymes like superoxide dismutase (SOD), catalase (CAT), glutathione peroxidases (GPx), and total glutathione (GSH) to protect organs from adverse effects (Chen et al., 2019).

3.3.2 Ingestion

Microplastics, owing to their diminutive size, can be consumed by lower-level aquatic organisms, and the repercussions on their health remain uncertain. In contrast, larger plastic items entangling and being ingested by vertebrates are widely recognized. Plastic ingestion is believed to affect over 250 marine species (Laist, 1997). A technical report assessing the impact of marine debris on biodiversity revealed that plastic was associated with over 80% of reported incidents between organisms and marine debris, while microplastics accounted for 11% of all reported encounters (GEF, 2012). Microplastics, with their extremely small size, are potentially accessible to a wide range of organisms, including low trophic suspension, filter, and deposit feeders, detritivores, and planktivores (Browne et al., 2008; Graham & Thompson, 2009; Murray & Cowie, 2011; Thompson et al., 2004). As a result, microplastics may accumulate within organisms, leading to physical harm such as internal abrasions and blockages. Moreover, ingestion of microplastics may introduce toxicity through the release of constituent contaminants like monomers and plastic additives, capable of causing carcinogenic and endocrine-disrupting effects (Oehlmann et al., 2009; Talsness et al., 2009). Microplastics can become heavily contaminated, sometimes up to six orders of magnitude greater than ambient seawater, with waterborne POPs (Hirai et al., 2011; Mato et al., 2001). This presents a potential pathway of exposure for marine organisms, allowing for bioaccumulation and biomagnification through the food chain.

The widespread use of plastic food containers, packaging, disposable cups, bottles, infant feeders, and plastic-coated paper cartons has raised concerns about direct contact and the release of plastic flakes in food and beverages. It is a common practice to store, transport, reheat, and eat food in plastic containers. People handle food and beverages in plastic containers without proper knowledge of the fact that plastic from these containers leaches into foods and beverages. Consequently, they end up directly consuming microplastics, which can harm their physiological systems and inhibit normal metabolic processes. Interaction between microplastics and human cells exacerbates cell damage and acute inflammation (Jadhav et al., 2021). Ingestion of microplastics in human beings can happen via food or drinking water intake. Globally, humans are expected to ingest 11,845 to 193,200 microplastics per year with the largest source being drinking water. In other words, on a global scale, individuals might potentially ingest an average of 0.1–5 g of microplastics per week. The quantity of ingested microplastics varies significantly among individuals and hinges on multiple factors. These factors encompass the characteristics of the microplastics, such as an individual's age, BMI, size, demographic profile, cultural background, geographical location, environmental surroundings, and lifestyle choices (Senathirajah et al., 2020). Microplastics have been detected in various species intended for human consumption, including invertebrates, crustaceans, and fish (Rochman, 2015; Van Cauwenberghe & Janssen, 2014). Estimates by researchers indicate that a high consumer of European shellfish may ingest approximately 11,000 plastic particles on an annual basis (Karami et al., 2017; Li et al., 2015). Furthermore, Van Cauwenberghe and Janssen (2014) revealed that farmed mussels had significantly higher concentrations of microplastics (178 microfibers) compared to their wild-caught counterparts (126 microfibers). Studies have documented the translocation of microplastic particles in various species, including zebrafish, European seabass, and common goby (*Pomatoschistus microps*) (de Sá et al., 2015; Lu et al., 2018). These findings confirm the presence of microplastics in seafood, although they do not address the specific chemical constituents. This suggests that the challenge of microplastic contamination in seafood could be widespread due to the ubiquity of microplastics in the environment and their potential to move within animals, potentially reaching parts typically consumed by humans.

Food and beverage packaging materials made from plastics have found widespread use in the food industry. They are highly valued for their ability to store fresh produce and other food items, control temperature and atmosphere within the packaging, their lightweight nature, and the ease with which they can be manufactured and processed (Bott et al., 2014). Beyond extending the shelf life of perishable goods, plastic packaging also serves to safeguard food against microbial and chemical contamination, as well as physical damage during distribution and marketing (Rydz et al., 2018). Microplastics have been detected in a diverse range of food and beverage products, including honey, sugar, table salt, beer, mineral and tap water, milk, and seafood such as bivalves, crustaceans, and fish (Liebezeit and Liebezeit, 2014, 2015; Iñiguez et al., 2017; Kosuth et al., 2018; Kutralam-Muniasamy et al., 2020; Van Cauwenberghe & Janssen, 2014; Devriese et al., 2015; Akhbarizadeh et al., 2018; Karami et al., 2018). These microplastics, originating from various sources, pose a growing concern as they are directly ingested by humans (Oßmann et al., 2018). They also contaminate drinking water sources, making them a significant route of microplastic exposure (Marsden et al., 2019). In 2019, Cox et al. estimated that the daily human consumption of microplastics through the air and commonly consumed foods ranges from 203 to 332 particles per person (2019). Research findings indicate that individuals may ingest between 37 million to as high as billions of plastic particles each year from various food products (Fadare et al., 2020). Microplastics in seafood products have been linked to marine plastic pollution, and they have even been detected in human stools, indicating their presence in the human digestive system due to food consumption (Schwabl et al., 2019). While scientific evidence regarding the direct or indirect health hazards resulting from human exposure to plastics or microplastics is currently lacking (Rist et al., 2018), there is a growing concern about potential health risks. Vigilance in monitoring these issues is therefore paramount.

In vitro testing has revealed the cytotoxic effects of environmental plastic particles (Furukuma and Fuji 2016). Choi et al. (2020) demonstrated that exposure to irregularly shaped polystyrene microplastics (PS MPs) significantly influenced the release of IL-6 and TNF-a, even at low concentrations of 100 μg/mL during 24-h experiments. Choi et al. (2021) reported that both spherical polyethylene (PE) and irregular low-density polyethylene (LDPE) MPs, at concentrations of 500 to 1000 μg/mL during 96-h experiments, had a significant effect on the same biomarkers. Choi et al. (2021) found that both spherical polyethylene (PE) and irregular low-density polyethylene (LDPE) MPs, at concentrations ranging from 500 to

1000 μg/mL, had a significant effect on biomarkers over 96 h. Hwang et al. (2020) found similar effects on these markers caused by spherical PS MPs ranging from 0.46 to 10 μm at a concentration of 500 μg/mL in both 4-h and 96-h exposures. Additionally, Liu et al. (2020) noted the impact on IL-8 and MCP-1 release induced by irregular PS MPs (0.404 μm) at a very low concentration of 20 μg/mL during 96-h tests. It is worth mentioning that Liu et al. (2020) was the sole study investigating MCP-1, while other studies focused on IL-8. Dong et al. (2020) reported that both IL-6 and IL-8 were affected by spherical PS MPs (1.72 μm) at concentrations of 10 and 1000 $\mu g/cm^2$ after 24-h exposures. Histamine release was assessed by four studies (Choi et al., 2021; Han et al., 2020; Hwang et al., 2019, 2020) using the HMC-1 cell model. They examined five different polymers and utilized two distinct tests (ELISA kit and histamine assay). Most of these studies (Choi et al., 2021; Hwang et al., 2020) tested two polymers, PE and PS, in sizes ranging from 5.5 to 100 μm and MP concentrations between 10 and 1000 μg/mL for PE and 0.46–100 μm and MP concentrations of 500 μg/mL for PS. It's important to note that all tested MPs were spherical in shape.

Wu et al. (2019) reported a significant increase in intracellular reactive oxygen species (ROS) production after exposure to spherical PS MPs, specifically those with sizes of 0.1 and 5 μm, at a concentration of 200 μg/mL using Caco-2 cells. Dong et al. (2020) observed a similar response after exposure to 1.72 μm spherical PS MPs using BEAS-2B cells at a concentration of 1000 $\mu g/cm^2$. Genotoxicity was assessed by Hesler et al., 2019, who used a p53 reporter and exposed Caco-2 cells to spherical 0.5 μm PS MPs (up to 10 μg/mL), but all results were nonsignificant. Liu et al. (2020) was the only study to investigate paracellular transport, assessing the expression of ZO-1 and Occludin using qPCR. They reported a significant down-regulation only after exposure to nanoparticles (NPs).

Microplastic uptake and internalization were the focus of seven studies, with most of them employing qualitative measures. Hesler et al. (2019) noted that spherical PS MPs (0.5 μm) were internalized by both cocultures they used after 24-h exposure, with translocation of MPs detected in the apical but not basolateral compartments of the models. Stock et al. (2021) exposed various MPs (PE, PP, PET, PVC) to a Caco-2 trans-well model and found intracellular uptake was evident only for spherical PE MPs (1–4 μm) via microscopic examination and fluorescence quantification of the cell membranes. Wu et al. (2019) reported that both 0.1 and 5 μm spherical PS MPs entered Caco-2 cells after a 12-h exposure. Furthermore, Goodman et al. (2022) confirmed the internalization of 1 μm spherical PS MPs during exposures ranging from 24 to 96 h using flow cytometry (Calcein AM and Ki67 assays) and phase-contrast microscopy with A549 cells. It's important to note that the specific routes and extent of exposure can vary depending on factors such as the type of plastic, chemical composition of plastic, environmental conditions, and individual behaviors.

Additionally, there is ongoing research to better understand the potential health effects of exposure to plastic-borne chemicals, especially concerning microplastics and their associated contaminants. Reducing plastic use, proper disposal, and regulatory measures are among the strategies being considered to mitigate potential risks associated with these exposure routes. A well-executed study conducted recently tracked the urinary excretion of phthalates in individuals who fasted for 48 h has revealed that diet is the primary pathway for exposures to DEHP, DiNP, and DiDP, whereas nonfood exposures are primarily associated with DMP, DEP, DiBP, DnBP, and BBzP (Koch et al., 2013). Phthalates have the potential to transfer into food products from PVC materials that have been softened, such as tubing commonly utilized during milking, lid seals, films used in food packaging, gloves worn during food preparation, and conveyor belts (Cao et al., 2010). Despite the limited number of studies conducted in India, recent comprehensive research from the country has uncovered that the primary exposure pathways for PAEs in humans are through dietary intake and inhalation (Mukhopadhyay et al., 2022; Chakraborty et al., 2021, 2022).

Polycystic Ovary Syndrome (PCOS) is a lifestyle-related condition with a notable prevalence rate, particularly among Indian women, estimated at around 10% (Bharali et al., 2022). In a pilot cross-sectional study carried out in Tamil Nadu, young adolescent females displayed a higher prevalence of 18% for PCOS (Balaji et al., 2015). Srnovršnik et al. (2023) have underscored the connection between EDCs and disturbances in reproductive and metabolic functions, including PCOS. They discovered elevated levels of phenolic EDCs such as BPA, triclosan, and nonyl phenol, which are known endocrine disruptors, in maternal blood plasma and amniotic fluid among pregnant women. This marks the first study in India to report on transplacental transfer (Shekhar et al., 2017). The timing of exposure to many EDCs is critical for their impact on reproductive health. Notably, prenatal and early postnatal exposure to BPA is believed to influence long-term ovarian functions, trigger the development of PCOS, and disrupt hormone profiles and metabolism. Environmental chemicals have the potential to disrupt the Hypothalamic-Pituitary-Ovarian (HPO) axis, lead to pregnancy-related insulin resistance, resulting in decreased glucose tolerance, and elevate insulin, triglycerides, glycerol, and leptin levels, affecting fetal growth and metabolic patterns (Barrett et al., 2014). In PCOS, androgens are known to inhibit Th1 differentiation by interfering with interferon-gamma (IFN-γ) and IL-12 production. This is associated with an increase in the Th1/Th2 ratio and Th17/Th2 ratio during the follicular phase in women with PCOS (Johnson & Laloraya, 2022).

3.3.3 Dermal Contact

Considering the prevalent occurrence of microplastics and plastic additives detected in indoor dust, along with the deposition of these particles from both indoor and outdoor air, the widespread utilization of microbeads in cosmetics, and the continuous degradation of microfibers from textiles, it is reasonable to consider dermal contact as a potential pathway for human exposure to these contaminants. Although human skin is recognized for its effective barrier against the penetration of larger particles, it has been proposed that only particles smaller than 100 nm (nanoplastics) may directly penetrate the dermal barrier. Nonetheless, there exist alternative avenues for larger particles to potentially penetrate the skin, such as through hair follicles, sweat glands, or open skin injuries. A recent study has indicated that human exposure to additives employed in microfibers, when inhaled, is estimated to range from 4.5 to 6440 μg/per person annually, while ingestion may result in exposure levels of 0.1 to 204 μg/per person. Interestingly, the release of additives from face masks through inhalation is estimated to be around 491 to 1820 μg/per person (Chen et al., 2022). Microplastic beads, typically measuring less than 1 mm in diameter, have found widespread application in dermal exfoliation and cleansing products, as well as in toothpaste and denture fillings (Anagnosti et al., 2021). Limited research has focused on estimating the per person consumption of microbeads stemming from the use of specific personal care items. For instance, a study examining various brands of facial scrubs in the UK revealed microplastic content ranging from 10 to 100 g per liter, resulting in per person daily consumption ranging from 40.5 to 215 mg (Napper et al., 2015). Additionally, Gouin et al. 2011 projected an average daily per person consumption of 2.4 mg of microplastics in the United States (US) population attributable to the use of liquid soap (Gouin et al., 2011). The US Microbead-Free Waters Act of 2015 prohibits the production, packaging, and distribution of rinse-off cosmetics using plastic microbeads. It also applies to both cosmetic and nonprescription items, such as toothpaste. Similarly, several articles have been published on the presence of plastic additives, specifically phthalates, in a wide variety of personal care products (Guo and Kannan 2013; Koniecki et al., 2011; Liao and Kannan 2014).

3.4 *Human, Societal, and Environmental Risks Associated with Plastic and Chemical Interlinkages*

Plastic and chemical interlinkages pose a range of human health, societal, and environmental risks, which are interconnected and can have far-reaching consequences.

3.4.1 Human Health Risks

All along the plastic lifecycle, humans are exposed to several toxic chemicals and microplastics through inhalation, ingestion, and direct skin contact. As discussed, prolonged exposure to plastics have several toxic effects on human health causing hormonal disruptions, reproductive issues, neurodegenerative diseases, certain cancers, cardiovascular diseases, chronic inflammation, auto-immune conditions, neurodegenerative diseases, and stroke. Studies indicate that, on average, individual may ingest 5 g of plastic per week (WWF Australia, 2019).

The plastic lifecycle stages include extraction/transport of raw materials, design and manufacture, packaging and distribution of consumer products, use and maintenance, and waste management (recycling, reuse, recovery, or final disposal). Workers involved in plastic production face heightened risks, such as leukemia, lymphoma, hepatic angiosarcoma, brain cancer, breast cancer, mesothelioma, neurotoxic injuries, and decreased fertility. Meanwhile, those producing plastic textiles experience elevated mortality rates due to bladder cancer, lung cancer, mesothelioma, and interstitial lung disease. Health impacts ripple throughout the plastic value chain, with plastic recycling workers exhibiting increased rates of cardiovascular disease, toxic metal poisoning, neuropathy, and lung cancer (Landrigan et al., 2023).

Macro-or microplastics that are released into the environment contaminate and build up in food systems through agricultural soils, terrestrial and aquatic food chains, and water supplies. As plastics degrade, they break down into smaller particles or microplastics (less than 5 mm in length), additives and monomers causing several health hazards in humans and animals.

It is crucial to understand the potential transfer of microplastics and associated toxic chemicals to crops, and human tissues. At present, there is a scarcity of comprehensive scientific research on the direct effects of plastic exposure on human health and the consequential diseases. This underscores the urgency for more extensive and rigorous studies to better understand and address these interconnected risks posed by plastics and chemicals to both human health and the environment (Davey, 2023).

3.4.2 Environmental Risks

Plastic is an emerging environmental threat due to its abundance, durability, and persistence. Plastic pollution stems from inadequate collection and waste disposal systems. Plastic waste, especially when not properly managed, can accumulate in ecosystems contributing to microplastic pollution in rivers, oceans, and soil. Accumulation of mismanaged plastic waste results in flooding in low-lying areas (by clogging of drainage systems) and also contributes to visual pollution. Microplastics generally emerge from industrial plastic pellets, synthetic textiles, road markings, and tire wear.

Only 9% of plastic is recycled or incinerated in waste-to-energy facilities (OECD, 2022). Globally, about 49% of it is disposed of in landfills, where it remains in the environment for almost a thousand years, and then decomposes and releases harmful chemicals into the soil and water (UNEP, 2021a). In India, 46% of plastic remains mismanaged or as uncollected litter. Research indicates that the impact of microplastic pollution in terrestrial ecosystems (soils, sediments, and freshwater) has long-term and higher-level impacts (about 4–23 times) than marine microplastic pollution. Also, terrestrial microplastic pollution has led to a decrease in the number of subsurface soil species (mites, larvae) with potential consequences on soil carbon and nutrient cycling (Dunmei et al., 2020).

An important source of microplastic pollution in water is through laundry wherein tiny cloth fibers (nylon, spandex, polyester) are transported to wastewater treatment plants or discharged into the open environment. About 80–90% of microplastic particles in sewage are from cloth fibers and it remains in the sludge A staggering 1.5 million trillion microscopic plastic fibers are present in the ocean, which are mistakenly ingested by marine animals impacting their health (Mishra et al., 2019). About 17% of species affected by entanglement and ingestion of plastics were listed as threatened or near threatened, as per IUCN list. Another plastic pollutant is the microbead, a solid plastic particle (10 μm to 1 mm), emerging from cosmetics and personal care products.

3.4.3 Societal Risks

The impacts of plastic pollution are extensive, affecting lives and livelihoods of communities, especially those living in vulnerable situations, including women, children, poor, migrants, indigenous people, and those with special needs. Some of the societal challenges of plastic pollution are discussed below.

- *Waste Management Challenges:* Increased plastics usage and disposal causes a growing waste management challenge. Poorly managed plastic waste can lead to litter, clogged drainage systems, increased municipal costs, and hence affect the overall quality of life of people. Government initiatives adopted earlier to reduce the consumption of single-use plastic were reversed during the COVID-19 pandemic, which lead to a big rise in the use of disposable plastic products (masks, gloves, sanitizer bottles, PPE suits, test kits, food takeout containers, delivery packaging, etc).
- *Resource Depletion:* The production of plastics relies heavily on fossil fuels and natural resources. The overextraction and consumption of these resources contribute to habitat destruction, land degradation, and resource depletion.
- *Loss of Ecosystem Services:* The presence of plastic and chemical waste in any ecosystem can disrupt the natural processes and reduce its ability to provide essential services such as clean water and climate regulation.
- *Resistance and Adaptation:* The overuse of chemicals, such as pesticides and antimicrobial agents in plastics, can lead to the development of resistance in pests and microorganisms, posing a big challenge.
- *Climate Change:* Greenhouse gas emissions result from the manufacture of chemicals and plastics, including the extraction and processing of raw materials. Being one of the most energy-intensive manufacturing processes, plastic production is closely related to the petrochemical industry, which is a substantial source of greenhouse gas emissions.
- *Economic Impact:* Local economies may be strained by the price of cleaning up plastic pollution, medical expenses related to chemical exposure, and lost tourism revenue. In terms of the value of benefits derived from marine ecosystem services, up to $2500 billion is lost each year due to marine plastic pollution (Beaumont et al., 2019).

- *Inequity and Environmental Justice:* Plastic pollution is an environmental injustice issue. It disproportionately affects underprivileged communities and those living near plastic production/waste sites (UNEP, 2021b). For example, the occupational hazards of about two million informal waste pickers (including women and children) in India is a big challenge. The societal impacts of plastic waste can be far-reaching: it includes deforestation and displacement of local communities living in or near oil extraction project sites, health issues of people living in the vicinity of such sites, as well as contamination of potable water by plastic-containing wastewater. Plastic waste threatens marine-dependent livelihoods, as well as causes health issues for people who consume seafood infested with toxic micro-and nanoplastics.

4 Conclusion

To manage the existing and projected impacts of plastic pollution, national and local governments may adopt regulatory solutions and restrict the use and import of specific plastics and related chemicals. To encourage reuse, plastic carry bags' allowed thickness increased from 50 to 75 microns by September 30, 2021, and to 120 microns by December 31, 2022. There's also a blanket ban on sachets using plastic for pan masala, gutkha, and tobacco. Additionally, the Ministry of Environment, Forest and Climate Change (MoEFCC) issued guidelines on Extended Producer Responsibility (EPR) for plastic packaging through the Plastic Waste Management (Amendment) Rules, 2022, specifying mandatory targets on EPR, plastic packaging recycling, reuse of rigid plastic packaging, and recycled plastic content usage. Efforts must be made to eliminate unnecessary packaging, implement extended producer responsibility (EPR) wherever possible, as well as develop and adopt alternatives to single-use plastics. Capacity building, education, and awareness are effective tools to support these actions by informing people of the consequences of plastic consumption and encouraging them to choose alternatives. Addressing the health, environmental, and societal risks associated with plastic and chemical interlinkages by adoption of such comprehensive strategies will pave way towards the achievement of Sustainable Development Goals (SDGs), especially SDG1 (no poverty), SDG2 (zero hunger), SDG3 (good health and well-being), SDG14 (protecting marine ecosystems), and SDG16 (providing access to justice for all). By embracing these concerted efforts, we can pave the way for a sustainable future that prioritizes human well-being, environmental integrity, and social equity.

Acknowledgments This research was supported by SRMIST's institutional fund given to the Centre for Research in Environment, Sustainability Advocacy and Climate Change (REACH), Directorate of Research the Royal Norwegian Embassy through the INOPOL Project (Project No. IND-19/0005 and Project No: IND-22/0007).

References

Akhbarizadeh, R., Moore, F., & Keshavarzi, B. (2018). Investigating a probable relationship between microplastics and potentially toxic elements in fish muscles from northeast of Persian Gulf. *Environmental Pollution, 232*, 154–163.

Anagnosti, L., Varvaresou, A., Pavlou, P., Protopapa, E., & Carayanni, V. (2021). Worldwide actions against plastic pollution from microbeads and microplastics in cosmetics focusing on European policies. Has the issue been handled effectively? *Marine Pollution Bulletin, 162*, 111883.

Bakir, A., O'Connor, I. A., Rowland, S. J., Hendriks, A. J., & Thompson, R. C. (2016). Relative importance of microplastics as a pathway for the transfer of hydrophobic organic chemicals to marine life. *Environmental Pollution, 219*, 56–65.

Balaji, S., Amadi, C., Prasad, S., et al. (2015). Urban rural comparisons of polycystic ovary syndrome burden among adolescent girls in a hospital setting in India. *BioMed Research International, 2015*, 158951. https://doi.org/10.1155/2015/158951

Bandow, N., Will, V., Wachtendorf, V., & Simon, F.-G. (2017). Contaminant release from aged microplastic. *Environmental Chemistry, 14*(6), 394–405.

Barboza, L. G. A., Vethaak, A. D., Lavorante, B. R., Lundebye, A.-K., & Guilhermino, L. (2018). Marine microplastic debris: An emerging issue for food security, food safety and human health. *Marine Pollution Bulletin, 133*, 336–348.

Barrett, E. S., & Sobolewski, M. (2014, May). Polycystic ovary syndrome: Do endocrine-disrupting chemicals play a role? *Seminars in Reproductive Medicine, 32*(03), 166–176.

Beaumont, N. J., et al. (2019). Global ecological, social and economic impacts of marine plastic. *Marine Pollution Bulletin, 142*, 189–195. https://doi.org/10.1016/j.marpolbul.2019.03.022. Epub 2019 Mar 27.

Beldi, G., Pastorelli, S., Franchini, F., & Simoneau, C. (2012). Time-and temperature-dependent migration studies of Irganox 1076 from plastics into foods and food simulants. *Food Additives & Contaminants: Part A, 29*(5), 836–845.

Bharali, M. D., Rajendran, R., Goswami, J., Singal, K., & Rajendran, V. (2022). Prevalence of polycystic ovarian syndrome in India: A systematic review and meta-analysis. *Cureus, 14*(12).

Bott, J., Störmer, A., & Franz, R. (2014). A model study into the migration potential of nanoparticles from plastics nanocomposites for food contact. *Food Packaging and Shelf Life, 2*(2), 73–80.

Boyer, R. (1951). Relation of tensile strength to brittle temperature in plasticized polymers. *Journal of Applied Physics, 22*(6), 723–729.

Bridson, J. H., Gaugler, E. C., Smith, D. A., Northcott, G. L., & Gaw, S. (2021). Leaching and extraction of additives from plastic pollution to inform environmental risk: A multidisciplinary review of analytical approaches. *Journal of Hazardous Materials, 414*, 125571.

Browne, M. A., Dissanayake, A., Galloway, T. S., Lowe, D. M., & Thompson, R. C. (2008). Ingested microscopic plastic translocates to the circulatory system of the mussel, Mytilus edulis (L.). *Environmental Science & Technology, 42*(13), 5026–5031.

Cao, X.-L. (2010). Phthalate esters in foods: Sources, occurrence, and analytical methods. *Comprehensive Reviews in Food Science and Food Safety, 9*, 21–43. https://doi.org/10.1111/j.1541-4337.2009.00093.x

Carter, K. R. (1995). Aryl fluoride monomers in nucleophilic aromatic substitution polymerization: Evaluation of monomer reactivity by 19F NMR spectroscopy. *Macromolecules, 28*(19), 6462–6470.

Chakraborty, P., Zhang, G., Li, J., Selvaraj, S., Breivik, K., & Jones, K. C. (2016). Soil concentrations, occurrence, sources and estimation of air–soil exchange of polychlorinated biphenyls in Indian cities. *Science of the Total Environment, 562*, 928–934.

Chakraborty, P., Selvaraj, S., Nakamura, M., Prithiviraj, B., Cincinelli, A., & Bang, J. J. (2018). PCBs and PCDD/fs in soil from informal e-waste recycling sites and open dumpsites in India: Levels, congener profiles and health risk assessment. *Science of the Total Environment, 621*, 930–938.

Chakraborty, P., Sampath, S., Mukhopadhyay, M., Selvaraj, S., Bharat, G. K., & Nizzetto, L. (2019). Baseline investigation on plasticizers, bisphenol a, polycyclic aromatic hydrocarbons and heavy metals in the surface soil of the informal electronic waste recycling workshops and nearby open dumpsites in Indian metropolitan cities. *Environmental Pollution, 248*, 1036–1045.

Chakraborty, P., Gadhavi, H., Prithiviraj, B., Mukhopadhyay, M., Khuman, S. N., Nakamura, M., & Spak, S. N. (2021). Passive air sampling of PCDD/Fs, PCBs, PAEs, DEHA, and PAHs from informal electronic waste recycling and allied sectors in Indian megacities. *Environmental Science & Technology, 55*(14), 9469–9478.

Chakraborty, P., Bharat, G. K., Gaonkar, O., Mukhopadhyay, M., Chandra, S., Steindal, E. H., & Nizzetto, L. (2022). Endocrine-disrupting chemicals used as common plastic additives: Levels, profiles, and human dietary exposure from the Indian food basket. *Science of the Total Environment, 810*, 152200.

Chen, Q., Allgeier, A., Yin, D., & Hollert, H. (2019). Leaching of endocrine disrupting chemicals from marine microplastics and mesoplastics under common life stress conditions. *Environment International, 130*, 104938.

Chen, Y., Chen, Q., Zhang, Q., Zuo, C., & Shi, H. (2022). An overview of chemical additives on (micro) plastic fibers: Occurrence, release, and health risks. *Reviews of Environmental Contamination and Toxicology, 260*(1), 22.

Choi, J. S., Hong, S. H., & Park, J. W. (2020). Evaluation of microplastic toxicity in accordance with different sizes and exposure times in the marine copepod Tigriopus japonicus. *Marine Environmental Research, 153*, 104838.

Choi, D., Hwang, J., Bang, J., Han, S., Kim, T., Oh, Y., et al. (2021). In vitro toxicity from a physical perspective of polyethylene microplastics based on statistical curvature change analysis. *Science of the Total Environment, 752*, 142242.

Covello, C., Di Vincenzo, F., Cammarota, G., & Pizzoferrato, M. (2024 Mar 21). Micro(nano) plastics and their potential impact on human gut health: A narrative review. *Current Issues in Molecular Biology, 46*(3), 2658–2677. https://doi.org/10.3390/cimb46030168. PMID: 38534784; PMCID: PMC10968954

Cox, K. D., Covernton, G. A., Davies, H. L., Dower, J. F., Juanes, F., & Dudas, S. E. (2019). Human consumption of microplastics. *Environmental Science & Technology, 53*(12), 7068–7074.

Cuadros-Rodríguez, L., Lazúen-Muros, M., Ruiz-Samblás, C., & Navas-Iglesias, N. (2020). Leachables from plastic materials in contact with drugs. State of the art and review of current analytical approaches. *International Journal of Pharmaceutics, 583*, 119332.

Davey. (2023). *The Guardian*. Plastics cause wide-ranging health issues from cancer to birth defects, landmark study finds. https://www.theguardian.com/environment/2023/mar/29/plastics-cause-wide-ranging-health-issues-from-cancer-to-birth-defects-landmark-study-finds

de Sá, L. C., Luís, L. G., & Guilhermino, L. (2015). Effects of microplastics on juveniles of the common goby (Pomatoschistus microps): Confusion with prey, reduction of the predatory performance and efficiency, and possible influence of developmental conditions. *Environmental Pollution, 196*, 359–362.

Devriese, L. I., Van der Meulen, M. D., Maes, T., Bekaert, K., Paul-Pont, I., Frère, L., et al. (2015). Microplastic contamination in brown shrimp (Crangon crangon, Linnaeus 1758) from coastal waters of the Southern North Sea and Channel area. *Marine Pollution Bulletin, 98*(1–2), 179–187.

Dong, C. D., Chen, C. W., Chen, Y. C., Chen, H. H., Lee, J. S., & Lin, C. H. (2020). Polystyrene microplastic particles: In vitro pulmonary toxicity assessment. *Journal of Hazardous Materials, 385*, 121575.

Dunmei, L., et al. (2020). Microplastics negatively affect soil fauna but stimulate microbial activity: Insights from a field-based microplastic addition experiment. *Proceedings of The Royal Society B, 287*, 2020126820201268. https://doi.org/10.1098/rspb.2020.1268

Fadare, O. O., Wan, B., Guo, L. H., & Zhao, L. (2020). Microplastics from consumer plastic food containers: are we consuming it? *Chemosphere, 253*, 126787.

Furukuma, S., & Fujii, N. (2016). In vitro cytotoxicity evaluation of plastic marine debris by colony-forming assay. *Japanese Journal of Environmental Toxicology, 19*(2), 71–81.

Gasperi, J., Wright, S. L., Dris, R., Collard, F., Mandin, C., Guerrouache, M., Langlois, V., Kelly, F. J., & Tassin, B. (2018). Microplastics in air: Are we breathing it in? *Current Opinion in Environmental Science & Health, 1*, 1–5.

Giuliani, D., Dickson, M. M., Espa, G., & Santi, F. (2020). Modelling and predicting the spatio-temporal spread of COVID-19 in Italy. *BMC Infectious Diseases, 20*(1), 1–10.

Goodman, K. E., Hua, T., & Sang, Q. X. A. (2022). Effects of polystyrene microplastics on human kidney and liver cell morphology, cellular proliferation, and metabolism. *ACS Omega, 7*(38), 34136–34153.

Gouin, T., Roche, N., Lohmann, R., & Hodges, G. (2011). A thermodynamic approach for assessing the environmental exposure of chemicals absorbed to microplastic. *Environmental Science & Technology, 45*(4), 1466–1472.

Graham, E. R., & Thompson, J. T. (2009). Deposit-and suspension-feeding sea cucumbers (Echinodermata) ingest plastic fragments. *Journal of Experimental Marine Biology and Ecology, 368*(1), 22–29.

Guo, Y., & Kannan, K. (2013). A survey of phthalates and parabens in personal care products from the United States and its implications for human exposure. *Environmental Science & Technology, 47*(24), 14442–14449.

Hesler, M., Aengenheister, L., Ellinger, B., Drexel, R., Straskraba, S., Jost, C., Wagner, S., Meier, F., von Briesen, H., Büchel, C., & Wick, P. (2019). Multi-endpoint toxicological assessment of polystyrene nano-and microparticles in different biological models in vitro. *Toxicology In Vitro, 61*, 104610.

Hirai, H., Takada, H., Ogata, Y., Yamashita, R., Mizukawa, K., Saha, M., Kwan, C., Moore, C., Gray, H., & Laursen, D. (2011). Organic micropollutants in marine plastics debris from the open ocean and remote and urban beaches. *Marine Pollution Bulletin, 62*(8), 1683–1692.

Horton, A. A., Walton, A., Spurgeon, D. J., Lahive, E., & Svendsen, C. (2017 May 15). Microplastics in freshwater and terrestrial environments: Evaluating the current understanding to identify the knowledge gaps and future research priorities. *Science Total Environment, 586*, 127–141. https://doi.org/10.1016/j.scitotenv.2017.01.190. Epub 2017 Feb 4. PMID: 28169032

Hwang, J., Choi, D., Han, S., Choi, J., & Hong, J. (2019). An assessment of the toxicity of polypropylene microplastics in human derived cells. *Science of the Total Environment, 684*, 657–669.

Hwang, J., Choi, D., Han, S., Jung, S. Y., Choi, J., & Hong, J. (2020). Potential toxicity of polystyrene microplastic particles. *Scientific Reports, 10*(1), 7391.

Iñiguez, M. E., Conesa, J. A., & Fullana, A. (2017). Microplastics in Spanish table salt. *Scientific Reports, 7*(1), 8620.

Jadhav, E. B., Sankhla, M. S., Bhat, R. A., & Bhagat, D. S. (2021). Microplastics from food packaging: An overview of human consumption, health threats, and alternative solutions. *Environmental Nanotechnology, Monitoring & Management, 16*, 100608. https://doi.org/10.1016/j.enmm.2021.100608

Jeong, C.-B., Won, E.-J., Kang, H.-M., Lee, M.-C., Hwang, D.-S., Hwang, U.-K., Zhou, B., Souissi, S., Lee, S.-J., & Lee, J.-S. (2016). Microplastic size-dependent toxicity, oxidative stress induction, and p-JNK and p-p38 activation in the monogonont rotifer (Brachionus koreanus). *Environmental Science & Technology, 50*(16), 8849–8857.

Johnson, B. S., & Laloraya, M. (2022). Polycystic ovary syndrome and immune deregulation: What do hormones want to say? *Exploration of Immunology, 2*(3), 393–413.

Karami, A., Golieskardi, A., Ho, Y. B., Larat, V., & Salamatinia, B. (2017). Microplastics in eviscerated flesh and excised organs of dried fish. *Scientific Reports, 7*(1), 5473.

Karami, A., Golieskardi, A., Choo, C. K., Larat, V., Karbalaei, S., & Salamatinia, B. (2018). Microplastic and mesoplastic contamination in canned sardines and sprats. *Science of the Total Environment, 612*, 1380–1386.

Koch, H. M., Lorber, M., Christensen, K. L., Pälmke, C., Koslitz, S., & Brüning, T. (2013). Identifying sources of phthalate exposure with human biomonitoring: results of a 48 h fasting study with urine collection and personal activity patterns. *International Journal of Hygiene and Environmental Health, 216*(6), 672–681.

Koniecki, D., Wang, R., Moody, R. P., & Zhu, J. (2011). Phthalates in cosmetic and personal care products: Concentrations and possible dermal exposure. *Environmental Research, 111*(3), 329–336.

Kosuth, M., Mason, S. A., & Wattenberg, E. V. (2018). Anthropogenic contamination of tap water, beer, and sea salt. *PLoS One, 13*(4), e0194970.

Kutralam-Muniasamy, G., Pérez-Guevara, F., Elizalde-Martínez, I., & Shruti, V. C. (2020). Branded milks—Are they immune from microplastics contamination? *Science of the Total Environment, 714*, 136823.

Laist, D. W. (1997). Impacts of marine debris: Entanglement of marine life in marine debris including a comprehensive list of species with entanglement and ingestion records. In *Marine debris: Sources, impacts, and solutions* (pp. 99–139). Springer.

Landrigan, et al. (2023). The Minderoo-Monaco commission on plastics and human health. *Annals of Global Health, 89*(1), 23. https://doi.org/10.5334/aogh.4056. PMID: 36969097; PMCID: PMC10038118.

Lau, O.-W., & Wong, S.-K. (2000). Contamination in food from packaging material. *Journal of Chromatography A, 882*(1-2), 255–270.

Li, J., Yang, D., Li, L., Jabeen, K., & Shi, H. (2015). Microplastics in commercial bivalves from China. *Environmental Pollution, 207*, 190–195.

Liao, C., & Kannan, K. (2014). A survey of alkylphenols, bisphenols, and triclosan in personal care products from China and the United States. *Archives of Environmental Contamination and Toxicology, 67*, 50–59.

Liebezeit, G., & Liebezeit, E. (2014). Synthetic particles as contaminants in German beers. *Food Additives & Contaminants: Part A, 31*(9), 1574–1578.

Liebezeit, G., & Liebezeit, E. (2015). Origin of synthetic particles in honeys. *Polish Journal of Food and Nutrition Sciences, 65*(2), 143–147.

Liu, K., Wang, X., Wei, N., Song, Z., & Li, D. (2019). Accurate quantification and transport estimation of suspended atmospheric microplastics in megacities: Implications for human health. *Environment International, 132*, 105127.

Liu, S., Wu, X., Gu, W., Yu, J., & Wu, B. (2020). Influence of the digestive process on intestinal toxicity of polystyrene microplastics as determined by in vitro Caco-2 models. *Chemosphere, 256*, 127204.

Lu, K., Qiao, R., An, H., & Zhang, Y. (2018). Influence of microplastics on the accumulation and chronic toxic effects of cadmium in zebrafish (Danio rerio). *Chemosphere, 202*, 514–520.

Marsden, P., Koelmans, A. A., Bourdon-Lacombe, J., Gouin, T., D'anglada, L., Cunliffe, D., Jarvis, P., Fawell, J., & De France, J. (2019). *Microplastics in drinking water*. World Health Organization. https://shorturl.at/auF0i

Mato, Y., Isobe, T., Takada, H., Kanehiro, H., Ohtake, C., & Kaminuma, T. (2001). Plastic resin pellets as a transport medium for toxic chemicals in the marine environment. *Environmental Science & Technology, 35*(2), 318–324.

Meeker, J. D., Sathyanarayana, S., & Swan, S. H. (2009). Phthalates and other additives in plastics: Human exposure and associated health outcomes. *Philosophical Transactions of the Royal Society B: Biological Sciences, 364*(1526), 2097–2113.

Mishra, S., Rath, C. C., & Das, A. P. (2019). Marine microfiber pollution: A review on present status and future challenges. *Marine Pollution Bulletin, 140*, 188–197. https://doi.org/10.1016/j.marpolbul.2019.01.039. Epub Jan 25. PMID: 30803634.

Mukhopadhyay, M., Jalal, M., Vignesh, G., Ziauddin, M., Sampath, S., Bharat, G. K., Nizzetto, L., & Chakraborty, P. (2022). Migration of plasticizers from polyethylene terephthalate and low-density polyethylene casing into bottled water: A case study from India. *Bulletin of Environmental Contamination and Toxicology, 109*(6), 949–955.

Murray, F., & Cowie, P. R. (2011). Plastic contamination in the decapod crustacean Nephrops norvegicus (Linnaeus, 1758). *Marine Pollution Bulletin, 62*(6), 1207–1217.

Napper, I. E., Bakir, A., Rowland, S. J., & Thompson, R. C. (2015). Characterisation, quantity and sorptive properties of microplastics extracted from cosmetics. *Marine Pollution Bulletin, 99*(1–2), 178–185.

Oßmann, B. E., Sarau, G., Holtmannspötter, H., Pischetsrieder, M., Christiansen, S. H., & Dicke, W. (2018). Small-sized microplastics and pigmented particles in bottled mineral water. *Water Research, 141*, 307–316.

OECD. (2022). *Plastic pollution is growing relentlessly as waste management and recycling fall short, says OECD*. OECD. https://www.oecd.org/environment/plastic-pollution-is-growing-relentlessly-as-waste-management-and-recycling-fall-short.htm

Oehlmann, J., Schulte-Oehlmann, U., Kloas, W., Jagnytsch, O., Lutz, I., Kusk, K. O., Wollenberger, L., Santos, E. M., Paull, G. C., & Van Look, K. J. (2009). A critical analysis of the biological impacts of plasticizers on wildlife. *Philosophical Transactions of the Royal Society B: Biological Sciences, 364*(1526), 2047–2062.

Pauly, J. L., Stegmeier, S. J., Allaart, H. A., Cheney, R. T., Zhang, P. J., Mayer, A. G., & Streck, R. J. (1998). Inhaled cellulosic and plastic fibers found in human lung tissue. *Cancer Epidemiology, Biomarkers & Prevention: A Publication of the American Association for Cancer Research, cosponsored by the American Society of Preventive Oncology, 7*(5), 419–428.

Peries, S. D., Sewwandi, M., Sandanayake, S., Kwon, H. H., & Vithanage, M. (2024 Apr 29). Airborne transboundary microplastics-a swirl around the globe. *Environmental Pollution, 353*, 124080. https://doi.org/10.1016/j.envpol.2024.124080. PMID: 38692389

Prata, J. C. (2018). Airborne microplastics: Consequences to human health? *Environmental Pollution, 234*, 115–126.

Rani, M., Shanker, U., & Jassal, V. (2017). Recent strategies for removal and degradation of persistent & toxic organochlorine pesticides using nanoparticles: A review. *Journal of Environmental Management, 190*, 208–222.

Rist, S., Almroth, B. C., Hartmann, N. B., & Karlsson, T. M. (2018). A critical perspective on early communications concerning human health aspects of microplastics. *Science of the Total Environment, 626*, 720–726.

Rochman, C. M. (2015). The complex mixture, fate and toxicity of chemicals associated with plastic debris in the marine environment. In *Marine anthropogenic litter* (pp. 117–140). Springer.

Rochman, C. M. (2018). Microplastics research—From sink to source. *Science, 360*(6384), 28–29.

Rydz, J., Musioł, M., Zawidlak-Węgrzyńska, B., & Sikorska, W. (2018). Present and future of biodegradable polymers for food packaging applications. In *Biopolymers for food design* (pp. 431–467).

Schwabl, P., Köppel, S., Königshofer, P., Bucsics, T., Trauner, M., Reiberger, T., & Liebmann, B. (2019). Detection of various microplastics in human stool: A prospective case series. *Annals of Internal Medicine, 171*(7), 453–457.

Secretariat of the Convention on Biological Diversity and the Scientific and Technical Advisory Panel—GEF. (2012). *Impacts of marine debris on biodiversity: Current status and potential solutions* (Montreal, Technical Series No. 67, 61 pages). https://www.cbd.int/doc/publications/cbd-ts-67-en.pdf

Senathirajah, Y., Kaufman, D. R., Cato, K. D., Borycki, E. M., Fawcett, J. A., & Kushniruk, A. W. (2020). Characterizing and visualizing display and task fragmentation in the electronic health record: Mixed methods design. *JMIR Human Factors, 7*(4), e18484.

Shekhar, S., Sood, S., Showkat, S., Lite, C., Chandrasekhar, A., Vairamani, M., Barathi, S., & Santosh, W. (2017). Detection of phenolic endocrine disrupting chemicals (EDCs) from maternal blood plasma and amniotic fluid in Indian population. *General and Comparative Endocrinology, 241*, 100–107.

Srnovršnik, T., Virant-Klun, I., & Pinter, B. (2023). Polycystic ovary syndrome and endocrine disruptors (bisphenols, parabens, and triclosan)—A systematic review. *Life, 13*(1), 138.

Stock, V., Laurisch, C., Franke, J., Dönmez, M. H., Voss, L., Böhmert, L., Braeuning, A., & Sieg, H. (2021). Uptake and cellular effects of PE, PP, PET and PVC microplastic particles. *Toxicology In Vitro, 70*, 105021.

Suhrhoff, T. J., & Scholz-Böttcher, B. M. (2016). Qualitative impact of salinity, UV radiation and turbulence on leaching of organic plastic additives from four common plastics—A lab experiment. *Marine Pollution Bulletin, 102*(1), 84–94.

Sun, J., Dai, X., Wang, Q., Van Loosdrecht, M. C., & Ni, B.-J. (2019). Microplastics in wastewater treatment plants: Detection, occurrence and removal. *Water Research, 152*, 21–37.

Talsness, C. E., Andrade, A. J., Kuriyama, S. N., Taylor, J. A., & Vom Saal, F. S. (2009). Components of plastic: Experimental studies in animals and relevance for human health. *Philosophical Transactions of the Royal Society B: Biological Sciences, 364*(1526), 2079–2096.

Teuten, E. L., Rowland, S. J., Galloway, T. S., & Thompson, R. C. (2007). Potential for plastics to transport hydrophobic contaminants. *Environmental Science & Technology, 41*(22), 7759–7764.

Teuten, E. L., Saquing, J. M., Knappe, D. R., Barlaz, M. A., Jonsson, S., Björn, A., Rowland, S. J., Thompson, R. C., Galloway, T. S., & Yamashita, R. (2009). Transport and release of chemicals from plastics to the environment and to wildlife. *Philosophical Transactions of the Royal Society B: Biological Sciences, 364*(1526), 2027–2045.

Thompson, R. C., Olsen, Y., Mitchell, R. P., Davis, A., Rowland, S. J., John, A. W., McGonigle, D., & Russell, A. E. (2004). Lost at sea: Where is all the plastic? *Science, 304*(5672), 838–838.

Treloar, L. (1974). The elasticity and related properties of rubbers. *Rubber Chemistry and Technology, 47*(3), 625–696.

Turner, A., & Lau, K. S. (2016). Elemental concentrations and bioaccessibilities in beached plastic foam litter, with particular reference to lead in polyurethane. *Marine Pollution Bulletin, 112*(1-2), 265–270.

UNEP. (2021a). *News & Stories—Plastic planet: How tiny plastic particles are polluting our soil.* UNEP. https://www.unep.org/news-and-stories/story/plastic-planet-how-tiny-plastic-particles-are-polluting-our-soil#:~:text=Very%20little%20of%20the%20plastic,into%20the%20soil%20and%20water

UNEP. (2021b). *Neglected: Environmental justice impacts of marine litter and plastic pollution.* UNEP. https://wedocs.unep.org/bitstream/handle/20.500.11822/35417/EJIPP.pdf

Van Cauwenberghe, L., & Janssen, C. R. (2014). Microplastics in bivalves cultured for human consumption. *Environmental Pollution, 193*, 65–70.

Wright, S. L., & Kelly, F. J. (2017). Plastic and human health: A micro issue? *Environmental Science & Technology, 51*(12), 6634–6647.

Wu, B., Wu, X., Liu, S., Wang, Z., & Chen, L. (2019). Size-dependent effects of polystyrene microplastics on cytotoxicity and efflux pump inhibition in human Caco-2 cells. *Chemosphere, 221*, 333–341.

WWF Australia. (June 2019). *Revealed: Plastic ingestion by people could be equating to a credit card a week.* https://wwf.org.au/news/2019/revealed-plastic-ingestion-by-people-could-be-equating-to-a-credit-card-a-week/

Ziccardi, L. M., Edgington, A., Hentz, K., Kulacki, K. J., & Kane Driscoll, S. (2016). Microplastics as vectors for bioaccumulation of hydrophobic organic chemicals in the marine environment: A state-of-the-science review. *Environmental Toxicology and Chemistry, 35*(7), 1667–1676.

Zimmerman, L., et al. (September 7, 2021). Plastic products leach chemicals that induce in vitro toxicity under realistic use conditions. *Environmental Science & Technology, 55*, 11814. https://doi.org/10.1021/acs.est.1c01103

Sampling and Analytical Techniques of Macro-and Microplastics

Girija Prasad, Himalaya S. Vardikar, Smita Mohanty, Baba Linkanmani Sahoo, Rachel Hurley, Nina Buenaventura, Marianne Olsen, and S. K. Nayak

1 Introduction

Plastics are ubiquitous materials used in manifold application sectors –including agriculture, automobile, aerospace, electronics, infrastructure, healthcare, packaging, and textiles. The annual global production of plastics in 2017 was about 350 MMT (million metric tonnes) (PlasticsEurope, 2018), while the virgin polymer consumption in India during 2016–2017 was 15,500 KT (PlastIndia, 2018). Plastic pollution has become a global concern due to the rapidly increasing production of disposable plastic products. Certain lightweight and smaller plastic products are not collected effectively in many countries, and accumulate in both terrestrial and aquatic/marine environments. Globally, the problem is being addressed by various means such as improving product design to reduce plastic use (and hence littering), organized plastic waste management and creating awareness among consumers and producers on the importance of efficient plastic waste management and recycling. The Central Pollution Control Board (CPCB) estimates that 25,940 tonnes of plastic waste is produced every day in India, of which about 60% is recycled and rest are discarded or disposed of in landfills or incinerated (MoEF&CC/Lok Sabha, 2019). Plastic waste contributes to an average of 7% of municipal solid waste (MSW) but varies from city to city. There are very few organized plastic waste management or

G. Prasad · H. S. Vardikar · S. Mohanty (✉) · B. L. Sahoo
Central Institute of Petrochemicals Engineering and Technology (CIPET): School for Advanced Research in Petrochemicals (SARP)-Laboratory for Advanced Research in Polymeric Materials (LARPM), Bhubaneswar, Odisha, India

R. Hurley · N. Buenaventura · M. Olsen
Norwegian Institute for Water Research (NIVA), Oslo, Norway

S. K. Nayak
Ravenshaw University, Cuttack, India

P. Chakraborty et al. (eds.), *Efficient and Enhanced Management of Plastic Waste in India*, https://doi.org/10.1007/978-3-031-89398-8_4

recycling facilities in India. Several participatory education programmes and cleanliness drives like beach clean-ups are arranged from time to time to raise public awareness on the harmful effects of plastic litter. Volunteers from various organizations like Project Mumbai, Beach Please, Malhar with eco-crusaders like Afroz Shah with a team of more than 350 volunteers have been involved in such cleanup activities across beach fronts.

Plastics enter the terrestrial environment through various pathways. Sewage sludge or biosolid application can introduce plastics directly into soil. In agriculture, plastic films are widely used to cover crops, and remnants of these films can degrade and mix with the soil. Littering is another common source, with discarded plastic waste accumulating in the environment over time. Moreover, atmospheric deposition can transport plastics from one area to another, further exacerbating the spread of these pollutants. Once in the environment, these plastics can break down into smaller fragments known as microplastics and persist for extended periods.

Microplastics (MPs) are defined as plastic particles smaller than 5 millimeters (mm) in size, while macroplastics are larger plastic items like bags, bottles, and packaging materials. These plastics can further degrade into smaller particles through physical, chemical, and biological processes, exacerbating their environmental impact. MPs have gained significant attention due to their potential to accumulate in organisms and transfer up the food chain, posing risks to ecosystem health and potentially human health.

1.1 Study of Macro-and Microplastics Abundance in India

In a study conducted on the quantification of macro-and MPs along the west coast of India, ten random beaches were selected in the states of Maharashtra, Karnataka, and Goa (Maharana et al., 2020). A total of 5,095 plastic debris items were collected, primarily ranging in size from 1 to 100 mm, with a few smaller than 1 mm. Of the three states, Maharashtra coast revealed to have a higher percentage of macroplastics followed by Karnataka and Goa. Improper waste management and tourism activities add to a majorfraction in macroplastic waste accumulation. Figure 1 shows the abundance of macroplastics in beaches of Maharashtra, Goa, and Karnataka (Maharana et al., 2020).

Research surveys done on river Ganga (Sarkar et al., 2019) indicated that plastic pollution in eastern parts of river Ganga is mainly due to Polyethylene terephthalate (PET, 39%) followed by polyethylene (PE, 30%). Another study (Baroth et al., 2019) found that polypropylene was the most abundant microplastic found in River Ganga followed by PE and polyvinyl aldehyde. The sedimental microplastic content was found to be higher in the middle zone of the river.

A survey conducted by National Institute of Oceanography (NIO) in six beaches of Goa revealed the presence of microplastics in fish, mussels, turtles, and sea birds (Fernandes, 2018). In another study conducted in Rameshwaram Island, Gulf of Mannar, India, large amounts of white, irregular-shaped plastic debris composed of

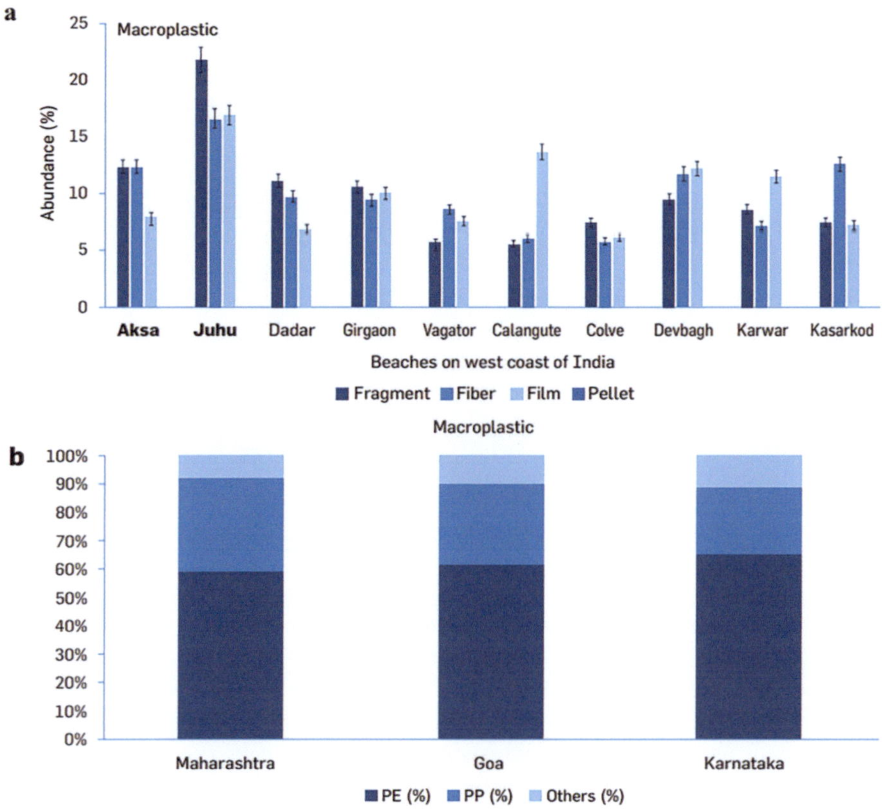

Fig. 1 Graphical representation of: (**a**) Abundance (%) of various macroplastics (5–100 mm) along the beaches of west coast of India during December 2016—February 2017. Data, expressed as mean ± SD, $n = 3$. The locations are beaches of Maharashtra, Goa, and Karnataka; (**b**) Relative abundance (%) of polymer types (polyethylene = PE, polypropylene = PP, others = unspecified polymers) of plastics collected from different beaches of the west coast of India based on the studies done by FTIR spectroscopy (Maharana et al., 2020)

polypropylene (major share), polyethylene, polystyrene, nylon, and polyvinyl chloride was found (Vidyasakar et al., 2018).

Another survey conducted across six beaches of Goa studied the spatial and seasonal variability in the distribution, composition, weathering pattern and possible sources of microplastic pellets (MPPs). A majority of the microplastics reach the coast during southwest monsoon, of which PE and PP were the dominant types. The probable sources include ocean-based activities like unintentional/accidental spills and tourism (Veerasingam et al., 2020). Figure 2 shows the microscopic image of the surface properties of microplastic pellets during the southwest and northwest monsoon.

Identifying and quantifying MPs and macroplastics in the environment is crucial for understanding their distribution, fate, and potential impacts. Various

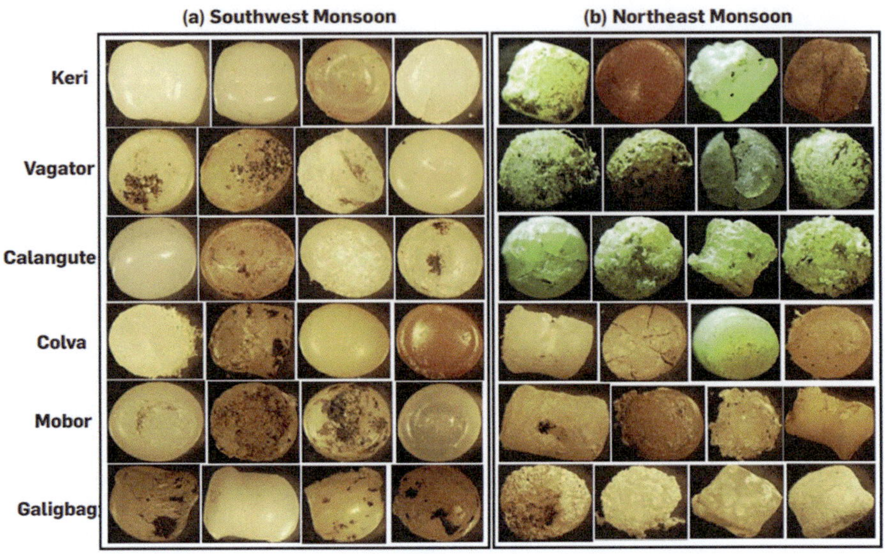

Fig. 2 Microscopic image showing the surface properties of microplastic pellets during the (**a**) southwest and (**b**) northwest monsoon (Veerasingam et al., 2016)

methodologies have been developed to address this challenge, including spectroscopic techniques like Fourier-transform infrared spectroscopy (FT-IR) and Raman spectroscopy, thermal analysis methods like pyrolysis-gas chromatography/mass spectrometry (Py-GC/MS), and imaging techniques like scanning electron microscopy (SEM) and micro-Fourier transform infrared (μFT-IR) imaging.

2 Sampling Techniques

2.1 Importance of Sampling

Sampling and analysis of macroplastics and MPs in different environmental compartments like water, soil, and air is of paramount importance to understand the scale and implications of this emerging pollutant. In aquatic systems, sampling water bodies and sediments allows quantification of microplastic loads, which can have detrimental effects on marine life through ingestion and bioaccumulation. Identifying microplastic hotspots in the oceans, rivers, and lakes enables prioritization of mitigation strategies and policies to curb plastic pollution. Soil sampling is crucial to assess the extent of microplastic contamination in terrestrial ecosystems, which can impact soil health, plant growth, and food chain contamination through uptake by organisms. Air sampling is also of grave importance as inhalation of airborne microplastics, especially fibers, can lead to respiratory issues in humans and animals upon prolonged exposure.

2.2 *Monitoring and Sampling of Macro-and Microplastics*

Macroplastics may be found floating in oceans and seas, in freshwaters and in rivers, stranded or trapped in vegetation along beaches and river shorelines, or settled in the sediments on the river or sea bed. Hence, monitoring of macro-plastics in various environmental matrices is required to be carried out by using different methods and scope.

Monitoring of stranded macroplastic relates to beach clean-ups. Based on standardized monitoring methods, visible litter items should be recorded and the registrations carried out within a defined shoreline (between backshore and waterline). Monitoring macroplastics in rivers, seas, and other aquatic environments is essential for understanding the distribution, sources, and impacts of plastic pollution. In rivers, macroplastics can be transported long distances, eventually reaching the sea and accumulating in coastal areas. Monitoring involves surveying riverbanks, estuaries, and coastal areas to detect visible plastic debris and assess its types, quantities, and sources. Drones or satellite imagery for large-scale monitoring may be used, while boat-based and underwater methods are used for nearshore and submerged debris. In open waters, monitoring includes using nets and trawls to collect floating debris, along with aerial and satellite surveys to capture large-scale distribution patterns. These monitoring efforts contribute to identifying pollution hotspots, understanding the movement of macroplastics through aquatic systems, and informing strategies for reducing plastic pollution and protecting marine life.

On the other hand, there is no uniform method to monitor floating macroplastics in the river or freshwater system. The methods used today can be roughly divided into five different methodologies.

- Tracking floating litter.
- Active sampling of litter; surface and water column.
- Passive sampling; surface and water column.
- Visual observations of floating litter.
- Advanced observations (such as UAV or spaceborne remote sensing).
- Citizen science.

Microplastics are small fragments of plastics that may vary in size, shape, color, specific density, and polymer type. They come in several shapes and buoyancy and thus they sink or float or accumulate in the oceans (Hidalgo-Ruz, 2012). Due to their different shape, small size, and attractive colors, they are often ingested by different organisms, leading to bioaccumulation, and finally entering the food chains migrating from the lowest members to the prime predators, humans. Ocean pollution with microplastics has spread widely enough to become a serious global concern, earning the term "plastic soup" for the world's oceans. Due to their small size, these particles manage to evade both conventional and municipal wastewater treatment and filtration systems. Microplastics are divided into primary and secondary categories based on where they come from. Primary microplastics are produced with a millimetric or submillimetric size range in mind. They are utilized in personal hygiene products (like toothpaste, exfoliating creams, and face cleansers), air-blasting media, soft toys, fillers in furniture, and even medications as drug vectors (Toxics Link, 2018). Primary microplastic may also originate from industrial facilities or runoff/effluent from the production of plastic products. The breakdown processes of bigger plastic wastes produce secondary microplastics. Because the structural integrity of plastic trash is reduced with time, the breakdown process can be physical, chemical, or biological. Almost all mismanaged plastic garbage will eventually generate microplastics thanks to this ongoing fragmentation process. An outline of the processes in the microplastic assessment method is shown in Fig. 3. Sampling, extraction, identification, and quantification are the primary procedures, with particular standards for evaluating water, sediment, and biota samples.

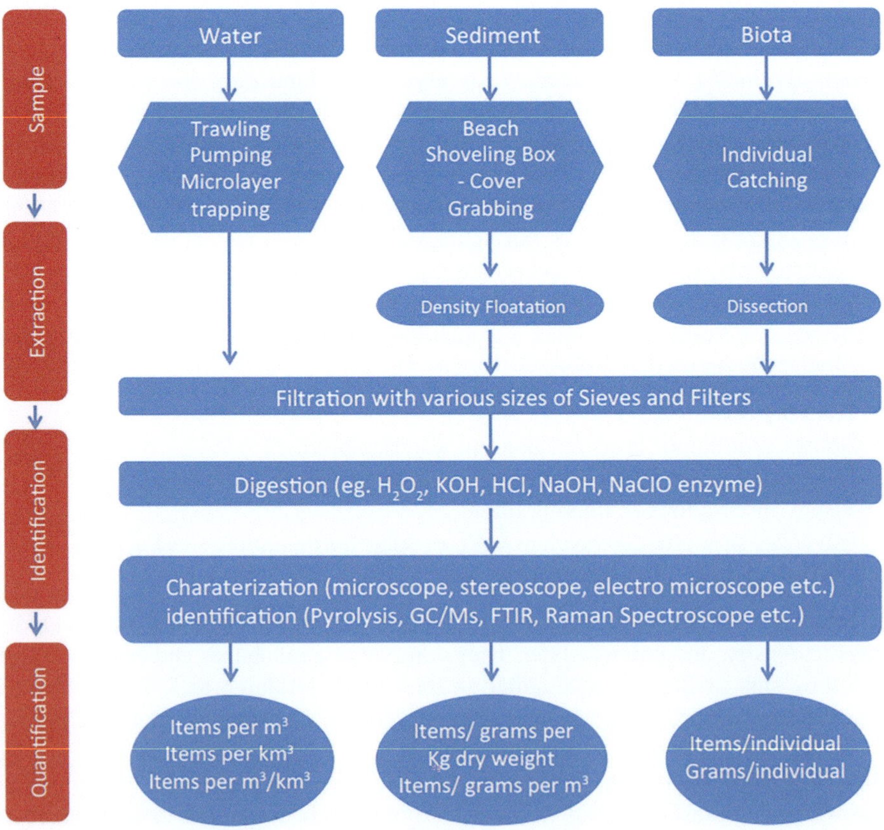

Fig. 3 Overall methodology of microplastics assessment (National Productivity Council 2020)

Sample collection is the first step of microplastic analysis. Accurate sample collection methods are essential to determine the types of collected, separated, and identified microplastics. The study location is crucial for the sampling method. There are three main methods of sampling:

(a) Selective sampling: Visually identifiable items are directly collected from the water surface or sediments. This method is applicable when different microplastics of similar morphology and size greater than 1 mm are present in the location under investigation, e.g. primary microplastics pellets and similarly shaped secondary microplastics. But the main disadvantage of this technique is the higher probability of selecting the wrong samples, particularly when they are mixed with other contaminants.
(b) Volume-reduced sampling: In this type of sampling, only the specific items of interest are considered for further analysis discarding most of the sample. This method is typically applicable to large water samples.

Table 1 Key methods of microplastics sampling/observations and their advantages and limitations

Method	Explanation	Advantage	Limitations	Sources
Net tows (manta trawl, Neuston net)	A fine-mesh net mounted on a large rectangular frame, designed for sampling surface and water column waters to collect plankton, insects, and other small organisms. A manta trawl equipped with floating wings to maintain its position on the surface	Small to large vessels can be deployed Use of flow meter to estimate volume and flow over time	Usage depends on weather. Care needed to minimize contamination from sampling vessel and tow ropes Can only estimate volume of water filtered when flow meter is used. To avoid clogging, towing speed and time must be limited	Viršek et a. (2016)
Pumping water sample	Pass a water sample through a 1 μm filter to measure all particles within the microplastic size range (1 μm to 5 mm)	This is well suited for use in a port environment due to its minimal clutter, which makes it possible to sample in confined spaces, such as between vessels and piers	More expensive, consumes more electricity, and is challenging to clean between sampling	Cutroneo et al. (2020)
Visual observation from a ship	A visual assessment of floating marine debris from a vessel's surface at sea can be conducted using either fixed-width transects (which assume all items are observed) or distance sampling (which accounts for reduced detection probability with increasing distance)	Easy to do from vessels of opportunity. Low cost, needs only binoculars (but ideally also a good quality Camera and lens)	Limited to water adjacent to the ship (up to 50 m typically). Biased against dark items and subsurface items, white and buoyant items easier to spot	Suaria et al. (2020)

(c) Bulk sampling: In this method, the entire volume of sample is taken without any reduction in the volume.

The most common equipment of sampling is to use a towed net such as manta trawl net or neuston with a fixed mesh size of 330 m, which might result in under-sampling of particles below this size. The details of different sampling methods are given in Table 1 (Cutroneo et al., 2020).

3 Extraction of Various Types of Micro-and Macroplastics

3.1 *Extraction of Macroplastics*

In accordance with OSPAR guidelines (GESAMP, 1996), a sampling unit refers to a designated section of beach spanning from the water's edge to the back of the beach, typically covering a 100-meter length. Macroplastics, often buried in beach or riverbank sediments, are seldom collected during standard monitoring. Nevertheless, buried plastics can be recovered by sifting sediment through sieves with a 10–20 mm mesh size. The sieve's mesh should be slightly smaller than the minimum size of macroplastic debris to enhance the chances of capturing irregularly shaped litter. The retained materials are then sorted by hand. The width of the transect is a compromise between obtaining representative data and the practicality of sieving large sediment volumes. The vertical sampling depth depends on factors such as the stability of the sediments, available personnel, and environmental dynamics like wave mixing or flood-influenced riverbanks.

3.2 *Extraction of Microplastics*

3.2.1 Extraction from Water Samples

Microplastics found in water consist of hard and soft plastics (e.g., foams), as well as films, fibers, sheets, and lines. To begin the analysis, solids collected using a 0.335 mm surface sampling net (such as a manta net for surface water) are filtered through 5.6-mm and/or 0.3-mm sieves to select particles of appropriate size. The sieved material is then dried to assess the mass of solids in the sample. Next, wet peroxide oxidation (WPO) is performed using an Fe(II) catalyst to break down easily degradable organic matter, leaving the plastic particles unchanged. The WPO-treated mixture undergoes density separation in NaCl solution, where the plastic debris is isolated through flotation. The floating plastic particles are separated from the denser mineral material using a density separator equipped with a custom 0.3-mm filter. After air-drying, the plastic material is collected, weighed, and the

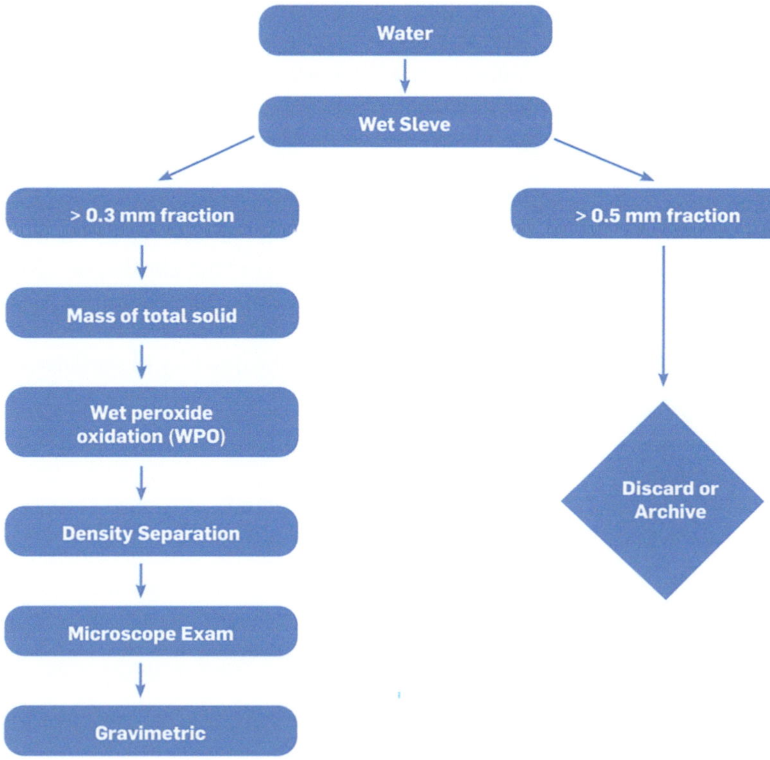

Fig. 4 The figure represents the protocol for the analysis of microplastics in the water samples (Masura et al., 2015; National Productivity Council. 2020)

concentration of microplastics is determined. Figure 4 represents the protocol for microplastic analysis.

The samples obtained through the sampling techniques are analyzed through:

- Pretreatment (removal of non-plastic materials aside from microplastics), sorting of larger plastic items, quantification and measurement, and identification of materials.
- Pretreatment process is selected based on purpose of the study.

Biological Digestion: To remove the organic substances, the collected samples are treated with enzymes or chemicals. The intention is to minimize the possibility of misidentifying plastic particles, with improved accuracy of the picking out process and overall work efficiency. If improperly conducted, however, it may lead to deterioration (deformation and/or weight reduction) of plastic particles from chemicals added or from heating. Different types of digestion methods include oxidative digestion (Masura et al., 2015), acid digestion (Claessens et al., 2013), alkaline digestion (Dehaut et al., 2016), and enzymatic digestion (Löder et al., 2017).

Density Separation: It is to fractionate between low-density and high-density natural particles of inorganic matter. Density separation is typically accomplished by combining samples into a solution that has a greater specific gravity (than that which is estimated) for the plastic particles that have been collected. This allows high-density inorganic materials to separate, recovering and fractionating the low-density floating plastic particles.

Picking Out Microplastics: This is crucial as it greatly impacts the accuracy of microplastic analysis. Two common methods adopted are size based on fractionation of plastic particles using a sieve and directly picking the plastic particles after filtering. Stereo microscopes are usually preferred to facilitate picking microplastics.

Counting and Measuring Sizes of Microplastics: The quantity of particles can be counted based on size via two different approaches:

- Individually measuring the maximum diameter (maximum Feret's diameter) of distinct particles;
- Determining the number of particles left in the sample after size fractionation using sieves with different mesh sizes.

The size of the microplastics can also be measured as Feret's diameter, which is defined as the distance between the two parallel planes restricting the object perpendicular to that direction, which means the longest diameter.

3.2.2 Extraction from Biota

The extraction and analysis of microplastics from biota samples is a crucial aspect of assessing the impact of these contaminants on marine organisms and ecosystems. The process typically involves several steps to ensure accurate and reliable results. Firstly, proper sampling methods and strategies are essential to obtain representative samples. Details such as the type of gear used, mesh size, location, depth, and time of sampling should be carefully recorded to ensure reproducibility. Additionally, appropriate sample sizes, typically 50 individuals per research unit, should be collected to provide statistically significant results.

Once the samples are collected, they must be carefully processed and stored to prevent contamination. Freezing at −20 °C or preserving in ethanol or formaldehyde is recommended, avoiding any handling or dissection prior to laboratory analysis. In the laboratory, rigorous measures must be taken to minimize contamination from airborne particles or equipment. This includes working in a clean air facility, such as a laminar flow cabinet or a clean room, and thoroughly washing and rinsing all materials used. To extract microplastics from biota samples, a digestion step is necessary to dissolve the organic matter without degrading the plastic particles. The most suitable method involves using a 10% potassium hydroxide (KOH) solution at room temperature for an extended period, typically 3 weeks. Alternatively, enzymatic digestion methods can be used for smaller organisms like bivalves or plankton species. Heating or drying the samples at high temperatures should be avoided to prevent deformation or clumping of the microplastic particles. After the digestion step, the samples are carefully inspected for the presence of microplastics.

3.2.3 Extraction from Sediments

The extraction of microplastics from sediment matrices is challenging due to the need to separate the plastics from the sediments. A study by Constant et al. (Constant 2021) investigated the efficiencies of 18 different combinations of six extracting solutions (ESs) and three isolation methods (IMs) for fine and coarse sediments, with low and high-density polymers. The ESs tested included oil, water, oil-in-water, NaCl, oil-in-NaCl, and NaI, while the IMs were hand stirring, centrifugation, and aeration. The authors found that the IMs (hand stirring, centrifugation, and aeration) and sediment features (coarse or fine) did not significantly affect the extraction efficiencies. However, the choice of ES played a crucial role in the extraction efficiency. Key findings regarding the ES efficiencies are the following:

- Fragments, pellets, and foams of low-density polymers (PE and PE-ABS) were well extracted (>80%) by all ESs except pure oil.
- For high-density polymers (PET, PA, and PES), the NaI solution is considered the best extraction efficiency (71 ± 17%).
- The addition of a few millilitres (ml) of oil into the aqueous solutions (oil-in-water and oil-in-NaCl) did not significantly improve the recovery of pristine microplastics.

4 Analysis of Microplastics

The analysis of microplastics typically involves several steps after the extraction process. Once the microplastics have been separated from the matrix, the next step is counting and identification. Counting can be done under a dissecting stereo microscope. Other characterization techniques involved are the following:

4.1 FTIR

Usually, a FTIR coupled with an attenuated total reflection (ATR) accessory is used to identify the compositional characteristics of microplastics. The ATR technique facilitates the acquisition of an infrared (IR) spectrum by merely placing the sample against a clear crystal, typically made of diamond. In this process, infrared light transmits through the crystal and into the sample, where it is absorbed, and then reflects into the crystal, resulting in the generation of a spectrum.

Table 2 Common methods used for microplastic characterization

Identification method	Advantages	Disadvantages
Microscopy	• Simple • Low cost • Color and morphological information	• No compositional information • High possibility of false positives • High possibility of missing small and transparent particles
Microscopy + FTIR spectroscopy	• No false positives • Reduction in false negatives • Nondestructive • The detection limit is 20 m particle size	• Using manual selection for particles may result in overlooking certain plastics. • Expensive • Demands significant effort and time for identifying of all particles • Demands proficiency in spectral interpretation • Contact analysis (ATR) • Transfer of particles is needed from filter paper to metal plate • Prerequisite removal of organic material
Microscopy + Raman spectroscopy	• No false positives—Confirmation of all plastic-like particles • Reduction in false negatives • 1 m particle detection limit • Nondestructive analysis • Noncontact analysis	• Manually selecting particles may result in the missing of certain plastics. • Expensive instrument • Demands significant effort and time for identifying of all particles • Demands proficiency in spectral interpretation • Interference by pigments • There is a risk of laser damage to the particles. • Prerequisite removal of organic material • Requires exact focusing

4.2 Raman Spectroscopy

Raman spectroscopy employs lasers with sub-micron wavelengths as the light source, enabling the resolution of particles down to 1 micron or smaller. These systems are constructed using high-quality white-light microscopes, allowing for straightforward observation of the particles. The laser in the Raman setup is concentrated on the sample, and the spectrum is obtained by collecting the light that is scattered.

4.3 Thermo-Analytical Methods

The common approaches include TGA, DSC, and Pyrolysis GC-MS.

Table 2 summarizes the common methods used for microplastic characterization including identification of polymer types.

References

Baroth, A., Pant, A., Sah, R., Hussain, S. A., & Chaudhary, P. (2019). Spatiotemporal trends of micro-plastics loading in the sediments of river ganga: First observation on occurrence, identification and quantification. In *Conference: SETAC Europe 29th annual meeting, Helsinki, Finland.*

Claessens, M., Van Cauwenberghe, L., Vandegehuchte, M., & Janssen, C. (2013). New techniques for the detection of microplastics in sediments and field collected organisms. *Marine Pollution Bulletin, 70*, 227. https://doi.org/10.1016/j.marpolbul.2013.03.009

Cutroneo, L., Reboa, A., Besio, G., Borgogno, F., Canesi, L., Canuto, S., & Lenoble, V. (2020). Microplastics in seawater: Sampling strategies, laboratory methodologies, and identification techniques applied to port environment. *Environmental Science and Pollution Research, 27*, 1–15.

Dehaut, A., Cassone, A.-L., et al. (2016). Microplastics in seafood: Benchmark protocol for their extraction and characterization. *Environmental Pollution, 215*, 223–233., ISSN 0269–7491. https://doi.org/10.1016/j.envpol.2016.05.018

Fernandes, Micropastics hits marine life, enters human food, The Times of India, https://timesofindia.indiatimes.com/city/goa/microplastic-hits-marine-life-enters-human-foodmicroplastics-in-fish-at-keri-galgibaga-experts/articleshow/65329255.cms (2018).

GESAMP (IMO/FAO/UNESCO-IOC/WMO/WHO/IAEA/UN/UNEP Joint Group of Experts on the Scientific Aspects of Marine Environmental Protection). (1996). *Report of the Twenty-sixth Session, Paris 25–29 March 1996* (Rep.Stud.GESAMP No. 60, 29 pp).

Hidalgo-Ruz, V., Gutow, L., Thompson, R. C., & Thiel, M. (2012). Microplastics in the marine environment: A review of the methods used for identification and quantification. *Environmental Science & Technology, 46*, 3060–3075.

Kershaw, P. J., Turra, A., & Galgani, F. (Eds.) (2019). Guidelines for the monitoring and assessment of plastic litter in the ocean. GESAMP Report and Studies No. 99, IMO/FAO/UNESCO-IOC/UNIDO/WMO/IAEA/UN/UNEP/UNDP/ISA Joint Group of Experts on the Scientific Aspects of Marine Environmental Protection, 130 pages.

Löder, M. G. J., Imhof, H. K., Ladehoff, M., Löschel, L. A., Lorenz, C., Mintenig, S. M., & Gerdts, G. (2017). Enzymatic purification of microplastics in environmental samples. *Environmental Science & Technology, 51*(24), 14283–14292. https://doi.org/10.1021/acs.est.7b03055

Maharana, D., Saha, M., Dar, J. Y., Rathore, C., Sreepada, R. A., Xu, X.-R., Koongolla, J. B., & Li, H.-X. (2020). Assessment of plastics along the west coast of India: Abundance, distribution, polymer type and toxicity. *Chemosphere, 246*, 125708. https://doi.org/10.1016/j.chemosphere.2019.125708

Masura, Laboratory methods for the analysis of microplastics in the marine environment: recommendations for quantifying synthetic particles in waters and sediments, https://doi.org/http://dx.doi.org/10.25607/OBP-604 (2015).

Sarkar, D. J., Sarkar, S. D., Das, B. K., Manna, R. K., Behera, B. K., & Samanta, S. (2019). Spatial distribution of meso and microplastics in the sediments of river Ganga at eastern India. *Science of the Total Environment, 694*, 133712. https://doi.org/10.1016/j.scitotenv.2019.133712

Suaria, G., Perold, V., Lee, J. R., Lebouard, F., Aliani, S., & Ryan, P. G. (2020). Floating macro- and microplastics around the Southern Ocean: Results from the Antarctic Circumnavigation Expedition. *Environment International, 136*, 105494. https://doi.org/10.1016/j.envint.2020.105494

Veerasingam, S., Ranjani, M., Venkatachalapathy, R., Bagaev, A., Mukhanov, V., Litvinyuk, D., Verzhevskaia, L., Guganathan, L., & Vethamony, P. (2020). Microplastics in different environmental compartments in India: Analytical methods, distribution, associated contaminants and research needs. *Trends in Analytical Chemistry, 133*, 116071. https://doi.org/10.1016/j.trac.2020.116071

Vidyasakar, A., Neelavannan, K., Krishnakumar, S., Prabaharan, G., Sathiyabama Alias Priyanka, T., Magesh, N. S., Godson, P. S., & Srinivasalu, S. (2018). Macrodebris and microplastic distribution in the beaches of Rameswaram Coral Island, Gulf of Mannar, Southeast coast of India: A first report. *Marine Pollution Bulletin, 137*, 610–616. https://doi.org/10.1016/j.marpolbul.2018.11.007

Environmental Monitoring: Global and Indian Perspectives

Rachel Hurley, Nina Buenaventura, Girija Prasad, Himalaya S. Vardikar, Smita Mohanty, Marianne Olsen, and S. K. Nayak

Abbreviations

ALDFG	Abandoned, lost, or otherwise discarded fishing gear
COBSEA	Coordinating Body on the Seas of East Asia
CPPS	Permanent Commission for the South Pacific
GESAMP	Group of Experts on the Scientific Aspects of Marine Environmental Protection
GPML	Global Partnership on Plastic Pollution and Marine Litter
G20	Group of Twenty
HELCOM	Baltic Marine Environment Protection Commission
ICES	International Council for the Exploration of the Sea
ILC	Interlaboratory calibration/comparison exercise
MSFD	Marine Strategy Framework Directive
NOAA	National Oceanic and Atmospheric Administration
OSPAR	Convention for the Protection of the Marine Environment of the North-East Atlantic
QA/QC	Quality assurance quality control
RAP	Reproducible analytical pipeline
SPREP	Secretariat of the Pacific Regional Environment Programme
TRL	Technological readiness level
UNEA	United National Environment Assembly

R. Hurley (✉) · N. Buenaventura · M. Olsen
Norwegian Institute for Water Research (NIVA), Oslo, Norway
e-mail: rachel.hurley@niva.no

G. Prasad · H. S. Vardikar · S. Mohanty · S. K. Nayak
Central Institute of Petrochemicals Engineering Technology, CIPET,
Bhubaneswar, Odisha, India

P. Chakraborty et al. (eds.), *Efficient and Enhanced Management of Plastic Waste in India*, https://doi.org/10.1007/978-3-031-89398-8_5

UNEP	United Nations Environment Program
UN	United Nations

1 Introduction

Environmental monitoring can provide baseline data from which trends can be observed, reveal spatial and temporal patterns in pollution, and point towards potential mechanisms or processes underpinning fate and impacts of pollution in the environment. The occurrence of plastic pollution was first observed during the second half of the twentieth century as plastic production increased and litter was first observed in the environment (Carpenter et al., 1972; Carpenter & Smith, 1972). Today, plastic pollution receives substantial scientific, media, and political attention and plastics are well-established as a global pollutant. Monitoring of plastic pollution is increasing in response to this, both through research activities and national or regional monitoring programmes.

India is an important player in global plastic production, consumption, and pollution (Sivadas et al., 2022). It is listed as the 12th largest contributor to marine plastic pollution globally by Jambeck et al. (2015), which has been linked to population growth and increasing urbanisation along the coastlines and a lag in plastic waste management to accommodate this (Das et al., 2020). The Ganga-Brahmaputra system is listed as the sixth highest in plastic waste releases to the marine environment by Schmidt et al. (2017), whilst Meijer et al. (2021) highlight the significance of small rivers across India, which rank second globally. Legislation addressing plastic pollution have been established in India, including the Plastic Waste Management Rules and subsequent amendments, but these typically focus on solid waste management as opposed to establishing a mandate for environment monitoring. Instead, plastic pollution monitoring in India has emerged primarily through scientific studies, beginning with post-monsoonal coastline monitoring in 1982 (Nigam, 1982) and expanding rapidly in the past decade (Sivadas et al., 2022).

This chapter aims to explore the evolving trends in international efforts to monitor plastic pollution across the globe. This will be compared to monitoring activities in India, including an assessment of some of the national approaches to monitoring and emerging perspectives. This includes a focus on two main categories of plastic pollution: macroplastic and microplastic, where macroplastic refers to larger pieces of litter (>25 mm) and microplastic includes smaller plastic particles (<5 mm). Additional size categories also exist, such as meso- or nanoplastic, but these are not specifically addressed here, due to a less prevalent focus on these in current monitoring perspective.

2 Global Macroplastic and Microplastic Monitoring

Monitoring of environmental plastic contamination is high on the international agenda. UN Member States, Regional Seas Conventions, and International Organisations are currently ratifying policy instruments and implementing monitoring programs in response to several international directives. This includes the ongoing development of a legally binding agreement on plastic pollution endorsed by 175 countries at the UN Environment Assembly (UNEA-5.2). Several regional, national, and local action plans for marine debris have now been developed under the UNEP Global Partnership on Marine Litter. Many of these action plans underline strategies for the mapping of sources, pathways, or patterns of litter in the marine environment, including different environmental matrices (e.g. water, sediment, biota). The terms 'litter' or 'debris are most commonly used and refer to a diverse suite of materials that constitute waste items discarded into the environment through a wide range of different potential channels. However, plastic is typically at the forefront of these initiatives, based on the significant proportion of waste items in the relevant environments that is composed of plastic and the high status of plastic and microplastic contamination on the global agenda.

Table 1 outlines the activities related to monitoring undertaken thus far by several international organisations, working groups, and initiatives. A majority of these address the marine environment and, in particular, floating litter, beach litter, and microplastics. A small number have directed activities towards freshwater environments, typically in the form of guidelines for monitoring riverine plastic litter (e.g. González et al., 2016; Barnardo & Ribbink, 2020). In addition to these, there are a range of national- and international-level monitoring programs and surveys that are commissioned by national environment agencies or conducted as part of collaborative research projects. These span a wide range of different environments (marine, riverine, lacustrine, terrestrial), matrices (water, sediment, biota, soils), and plastic types (microplastic, macroplastic, litter).

Due to the relative infancy of environmental monitoring efforts related to plastics and a lack of harmonised methodologies or standards, much of the current activity has taken the form of establishing technical guidelines or undertaking baseline or 'one-off' surveys. As such, both the methods used for environmental monitoring and the data collected from monitoring surveys vary in regard to resolution, focus, quality, and analytical approach. This hampers the ability to compare data from different monitoring activities, which in turn complicates any international efforts to set definitions or targets for different types of plastic contamination (e.g. macro- or microplastic). There has also been, thus far, a lack of monitoring activities that extend across large spatial or temporal scales. It is therefore difficult to conclude about the relative contributions from different sources of environmental contamination, and the different controls on the accumulation or transport of plastic, such as hydrological or meteorological conditions.

Included in Table 1 are international bodies that have started or completed activities related to the monitoring of environmental plastic contamination. Table 2 lists

Table 1 International organisations, groups, and initiatives that have undertaken activities related to environmental monitoring of plastics

Organisation/Group/Initiative	Geographic scope	Activities (related to environmental monitoring)
UNEP global partnership on marine litter (UNEP-GPML)	International (UN)	Oversee development and monitoring activities of international, regional, national, and local marine litter action plans Guidelines for monitoring marine litter
GESAMP working group 40 on plastic pollution	International (UN)	Guidelines for plastic (macro, micro) litter monitoring in the ocean
Ministry of the Environment—Japan; G20 action plan on marine litter	International (G20)	Harmonisation of methods for monitoring floating microplastics in marine waters
ICES working group on marine litter (ICES WG ML)	International (20 member nations)	Guidelines/protocols for monitoring seafloor litter and microplastics
EU marine strategy framework directive technical subgroup on marine litter (MSFD TSG-ML)	Europe	Guidelines for monitoring marine litter: Floating litter, beach litter, seafloor litter, litter in biota, and microlitter Guidelines for monitoring riverine litter
OSPAR intersessional correspondence group on marine litter (ICG ML)	Europe	Monitoring of macroplastic litter on beaches, the seabed, and in biota
COBSEA regional action plan on marine litter	Asia	Training of methods for monitoring marine litter and microplastics
African marine waste network	Africa	Guidelines for monitoring litter: Along shorelines, in mangroves, in rivers/estuaries, and on land
Western Indian Ocean marine science association	Africa	Monitoring of beach litter
SPREP Pacific regional action plan on marine litter	Pacific region	Monitoring of marine litter through coastal clean-up activities
CPPS regional Programme on marine litter in the Southeast Pacific	Southeast Pacific	Monitoring of microplastics in marine waters, sand, and fish.
NOAA marine debris program	USA and Canada	Monitoring of plastic (macro, micro) in freshwaters and marine environment. Guidelines for microplastic analysis
HELCOM expert network on marine litter	Baltic	Monitoring of beach litter, seafloor litter, and microlitter Guidelines for monitoring beach litter

examples of regional Action Plans or Strategies related to plastic pollution that include a specific monitoring component. The extent of the plans outlined for monitoring activities varies, ranging from promoting scientific monitoring by universities or other institutions or developing monitoring guidelines to establishing or maintaining national monitoring plans for plastic or microplastics.

Table 2 Examples of national or regional action plans or strategies related to plastic pollution, which include measures addressing environmental monitoring

Country or region	Action plans or strategies	Measures that address monitoring
Arctic region	Regional Action Plan on Marine Litter in the Arctic (2021)	Promote harmonised approaches to detect, monitor, characterise, and assess marine litter in the Arctic environment
ASEAN region	Regional Action Plan for Combating Marine Debris in the ASEAN Member States (2021)	Establish guidelines on monitoring methods and conduct a regional study on microplastic occurrence
Australia	National Plastics Plan (2021)	Establishment of a national monitoring protocol and database for plastic pollution monitoring data
Baltic Sea region	HELCOM Regional Action Plan on Marine Litter (2021)	Establish coordinated monitoring programmes for beach litter, seafloor litter, and microlitter
Belize	Marine Litter Action Plan (2019)	Strengthen institutional and financial capacity to conduct scientific monitoring programme in Belize and contribute to regional data sets
Black Sea region	Black Sea Marine Litter Regional Action Plan (2018)	Prepare guidelines on monitoring methods and implement a regional marine litter monitoring programme
Brazil	National Plan to Combat Marine Litter (2019)	Mapping of marine litter and creation of a unified database with the goal of better understanding dynamics of marine litter
Canada	Canada-wide Action Plan on Zero Plastic Waste (2019, 2020)	Identify opportunities and develop guidance for Canada-wide monitoring to achieve consistent data gathering on plastics in the environment
Chile	National Strategy for the Management of Marine Litter and Microplastics (2021)	Sponsor and encourage the development of research towards monitoring including the refinements of methods
China	Opinions on Further Strengthening Plastic Pollution Control (2020)	Establish a national marine litter and microplastics monitoring programme
Costa Rica	National Marine Residues Plan Costa Rica (2021)	Establish a national monitoring programme
EU	Action plan for the marine environment (marine strategy framework directive) (2008)	Funding for monitoring and research through the EU research and innovation Programme, European maritime and fisheries fund, and Regional Sea conventions, development of monitoring guidelines
Finland	Marine Strategy (2018)	Continue beach litter monitoring and establish national monitoring for marine litter and microplastics
France	National Roadmap against Marine Litter (2019)	Partnering with national agencies and research centres to monitor plastic sources and flows

(continued)

Table 2 (continued)

Country or region	Action plans or strategies	Measures that address monitoring
Great Lakes region	Great Lakes Marine Debris Action Plan (2014, 2015, 2016, 2017, 2018, 2020)	Undertake monitoring and research following several identified approaches or thematic topics, share research and monitoring outcomes, update and maintain a collaborative portal for sharing data
Indonesia	National Action Plan on Marine Debris Management (2018)	Research and monitoring of microplastic and land-based sources of marine plastic through government agencies
Japan	National Action Plan for Marine Plastic Litter (2019)	Establishing guidelines on harmonisation of monitoring methods and encouraging harmonisation of methods and data reporting internationally through funded initiatives
Kenya	National Action Plan on Marine Plastic Litter from Sea-based Sources (2022)	Facilitate collaborative research to estimate quantities of marine plastic litter
Malaysia	National Marine Litter Policy and Action Plan (2021)	Compilation of a national database on the status of marine litter and strengthening national and local capacity to harmonise monitoring of marine litter
Northwest Pacific region	NOWPAP Regional Action Plan on Marine Litter (2008)	Establish a regional marine litter monitoring programme and maintain a regional database on marine litter distribution,
Norway	Norwegian Plastics Strategy (2017)	Establish a national monitoring programme for microplastic
Pacific Island countries and territories	Pacific Regional Action Plan: Marine Litter (2018)	Implement a standardised marine litter and microplastics data collection system and app for the Pacific
Philippines	National Plan of Action for the Prevention, Reduction and Management of marine Litter (2021)	Establish a baseline for marine litter, including the establishment of robust monitoring
Republic of Korea	National action plan on marine litter and marine contaminated sediment (2021)	Maintain a national marine litter monitoring programme and include microplastic monitoring in this plan
Republic of Maldives	National Strategic Action Plan (2019)	Establish a microplastic testing lab for environmental monitoring
Saudi Arabia	National action plan for sustainable management of Marine Litter in the Red Sea coast of the Kingdom of Saudi Arabia (2018)	Regularly re-survey beaches that have baseline data and encourage and support local marine institutes and universities to undertake research
Singapore	National Action Strategy on Marine Litter (2022)	Establish a marine debris research project to collect baseline data on marine debris along shorelines and organise capacity building and regional training programmes on marine litter monitoring

(continued)

Table 2 (continued)

Country or region	Action plans or strategies	Measures that address monitoring
Spain	Program of Measures on Marine Litter (2018)	Monitoring (macrolitter and microplastics) on beaches, floating litter, seabed litter, and also biota
Turkey	Integrated Marine Pollution Monitoring Programme (2013)	Include microplastics and marine and coastal litter within the national marine pollution monitoring programme
UK	UK Marine Strategy (2019, 2021) Marine Litter Action Plan for Wales (2017, 2020) Marine Litter Strategy for Scotland (2021) Northern Ireland Marine Litter Strategy (2013)	Implementation of a monitoring programme, including monitoring of seafloor litter bycatch in fisheries surveys, beach litter, and microplastics
USA	U.S. Marine Debris Act (2006). Save our Seas Act (2018)	Establish and maintain a marine debris monitoring and assessment programme
Vietnam	National action plan for Management of Marine Litter by 2030 (2020)	Annual monitoring of marine litter in estuaries and drainage basins
West Indian Ocean	Action Plan on Marine Litter and Microplastics (2018)	Development of standard methods for monitoring marine litter and microplastics in the region and undertaking surveys to update baseline data on marine litter at regional and national levels

2.1 *Macroplastic*

Global Activities Related to Macroplastic Monitoring

Early monitoring efforts focused largely on plastic contamination in the ‘macro’ range (>25 mm), under the remit of surveying marine litter. This stemmed from reports of plastic waste in the ocean that entangled or was ingested by marine organisms in the 1960s and 70s, which prompted the first monitoring efforts to record this problem (Ryan, 2015). These studies were intended to survey amounts of different plastic waste types in the marine environment, and to unpick potential sources or transport mechanisms (e.g. Scott, 1972; Cundell, 1973; Dixon & Cooke, 1977). Since then, macroplastic monitoring has been conducted or planned for by several international organisations (Table 1) or within a total of 19 marine debris action plans across the globe (Table 1). Much of this work has, thus far, centered on the development of technical guidelines for undertaking plastic monitoring (Barnardo & Ribbink, 2020; Cheshire et al., 2009; GESAMP, 2019; González et al., 2016); however, several international programmes for monitoring macroplastic are already implemented and ongoing.

Many of the current macroplastic monitoring efforts by international organisations focus on the marine environment (Table 1). In particular, beach litter is the most commonly monitored component of environmental macroplastic contamination. Large efforts have been undertaken since 2000 by OSPAR to harmonise methods for recording plastic waste collected during beach cleans for the North East Atlantic region (OSPAR Commission, 2010, 2007). Beach monitoring has also been performed in many regions of the globe, under different local, national, and international schemes. In addition to beach litter, macroplastics floating in the open ocean and the Great Lakes of North America have also been surveyed within the scope of international organisations (Table 1). Macroplastic contamination in the environment has also been addressed by many national or local monitoring programmes in countries across the globe.

Table 3 presents a list of internationally funded collaborative research projects that have undertaken, or plan to undertake, work related to the monitoring of macro-, micro-, or nanoplastics in the environment. This includes conducting monitoring surveys or developing methods or guidelines for monitoring approaches. Projects initially focussed primarily on macroplastic contamination (2013–2016), except CLEANSEA, which added some microplastic monitoring in the later stages of the project. More recently, however, macroplastic monitoring has dropped from the focus of these international research projects. Despite this, the need for further work in developing, refining, and harmonising methods for macroplastic sampling and analysis is reflected in the work programme of the EUROqCHARM project, which addressed a specific European Commission H2020 call related to the improvement and standardisation of methods across the full-size spectrum of environmental plastic contamination.

Table 3 Examples of international collaborative research projects that have undertaken activities related to environmental monitoring of plastic pollution

Project	Funding agency/ programme	Timeframe	Activities (with regards to monitoring activities)
CLEANSEA *Towards a clean, litter-free European marine environment through scientific evidence, innovative tools and good governance*	European Commission seventh framework Programme (FP7)	2013–2015	Monitoring of macro- and microplastics in sediments, biota, water, on beaches, and the seabed
RIMMEL *RIverine and marine floating macro litter monitoring and modelling of environmental loading*	European Commission Joint Research Centre	2015–2017	Development of methods for monitoring floating macrolitter in rivers and marine waters, including a data collection mobile application; monitoring of floating macrolitter in European rivers
BLASTIC *Tackling plastic litter in the Baltic Sea*	European regional development fund	2016–2018	Development of guidelines for monitoring river plastic litter; monitoring of floating plastic litter in case study rivers

(continued)

Table 3 (continued)

Project	Funding agency/ programme	Timeframe	Activities (with regards to monitoring activities)
BASEMAN *Defining the baselines and standards for microplastic analyses in European waters*	Joint programming initiative (JPI) oceans	2016–2018	Development of analytical methodologies for microplastic samples; development of harmonised methodologies for microplastic monitoring
IMPASSE *Impact of MicroPlastics in AgroSystems and stream environments*	Water joint programming initiative (JPI)	2017–2020	Monitoring of microplastics in agricultural soils, stream water, stream sediments; development of methodology for monitoring microplastics in soil and sludge
LimnoPlast *Microplastics in Europe's freshwater ecosystems: From sources to solutions*	European Commission Horizon 2020	2019–2023	Monitoring of microplastic releases and distributions in selected case studies
ANDROMEDA *Analysis techniques for quantifying nano-and microplastic particles and their degradation in the marine environment*	Joint programming initiative (JPI) oceans	2020–2023	Development of monitoring methodologies for small microplastics (<10 μm); development of cost-effective monitoring techniques for microplastic
FACTS *Fluxes and fate of microplastics in northern European waters*	Joint programming initiative (JPI) oceans	2020–2023	Monitoring of small microplastics (<100 μm) in northern European marine waters
resPonse *Toward a risk-based assessment of microplastic pollution in marine ecosystems*	Joint programming initiative (JPI) oceans	2020–2023	Monitoring of marine microplastics across spatial and temporal scales
EUROqCHARM *EUROpean quality controlled harmonisation assuring reproducible monitoring and assessment of plastic pollution*	European Commission Horizon 2020	2020–2023	Development and harmonisation of protocols for environmental monitoring of plastic (nano, micro, macro) in water, soil/ sediment, air, biota
NAUTILOS *New approach to underwater Technologies for Innovative, Low-Cost Ocean obServation*	European Commission Horizon 2020	2020–2024	Complement and expand existing observation tools for marine monitoring, including samplers and sensors for micro- and macroplastic debris

The shift away from macroplastics as the focus of large research projects during the period 2016–2020 likely reflected the increased attention paid to microplastics as an emerging contaminant of concern. More recently, macroplastic has received renewed focus in the international research agenda as multiple papers have now estimated that significant volumes of plastic waste released from rivers into the ocean (Jambeck et al., 2015; Lebreton et al., 2017; Schmidt et al., 2017). Several recent papers have begun to monitor riverine transport of macroplastics (Geraeds et al., 2019; Lahens et al., 2018; van Emmerik et al., 2019a, 2019b, 2019c) and develop methods for measuring plastic flows (Geraeds et al., 2019; van Emmerik et al., 2018). This work must now begin to identify and quantify source inputs and establish the controls on macroplastic transport and accumulation. It is important to note that rivers are not simply conduits of plastic waste to the ocean but are highly complex and dynamic environments that also deserve protection from the potential negative consequences of plastic contamination.

A facet of the internationally funded collaborative research projects on plastic contamination is that they largely focus on European and North American environments, which reflects the geographic scope of the funding agencies that typically grant these types of projects. The monitoring efforts summarised in Table 1 span the globe, but there is limited coordination between them in regard to developing or defining approaches to monitoring. They may, therefore, establish different methods or techniques which reflect local or regional environmental settings. Further work is required to address the harmonisation of monitoring methods established for different environments or matrices across the globe.

2.2 Microplastic

Global Activities Related to Microplastic Monitoring

Table 1 indicates the role of microplastics as an important component in the 'marine litter' or 'plastic' definition for several international organisations and working groups. This is evidenced through the central role of microplastics in many of the monitoring activities that have been undertaken thus far. Many guidelines for microplastics (or 'microlitter') monitoring have been developed by these groups (e.g. Hanke et al., 2013; GESAMP, 2019; Michida et al., 2019) and monitoring surveys have been already undertaken by NOAA, HELCOM, and CPPS. This is also supplemented by a suite of monitoring programmes conducted across the world, operated by national research councils, environmental protection agencies, or non-governmental organisations. Many programmes focus primarily on mapping microplastic distributions; however, some are now beginning to extend beyond this. For example, the San Francisco Bay Area has a long record of environmental monitoring of plastic contamination extending back over the previous three decades (Gordon, 2006). The more recent strategies place greater focus on the microplastic fraction of marine debris and plan to characterise important pathways for microplastic releases from land and to incorporate this data into existing particle tracking models for the Bay Area and the wider regional ocean system (Sutton et al., 2019; Sutton & Sedlak, 2017). Similar developments should be considered in other monitoring strategies to better address questions related to the controls on microplastic distribution, and progress towards solutions for reducing releases or mitigating contamination.

Microplastics have dominated international collaborative research projects related to environmental monitoring of plastic contamination (Table 2). The work was initially focussed on establishing baseline assessments and developing sampling and analytical methodologies (e.g. BASEMAN, BLASTIC) and has progressed to include a wider range of environments (e.g. terrestrial environments covered in IMPASSE), smaller particle sizes (e.g. ANDROMEDA, FACTS), and greater harmonisation of monitoring approaches (e.g. EUROqCHARM). As observed for macroplastic monitoring, these international collaborative research projects remain largely European in focus and the findings from these projects should be adapted for wider geographic contexts, including specific regional settings. However, one-off or baseline surveys of microplastic contamination presented in the scientific literature do exist for a wider range of environments across the globe. For example, river contamination has now been documented in a large number of catchments outside of Europe and North America (e.g. Dahms et al., 2020; Gerolin et al., 2020; He et al., 2020; Lestari et al., 2020; Wang et al., 2020). Further work is required to continue to increase knowledge about riverine microplastic contamination in a more representative array of global environments. The scientific literature is also expanding to consider more advanced research questions within the monitoring of riverine microplastic contamination. This includes: (1) the consideration or quantification of potential sources (e.g. Kataoka et al., 2018; Campanale et al., 2020); (2) the investigation of fate and transport processes (e.g. Hurley et al., 2018; Xiong et al., 2019); (3) the evaluation of spatiotemporal trends (e.g. Eo et al., 2019; Stanton et al., 2020; Wong et al., 2020); and (4) the development of models to estimate fate and transport (e.g. Nizzetto et al., 2016) or export to the marine environment (van Wijnen et al., 2019). These developments currently exist for isolated examples. A holistic approach to future riverine microplastic monitoring should address several of these points within a single coordinated programme.

The need for harmonised definitions, methods, and materials for microplastic analysis has now been highlighted by several international working groups and researchers (e.g. Rochman et al., 2017; GESAMP, 2019; Hartmann et al., 2019; Michida et al., 2019). Interlaboratory comparison/calibration exercises (ILCs) represent one tool to achieve this goal, addressing the analytical component of microplastic studies. Thus far, five large international ILCs have been initiated and/or completed (Table 4). These involve the preparation of standard reference materials for microplastics and the spiking of different environmental matrices with reference particles. Materials and spiked samples are sent to participating laboratories and analyse them following established protocols or their internal methods. Disparities in the reported results between laboratories are used to assess individual laboratory performance, assess typical interlaboratory variability associated with established methods, and identify strengths and shortcomings associated with different analytical methods. Many of the international ILCs for microplastics are still ongoing, but preliminary results confirm the need for chemical characterisation methods for small particles and demonstrate under- and overestimation of microplastic particles across the size spectrum (Isobe et al., 2019). To maintain good QA/QC practices and ensure harmonisation between different laboratories, ILCs should become a

Table 4 International interlaboratory calibration exercises for microplastic analysis

Coordinating organisation(s)	Participating laboratories	Matrices
Ministry of Environment—Japan	12	Water
QUASIMEME, Vrije Universiteit Amsterdam, Norwegian Institute for Water Research, Norman network	30	Water, various complex environmental matrices
Southern California coastal water research project (SCCRWP)	30	Water, sediment, biota
EC Joint Research Centre (JRC)	130	Water
EUROqCHARM project	98	Sediment, biota

regular feature of projects and international initiatives to review detection efficiency and comparability of results.

3 Indian Macroplastic and Microplastic Monitoring

An entry for India is absent from Table 2 due to the lack of a national Action Plan for marine litter or plastics that specifically includes plans for monitoring activities. Yet, India already has an established history of plastic pollution monitoring. India has established policies that aim to reduce environmental or marine pollution, although none of these specifically mandate monitoring of plastic in the environment (Sivadas et al., 2022). India participated as a member of the South Asia Seas Region in the South Asia Co-operative Environment Programme; however, the Regional Action Plan developed by this group in 2019 does not include specific plans for implementing plastic pollution monitoring and comments on the lack of a common baseline for plastic pollution across the country and in India.

Instead, monitoring of plastic pollution in India has largely been performed by universities or institutes. This constitutes a large and growing body of literature reporting plastic pollution data: over 150 scientific articles have now been published. The earliest studies focused on plastic pellets or plastic litter occurring along beaches; however, since the mid-2010s this has shifted to a dominant focus on microplastic monitoring and activities undertaken in a wider variety of environmental compartments or addressing processes or mechanisms of pollution in addition to assessment of occurrence. Figure 1 presents the geographic distribution of micro- and macroplastic monitoring locations in India.

3.1 Macroplastic

Macroplastic monitoring in India predominately takes the form of beach litter observations, represented by the green circles covering much of the western and southern coastline of India in Fig. 1. Tamil Nadu, Kerala, and Karnataka states have

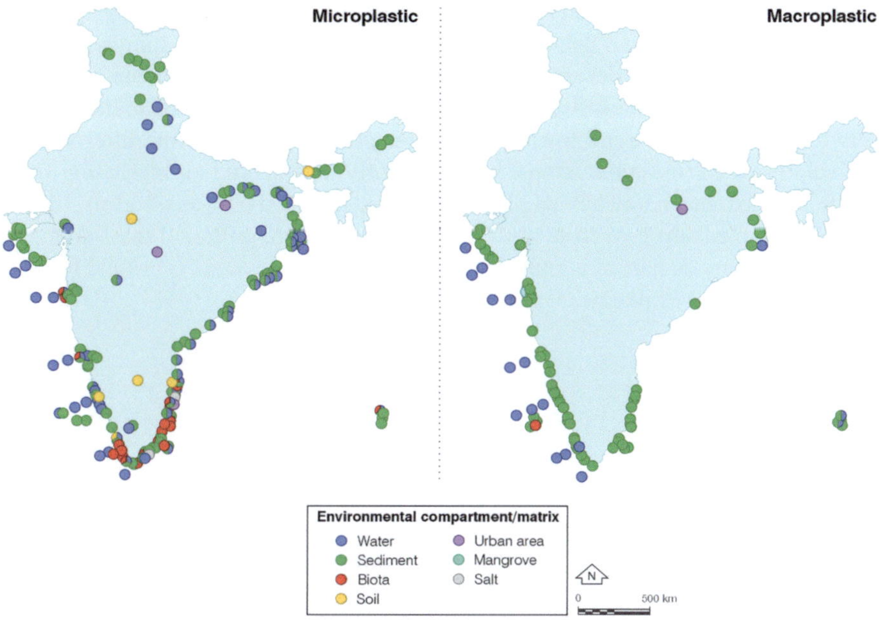

Fig. 1 Locations of environmental monitoring for micro- and macroplastic pollution in India, classified by the environmental matrix investigated

experienced the most intensive macroplastic monitoring in terms of the density of monitoring locations, with additional clusters associated with Goa and Mumbai. In contrast, the eastern coast of India remains relatively poorly monitored for macroplastic pollution, except in the state of Tamil Nadu. The majority of publications follow international standardised or harmonised guidelines for beach litter monitoring and classification of litter, such as (Cheshire et al., 2009).

Offshore monitoring of marine litter has been performed along the western coast of India by Naidu et al. (2021), revealing higher abundances in the Central and South Eastern Arabian Seas. Macroplastic represented 10% of the total collected plastic items by number, whereas microplastic instead dominated the count-based data. Only a single study has specifically monitored the interaction between macroplastic and biota, in an investigation of sea bird entanglement on the Kerala coastline and the Lakshadweep islands (Nisanth & Kumar, 2019).

The only inland assessments of macroplastic pollution in the published scientific literature comprise riverbank monitoring of abandoned, lost, or otherwise discarded fishing gear (ALDFG) in the river Ganga (Nelms et al., 2021) and quantification in litter aggregation pathways, such as along walkways or in open gutters, in the city of Varanasi along the Ganga river (Youngblood et al., 2022). In the Ganga, ADLFG was observed to increase with proximity to the river mouth, where nets made up the vast majority of observed ALDFG by volume (Nelms et al., 2021). In Varanasi, higher densities of litter were observed in areas with lower population, which the authors linked to potential socioeconomic factors relevant to those parts of the city and the potentially higher degree of street cleaning activities occurring in higher population density areas (Youngblood et al., 2022). This limited number of published scientific articles describing macroplastic pollution in freshwater or terrestrial environments reveals a strong marine focus on Indian macroplastic monitoring. This follows a global trend towards a greater prevalence of marine litter studies, particularly in earlier publications, but apparently lags behind an increasing focus on the sources and pathways of plastic litter on land (Al-Zawaidah et al., 2021). New studies measuring riverine or terrestrial macroplastic pollution may emerge in the coming years as part of a greater global focus on these areas is critical for controlling the release of plastic.

3.2 *Microplastic*

Microplastic monitoring in India has largely taken place following 2015, after which there has been an exponential increase in the number of published studies and a rapid increase in the geographic scope and variety of environmental matrices tested for microplastic occurrence. This includes a higher propensity for studies examining multiple matrices in a single monitoring campaign, demonstrated by the greater number of circles with multiple colours in Fig. 1. The coastlines of Tamil Nadu and Kerala have been subject to the most intensive monitoring activities, based on the density of monitoring locations and several individual studies. This corresponds with a higher density of monitoring locations observed in southern India for macroplastic; however, microplastic monitoring has been substantially more varied in terms of the different environmental matrices tested. This includes several studies monitoring microplastics in biota (e.g. Anandhan et al., 2022; Joshy et al., 2022; Karthik et al., 2018), urban dust (e.g. Narmadha et al., 2020; Pandey et al., 2022; Patchaiyappan et al., 2021a) soils (e.g. Reethu et al., 2023; Singh et al., 2023; Tun et al., 2022), and sea salt (Ravikumar et al., 2023; Selvam et al., 2020). In addition, there has been more monitoring of the eastern coastline of India compared to macroplastic monitoring activities, such as beach monitoring studies of Odisha and Andhra Pradesh (Patchaiyappan et al., 2021b; Raju et al., 2023) and the monitoring of coastal waters and shelf sediments along the central east coast undertaken by Sambandam et al. (2022).

The geographic distribution of microplastic monitoring locations also shows a larger number of studies investigating inland areas, including the Ganga River (e.g. Napper et al., 2021; Neelavannan et al., 2023; Sarkar et al., 2019), the Brahmaputra and Indus rivers of the Indian Himalayas (Tsering et al., 2021), and the urban areas of Varanasi, Nagpur, and Chennai (Narmadha et al., 2020; Pandey et al., 2022; Patchaiyappan et al., 2021a). Since 2022, studies have begun to emerge monitoring microplastic levels in Indian soils (Kannankai et al., 2022; Reethu et al., 2023;

Sarkar et al., 2022; Singh et al., 2023; Tun et al., 2022). These primarily focus on urban locations, including urban mangroves or dumpsites. A study comparing India with five other Asian countries found that soil microplastic pollution associated with dumpsites was lowest in India, which was investigated in Bangalore, Chennai, and Perungudi (Tun et al., 2022). It is likely that monitoring of microplastic in soil will increase in the coming years, following the emerging global trend and an increasing focus on the prevalence of microplastic pollution arising from agricultural practices.

3.3 Harmonisation in Indian Plastic Monitoring

A common factor of plastic monitoring globally is a lack of harmonisation between studies, hindering the potential for comparing datasets. This is also a challenge in Indian plastic pollution monitoring. Whilst some monitoring activities—such as beach litter monitoring—tend to follow international guidelines for methodological approaches and data reporting, many studies use adapted or wholly different methods for sampling and analysing macro- and microplastic pollution (Veerasingam et al., 2020). These differences yield results that are potentially poorly comparable on the basis that they are typically measuring different things, such as different size classes of plastic or different components of the environment. This is also characterised by diversity in data reporting, apparent in the wide variety of different reporting units that are used to report Indian monitoring data across different published studies (Sivadas et al., 2022). In some cases, methods are insufficiently reported, hindering the potential to understand how comparable the data are to other published studies.

The upshot of this diversity of different methodological and data reporting approaches is that it is difficult to conclude on the distribution of plastic pollution across India. Hotspots or areas with higher levels of contamination reported in one study cannot be reliably compared to the data reported in another (Veerasingam et al., 2020); thus, there is a challenge in establishing a national picture of the scales of plastic pollution. There is a need for coordination of monitoring activities at both the national and international levels to converge ongoing and future monitoring activities towards a more harmonised approach (Shaikh & Shaikh, 2021). This could comprise the following recommendations for reporting methodological details and reporting data in multiple units to facilitate a greater understanding of what the data represent and increase the potential utilisation of the data by different end users (Hurley et al., 2023). The use of reproducible analytical pipelines (RAPs) and assessments of the technological readiness levels (TRLs) of different monitoring methods could also accelerate national and international efforts toward harmonisation (Aliani et al., 2023).

4 Summary

Internationally, monitoring of plastic pollution had an early marine focus but in recent years there has been increasing attention paid to freshwater or terrestrial environments following their recognition as important domains for environmental releases of plastic. Monitoring is now underway in these settings to identify dominant sources of plastic pollution and better understand the environmental fate of plastics to tailor effective mitigation or management solutions. This early marine focus largely comprised a focus on marine litter; namely, macroplastics. However, this has evolved into a dominance of microplastic monitoring, in response to increasing scientific concern around this emerging contaminant.

In India, there remains a lack of a national action plan for plastic pollution like those prepared by other countries globally. India participates in a South Asian Seas Regional Action Plan, but this does not specifically mandate micro- or macroplastic monitoring beyond a list of recommendations. Legislation is emerging to address plastic pollution but these do not yet include monitoring as a component. Yet, there is an increasing abundance of monitoring data in India conducted instead by universities, NGOs, and other institutes and available in the scientific literature. Over 150 studies report micro- or macroplastic monitoring data for a range of different environmental compartments or matrices. Within these data, a corresponding early focus macroplastic has been replaced by an exponential increase in microplastic monitoring from the mid-2010s. There remains a marine focus in Indian plastic monitoring but new scientific articles published since 2022 reveal the emergence of a trend towards freshwater and terrestrial monitoring.

Harmonisation of monitoring methods and data reporting remains a challenge both internationally and nationally and with respect to both macro- and microplastics. There is a need for a convergence of definitions, methods, and data reporting approaches globally towards facilitating greater comparability and utilisation of end data. There is a need to harmonise monitoring efforts within India to allow an understanding of pollution trends to be developed but this should occur alongside and in collaboration with ongoing international efforts towards harmonisation to ensure that national data can be compared regionally or globally.

Acknowledgments This research was supported by the Royal Norwegian Embassy in New Delhi through the INOPOL Project (Project No: IND-22/0007).

References

Aliani, S., Lusher, A., Galgani, F., Herzke, D., Nikiforov, V., Primpke, S., Roscher, L., da Silva, V. H., Strand, J., Suaria, G., Vanavermaete, D., Verlé, K., De Witte, B., & van Bavel, B. (2023). Reproducible pipelines and readiness levels in plastic monitoring. *Nature Reviews Earth and Environment, 4*, 290–291. https://doi.org/10.1038/s43017-023-00405-0

Al-Zawaidah, H., Ravazzolo, D., & Friedrich, H. (2021). Macroplastics in rivers: Present knowledge, issues and challenges. *Environmental Science: Processes and Impacts, 23*, 535–552. https://doi.org/10.1039/D0EM00517G

Anandhan, K., Tharini, K., Thangal, S. H., Yogeshwaran, A., & Muralisankar, T. (2022). Occurrence of microplastics in the gastrointestinal tracts of edible fishes from south Indian rivers. *Bulletin of Environmental Contamination and Toxicology, 109*, 1023–1028. https://doi.org/10.1007/s00128-022-03595-3

Barnardo, T., & Ribbink, A. (2020). *African marine litter monitoring manual*. African Marine Waste Network, Sustainable Seas Trust.

Campanale, C., Stock, F., Massarelli, C., Kochleus, C., Bagnuolo, G., Reifferscheid, G., & Uricchio, V. F. (2020). Microplastics and their possible sources: The example of Ofanto river in Southeast Italy. *Environmental Pollution, 258*, 113284.

Carpenter, E. J., & Smith, K. L. (1972). Plastics on the Sargasso sea surface. *Science, 175*, 1240–1241. https://doi.org/10.1126/science.175.4027.1240

Carpenter, E. J., Anderson, S. J., Harvey, G. R., Miklas, H. P., & Peck, B. B. (1972). Polystyrene spherules in coastal waters. *Science, 178*, 749–750. https://doi.org/10.1126/science.178.4062.749

Cheshire, A., Adler, E., & Barbière, J. (2009). *UNEP/IOC guidelines on survey and monitoring of marine litter, Regional Seas reports and studies*. United Nations Environment Programme, Regional Seas Programme; Intergovernmental Oceanographic Commission, Integrated Coastal Area Management and Regional Programme. https://wedocs.unep.org/bitstream/handle/20.500.11822/13604/rsrs186.pdf?sequence=1&amp%3BisAllowed=

Cundell, A. M. (1973). Plastic materials accumulating in Narragansett Bay. *Marine Pollution Bulletin, 4*, 187–188. https://doi.org/10.1016/0025-326X(73)90226-9

Dahms, H. T., van Rensburg, G. J., & Greenfield, R. (2020). The microplastic profile of an urban African stream. *Science of the Total Environment, 731*, 138893.

Das, S., Jha, P., & Chatterjee, A. (2020). *Assessing marine plastic pollution in India*. Institute of Economic Growth, University Enclave, University of Delhi. https://iegindia.org/upload/profile_publication/doc-160420_154550wp389.pdf

Dixon, T. R., & Cooke, A. J. (1977). Discarded containers on a Kent beach. *Marine Pollution Bulletin, 8*, 105–109. https://doi.org/10.1016/0025-326X(77)90132-1

Eo, S., Hong, S. H., Song, Y. K., Han, G. M., & Shim, W. J. (2019). Spatiotemporal distribution and annual load of microplastics in the Nakdong River, South Korea. *Water Research, 160*, 228–237. https://doi.org/10.1016/j.watres.2019.05.053

Geraeds, M., van Emmerik, T., de Vries, R., & bin Ab Razak, M. S. (2019). Riverine plastic litter monitoring using unmanned aerial vehicles (UAVs). *Remote Sensing, 11*, 2045. https://doi.org/10.3390/rs11172045

Gerolin, C. R., Pupim, F. N., Sawakuchi, A. O., Grohmann, C. H., Labuto, G., & Semensatto, D. (2020). Microplastics in sediments from Amazon rivers, Brazil. *Science of the Total Environment, 749*, 141604.

GESAMP. (2019). *Guidelines for the monitoring and assessment of plastic litter in the ocean* (No. 2019 #99). http://www.gesamp.org/site/assets/files/2002/rs99e.pdfIMO

González, D., Hanke, G., Tweehuysen, G., Bellert, B., Holzhauer, M., Palatinus, A., Hohenblum, P., & Oosterbaan, L. (2016). *Riverine litter monitoring-options and recommendations* (MSFD GES TG Marine Litter Thematic Report; No. EUR 28307). Joint Research Centre (JRC). https://publications.jrc.ec.europa.eu/repository/bitstream/JRC104194/lb-na-28307-en-n%20.pdf

Gordon, M. (2006). *Eliminating land-based discharges of marine debris in California: A plan of action from the Plastic Debris Project*. California Coastal Commission. https://www.yumpu.com/en/document/view/42231824/eliminating-land-based-discharges-of-marine-debris-in-california-

Hanke, G., Galgani, F., Werner, S., Oosterbaan, L., Nilsson, P., Fleet, D., Kinsey, S., Thompson, R., Palatinus, A., Van Franeker, J. A., Vlachogianni, T., Scoullos, M., Veiga, J. M., Matiddi, M., Alcaro, L., Maes, T., Korpinen, S., Budziak, A., Leslie, H., Gago, J., & Liebezeit, G. (2013). *Guidance on Monitoring of Marine Litter in European Seas (EUR - Scientific and technical research reports)*. Publications Office of the European Union. https://doi.org/10.2788/99475

Hartmann, N. B., Hüffer, T., Thompson, R. C., Hassellöv, M., Verschoor, A., Daugaard, A. E., Rist, S., Karlsson, T., Brennholt, N., Cole, M., Herrling, M. P., Hess, M. C., Ivleva, N. P., Lusher, A. L., & Wagner, M. (2019). Are we speaking the same language? Recommendations for a definition and categorization framework for plastic debris. *Environmental Science & Technology, 53*(3), 1039–1047. https://doi.org/10.1021/acs.est.8b05297

He, B., Goonetilleke, A., Ayoko, G. A., & Rintoul, L. (2020). Abundance, distribution patterns, and identification of microplastics in Brisbane River sediments, Australia. *Science of the Total Environment, 700*, 134467.

Hurley, R., Woodward, J., & Rothwell, J. J. (2018). Microplastic contamination of river beds is significantly reduced by catchment-wide flooding. *Nature Geoscience, 11*, 251. https://doi.org/10.1038/s41561-018-0080-1

Hurley, R., Braaten, H. F. V., Nizzetto, L., Steindal, E. H., Lin, Y., Clayer, F., van Emmerik, T., Buenaventura, N. T., Eidsvoll, D. P., Økelsrud, A., Norling, M., Adam, H. N., & Olsen, M. (2023). Measuring riverine macroplastic: Methods, harmonisation, and quality control. *Water Research, 235*, 119902. https://doi.org/10.1016/j.watres.2023.119902

Isobe, A., Buenaventura, N. T., Chastain, S., Chavanich, S., Cózar, A., DeLorenzo, M., Hagmann, P., Hinata, H., Kozlovskii, N., Lusher, A. L., Martí, E., Michida, Y., Mu, J., Ohno, M., Potter, G., Ross, P. S., Sagawa, N., Shim, W. J., Song, Y. K., … Zhang, W. (2019). An interlaboratory comparison exercise for the determination of microplastics in standard sample bottles. *Marine Pollution Bulletin, 146*, 831–837. https://doi.org/10.1016/j.marpolbul.2019.07.033.

Jambeck, J. R., Geyer, R., Wilcox, C., Siegler, T. R., Perryman, M., Andrady, A., Narayan, R., & Law, K. L. (2015). Plastic waste inputs from land into the ocean. *Science, 347*, 768–771. https://doi.org/10.1126/science.1260352

Joshy, A., Krupesha Sharma, S. R., & Mini, K. G. (2022). Microplastic contamination in commercially important bivalves from the southwest coast of India. *Environmental Pollution, 305*, 119250. https://doi.org/10.1016/j.envpol.2022.119250

Kannankai, M. P., Alex, R. K., Muralidharan, V. V., Nazeerkhan, N. P., Radhakrishnan, A., & Devipriya, S. P. (2022). Urban mangrove ecosystems are under severe threat from microplastic pollution: A case study from Mangalavanam, Kerala, India. *Environmental Science and Pollution Research International, 29*, 80568–80580. https://doi.org/10.1007/s11356-022-21530-1

Karthik, R., Robin, R. S., Purvaja, R., Ganguly, D., Anandavelu, I., Raghuraman, R., Hariharan, G., Ramakrishna, A., & Ramesh, R. (2018). Microplastics along the beaches of southeast coast of India. *Science of the Total Environment, 645*, 1388–1399.

Kataoka, T., Nihei, Y., Kudou, K., & Hinata, H. (2018). Assessment of the sources and inflow processes of microplastics in the river environments of Japan. *Environmental Pollution, 244*, 958–965. https://doi.org/10.1016/j.envpol.2018.10.111

Lahens, L., Strady, E., Kieu-Le, T.-C., Dris, R., Boukerma, K., Rinnert, E., Gasperi, J., & Tassin, B. (2018). Macroplastic and microplastic contamination assessment of a tropical river (Saigon River, Vietnam) transversed by a developing megacity. *Environmental Pollution, 236*, 661–671.

Lebreton, L. C. M., Van der Zwet, J., Damsteeg, J.-W., Slat, B., Andrady, A., & Reisser, J. (2017). River plastic emissions to the world's oceans. *Nature Communications, 8*, 15611. https://doi.org/10.1038/ncomms15611

Lestari, P., Trihadiningrum, Y., Wijaya, B. A., Yunus, K. A., & Firdaus, M. (2020). Distribution of microplastics in Surabaya River, Indonesia. *Science of the Total Environment, 726*, 138560.

Meijer, L. J. J., van Emmerik, T., van der Ent, R., Schmidt, C., & Lebreton, L. (2021). More than 1000 rivers account for 80% of global riverine plastic emissions into the ocean. *Science Advances, 7*, eaaz5803. https://doi.org/10.1126/sciadv.aaz5803

Michida, Y., Chavanich, S., Chiba, S., Cordova, M. R., Cozsar Cabanas, A., Glagani, F., Hagmann, P., Hinata, H., Isobe, A., Kershaw, P. (2019). *Guidelines for harmonizing ocean surface micro plastic monitoring methods*. Version 1.1. https://repository.oceanbestpractices.org/handle/11329/1361

Naidu, S. A., Mawii, L., Ranga Rao, V., Anitha, G., Mishra, P., Narayanaswamy, B. E., Anil Kumar, V., Ramana Murthy, M. V., & Gvm, G. (2021). Characterization of plastic debris from surface waters of the eastern Arabian Sea–Indian Ocean. *Marine Pollution Bulletin, 169*, 112468. https://doi.org/10.1016/j.marpolbul.2021.112468

Napper, I. E., Baroth, A., Barrett, A. C., Bhola, S., Chowdhury, G. W., Davies, B. F. R., Duncan, E. M., Kumar, S., Nelms, S. E., Hasan Niloy, M. N., Nishat, B., Maddalene, T., Thompson, R. C., & Koldewey, H. (2021). The abundance and characteristics of microplastics in surface water in the transboundary Ganges River. *Environmental Pollution, 274*, 116348. https://doi.org/10.1016/j.envpol.2020.116348

Narmadha, V. V., Jose, J., Patil, S., Farooqui, M. O., Srimuruganandam, B., Saravanadevi, S., & Krishnamurthi, K. (2020). Assessment of microplastics in roadside suspended dust from urban and rural environment of Nagpur, India. *International Journal of Environmental Research, 14*, 629–640. https://doi.org/10.1007/s41742-020-00283-0

Neelavannan, K., Sen, I. S., Sinha, N., Thakur, A. K., & Misra, S. (2023). Microplastics in the ganga-Brahmaputra delta: Sources and pathways to the Sundarbans biosphere reserve—an UNESCO world heritage Centre. *Environmental Advances, 11*, 100350. https://doi.org/10.1016/j.envadv.2023.100350

Nelms, S. E., Duncan, E. M., Patel, S., Badola, R., Bhola, S., Chakma, S., Chowdhury, G. W., Godley, B. J., Haque, A. B., Johnson, J. A., Khatoon, H., Kumar, S., Napper, I. E., Niloy, Md. N. H., Akter, T., Badola, S., Dev, A., Rawat, S., Santillo, D., … Koldewey, H. (2021). Riverine plastic pollution from fisheries: Insights from the Ganges River system. *Science of the Total Environment, 756*, 143305. https://doi.org/10.1016/j.scitotenv.2020.143305.

Nigam, R. (1982). Plastic pellets on the Caranzalem beach sands, Goa, India. *Mahasagar Bulletin of the National Institute of Oceanography, 15*, 125–127.

Nisanth, H. P., & Kumar, B. (2019). Observations on the entanglement of plastic debris in seabirds of the family Laridae along Kerala coast, India. *Journal of Aquatic Biology & Fisheries, 7*, 115–119.

Nizzetto, L., Bussi, G., Futter, M. N., Butterfield, D., & Whitehead, P. G. (2016). A theoretical assessment of microplastic transport in river catchments and their retention by soils and river sediments. *Environmental Science: Processes and Impacts, 18*, 1050–1059. https://doi.org/10.1039/C6EM00206D

OSPAR Commission. (2007). *OSPAR pilot project on monitoring marine beach litter*. Monitoring of marine litter in the OSPAR region. https://qsr2010.ospar.org/media/assessments/p00306_Litter_Report.pdf

OSPAR Commission. (2010). *Guideline for monitoring marine litter on the beaches in the OSPAR maritime area*. https://www.ospar.org/ospar-data/10-02e_beachlitter%20guideline_english%20only.pdf

Pandey, D., Banerjee, T., Badola, N., & Chauhan, J. S. (2022). Evidences of microplastics in aerosols and street dust: A case study of Varanasi City, India. *Environmental Science and Pollution Research International, 29*, 82006–82013. https://doi.org/10.1007/s11356-022-21514-1

Patchaiyappan, A., Dowarah, K., Zaki Ahmed, S., Prabakaran, M., Jayakumar, S., Thirunavukkarasu, C., & Devipriya, S. P. (2021a). Prevalence and characteristics of microplastics present in the street dust collected from Chennai metropolitan city, India. *Chemosphere, 269*, 128757. https://doi.org/10.1016/j.chemosphere.2020.128757

Patchaiyappan, A., ZakiAhmed, S., Dowarah, K., Khadanga, S. S., Singh, T., Jayakumar, S., Thirunavukkarasu, C., & Devipriya, S. P. (2021b). Prevalence of microplastics in the sediments of Odisha beaches, southeastern coast of India. *Marine Pollution Bulletin, 167*, 112265. https://doi.org/10.1016/j.marpolbul.2021.112265

Raju, M. P., Veerasingam, S., Suneel, V., Saha, M., Rathore, C., Naik, A., Suneetha, P., & Ramakrishna, S. S. V. S. (2023). Seasonal variation and spatial distribution of microplastic pellets and their associated contaminants along the central east coast of India. *Environmental*

Science and Pollution Research International, 30, 68489–68503. https://doi.org/10.1007/s11356-023-27100-3

Ravikumar, S., Jeyameenakshi, A., Syed Ali, M., & Ebenezer, K. S. (2023). Assessment of microplastics in edible salts from solar saltpans and commercial salts. *Total Environment Research Themes, 6*, 100032. https://doi.org/10.1016/j.totert.2023.100032

Reethu, M., Biswajit, R., Aravind, G. H., Rafaz, A. K., Sandeep, K., Sijinkumar, A. V., & Warrier, A. K. (2023). A first report on the spatial and temporal variability of microplastics in coastal soils of an urban town in South-Western India: Pre- and post-COVID scenario. *Marine Pollution Bulletin, 190*, 114888. https://doi.org/10.1016/j.marpolbul.2023.114888

Rochman, C. M., Regan, F., & Thompson, R. C. (2017). On the harmonization of methods for measuring the occurrence, fate and effects of microplastics. *Analytical Methods, 9*, 1324–1325. https://doi.org/10.1039/C7AY90014G

Ryan, P. G. (2015). *A brief history of marine litter research, in: Marine anthropogenic litter* (pp. 1–25). Springer. https://doi.org/10.1007/978-3-319-16510-3_1

Sambandam, M., Dhineka, K., Sivadas, S. K., Kaviarasan, T., Begum, M., Hoehn, D., Sivyer, D., Mishra, P., & Murthy, M. V. R. (2022). Occurrence, characterization, and source delineation of microplastics in the coastal waters and shelf sediments of the central east coast of India, Bay of Bengal. *Chemosphere, 303*, 135135. https://doi.org/10.1016/j.chemosphere.2022.135135

Sarkar, D. J., Das Sarkar, S., Das, B. K., Manna, R. K., Behera, B. K., & Samanta, S. (2019). Spatial distribution of meso and microplastics in the sediments of river ganga at eastern India. *Science of the Total Environment, 694*, 133712. https://doi.org/10.1016/j.scitotenv.2019.133712

Sarkar, A., Deb, S., Ghosh, S., Mandal, S., Quazi, S. A., Kushwaha, A., Hoque, A., & Choudhury, A. (2022). Impact of anthropogenic pollution on soil properties in and around a town in eastern India. *Geoderma Regional, 28*, e00462. https://doi.org/10.1016/j.geodrs.2021.e00462

Schmidt, C., Krauth, T., & Wagner, S. (2017). Export of plastic debris by Rivers into the sea. *Environmental Science & Technology, 51*, 12246–12253. https://doi.org/10.1021/acs.est.7b02368

Scott, P. G. (1972). Plastics packaging and coastal pollution. *International Journal of Environmental Studies, 3*, 35–36. https://doi.org/10.1080/00207237208709489

Selvam, S., Manisha, A., Venkatramanan, S., Chung, S. Y., Paramasivam, C. R., & Singaraja, C. (2020). Microplastic presence in commercial marine sea salts: A baseline study along Tuticorin coastal salt pan stations, gulf of Mannar, South India. *Marine Pollution Bulletin, 150*, 110675. https://doi.org/10.1016/j.marpolbul.2019.110675

Shaikh, I. V., & Shaikh, V. A. E. (2021). A comprehensive review on assessment of plastic debris in aquatic environment and its prevalence in fishes and other aquatic animals in India. *Science of the Total Environment, 779*, 146421. https://doi.org/10.1016/j.scitotenv.2021.146421

Singh, S., Chakma, S., Alawa, B., Kalyanasundaram, M., & Diwan, V. (2023). Identification, characterization, and implications of microplastics in soil—A case study of Bhopal, Central India. *Journal of Hazardous Materials Advances, 9*, 100225. https://doi.org/10.1016/j.hazadv.2022.100225

Sivadas, S. K., Mishra, P., Kaviarasan, T., Sambandam, M., Dhineka, K., Murthy, M. V. R., Nayak, S., Sivyer, D., & Hoehn, D. (2022). Litter and plastic monitoring in the Indian marine environment: A review of current research, policies, waste management, and a roadmap for multidisciplinary action. *Marine Pollution Bulletin, 176*, 113424. https://doi.org/10.1016/j.marpolbul.2022.113424

Stanton, T., Johnson, M., Nathanail, P., MacNaughtan, W., & Gomes, R. L. (2020). Freshwater microplastic concentrations vary through both space and time. *Environmental Pollution, 263*(Pt B), 114481.

Sutton, R., & Sedlak, M. (2017). *Microplastic monitoring and science strategy*. San Francisco Estuary Institute. https://www.sfei.org/documents/microplastic-monitoring-and-science-strategy-san-francisco-bay

Sutton, R., Franz, A., Gilbreath, A., Lin, D., Miller, L., Sedlak, M., Wong, A., Box, C., Holleman, R., Munno, K., Zhu, X., & Rochman, C. (2019). Understanding microplastic levels. In

Pathways, and Transport in the San Francisco Bay Region (No. SFEI-ASC Publication #950). San Francisco Estuary Institute.

Tsering, T., Sillanpää, M., Sillanpää, M., Viitala, M., & Reinikainen, S.-P. (2021). Microplastics pollution in the Brahmaputra River and the Indus River of the Indian Himalaya. *Science of the Total Environment, 789*, 147968. https://doi.org/10.1016/j.scitotenv.2021.147968

Tun, T. Z., Kunisue, T., Tanabe, S., Prudente, M., Subramanian, A., Sudaryanto, A., Viet, P. H., & Nakata, H. (2022). Microplastics in dumping site soils from six Asian countries as a source of plastic additives. *Science of the Total Environment, 806*, 150912. https://doi.org/10.1016/j.scitotenv.2021.150912

van Emmerik, T., Kieu-Le, T.-C., Loozen, M., van Oeveren, K., Strady, E., Bui, X.-T., Egger, M., Gasperi, J., Lebreton, L., Nguyen, P.-D., Schwarz, A., Slat, B., & Tassin, B. (2018). A methodology to characterize riverine macroplastic emission into the ocean. *Frontiers in Marine Science, 5*. https://doi.org/10.3389/fmars.2018.00372

van Emmerik, T., Loozen, M., van Oeveren, K., Buschman, F., & Prinsen, G. (2019a). Riverine plastic emission from Jakarta into the ocean. *Environmental Research Letters, 14*, 084033. https://doi.org/10.1088/1748-9326/ab30e8

van Emmerik, T., Strady, E., Kieu-Le, T.-C., Nguyen, L., & Gratiot, N. (2019b). Seasonality of riverine macroplastic transport. *Scientific Reports, 9*, 1–9. https://doi.org/10.1038/s41598-019-50096-1

van Emmerik, T., Tramoy, R., van Calcar, C., Alligant, S., Treilles, R., Tassin, B., & Gasperi, J. (2019c). Seine plastic debris transport Tenfolded during increased river discharge. *Frontiers in Marine Science, 6*. https://doi.org/10.3389/fmars.2019.00642

van Wijnen, J., Ragas, A. M. J., & Kroeze, C. (2019). Modelling global river export of microplastics to the marine environment: Sources and future trends. *Science of the Total Environment, 673*, 392–401. https://doi.org/10.1016/j.scitotenv.2019.04.078

Veerasingam, S., Ranjani, M., Venkatachalapathy, R., Bagaev, A., Mukhanov, V., Litvinyuk, D., Verzhevskaia, L., Guganathan, L., & Vethamony, P. (2020). Microplastics in different environmental compartments in India: Analytical methods, distribution, associated contaminants and research needs. *TrAC Trends in Analytical Chemistry, 133*, 116071. https://doi.org/10.1016/j.trac.2020.116071

Wang, C., Xing, R., Sun, M., Ling, W., Shi, W., Cui, S., & An, L. (2020). Microplastics profile in a typical urban river in Beijing. *Science of the Total Environment, 743*, 140708.

Wong, G., Löwemark, L., & Kunz, A. (2020). Microplastic pollution of the Tamsui River and its tributaries in northern Taiwan: Spatial heterogeneity and correlation with precipitation. *Environmental Pollution, 260*, 113935.

Xiong, X., Wu, C., Elser, J. J., Mei, Z., & Hao, Y. (2019). Occurrence and fate of microplastic debris in middle and lower reaches of the Yangtze River–from inland to the sea. *Science of the Total Environment, 659*, 66–73. https://doi.org/10.1016/j.scitotenv.2018.12.313

Youngblood, K., Brooks, A., Das, N., Singh, A., Sultana, M., Verma, G., Zakir, T., Chowdhury, G. W., Duncan, E., Khatoon, H., Maddalene, T., Napper, I., Nelms, S., Patel, S., Sturges, V., & Jambeck, J. R. (2022). Rapid characterization of macroplastic input and leakage in the Ganges River basin. *Environmental Science & Technology, 56*, 4029–4038. https://doi.org/10.1021/acs.est.1c04781

Good Practices in Plastic Waste Management: Indian and Global Case Studies

Mary Abraham, Abdullah Atiq, Suneel Pandey, Girija Prasad, and Nathaniel Bhakupar Dkhar

1 Overview of Relevant Plastic Waste Collection Practices

This chapter explores the best practices in plastic waste service delivery, ranging from collection and recycling to concepts on managing plastic at a national, regional, and area level. The chapter also documents international best practices relevant to the Indian waste management scenario.

Waste collection is categorized as among the first and foremost waste management services. The following is a collation of innovative approaches to waste collection in the Indian and global contexts.

1.1 Indian Context

In India, accelerated urbanization has posed serious challenges to waste management. Many municipal corporations in the country are still grappling with identifying the right approach to their sustenance. One of the major problems has been to

M. Abraham · N. B. Dkhar (✉)
Mu Gamma Consultants, Gurgaon, Haryana, India
e-mail: nathaniel@mugammaconsultants.com

A. Atiq · S. Pandey
Environment and Waste Management Division, TERI, New Delhi, Delhi, India

G. Prasad
Central Institute of Petrochemicals Engineering & Technology, CIPET, Bhubaneswar, Odisha, India

P. Chakraborty et al. (eds.), *Efficient and Enhanced Management of Plastic Waste in India*, https://doi.org/10.1007/978-3-031-89398-8_6

operationalize source segregation, which is the most obvious yet pertinent step towards handling waste effectively. Its implementation has been challenging and limited to a few cities considering the vast expanse of area, population, and social diversity. A few best practices for the collection of waste documented from different Indian cities are presented below.

1.1.1 Panaji, Goa

Panaji, the capital of Goa, is one of the cities that have successfully adopted source segregation. The Corporation of the City of Panjim (CCP) initiated a Door-to-Door Collection (DTDC) system for segregated waste in 2003. All generators within the jurisdiction of CCP are needed to segregate waste into two primary fractions: biodegradable/organic waste (wet waste) and nonbiodegradable/inorganic waste (dry waste). In 2008, CCP divided the city into 12 waste management zones to enable better on-ground monitoring and thereby strengthen the implementation of its waste management services across the city. Each zone has dedicated sanitation workers and a supervisor assigned to monitor municipal solid waste management (MSWM) in the respective zone.

In September 2011, CCP initiated a “five-way segregation” system primarily covering large establishments with bulk waste generation, as well as housing societies and apartments (covering at least 60% of households across the city) to enhance recovery and recycling levels. Waste from residential establishments is collected separately, as dry waste and wet waste. The dry waste is then further sorted into four streams at the society level by CCP staff. The total five streams of waste are as follows (Fig. 1):

- Wet waste (Green bin).
- Glass and metals (Black or grey bin).
- Paper and cartons (Pink bin).
- Plastics (Orange bin).
- Nonrecyclables (White bin).

For implementing its waste collection services, CCP has a separate collection schedule for residents, commercial establishments, and hotels/restaurants. While dry waste is collected twice a week, wet waste is collected daily from

Fig. 1 Five-Way Segregation

both—residents (societies or neighborhood communities/individual houses) and commercial establishments (shops/offices); hotel waste (both dry and wet waste components) is collected daily.

Residents from housing colonies/societies hand over their waste (that has been segregated into two fractions) to a waste collector who is employed by the neighborhood community, and for single dwelling units (individual houses/bungalows), it is collected by CCP staff. Waste is collected door-to-door in the first half of the day, where the worker blows a whistle informing the residents about the handing over of waste (Fig. 2). In some neighborhood communities, domestic help assists in delivering the segregated waste to a single point within the colony. The CCP worker further segregates the dry waste into four fractions at the colony (for colony residents) and zonal level (for single dwelling units), respectively. For colonies, it is segregated within the colony gate, and for individual houses, it is taken to the zonal-level sorting units for further segregation.

In the case of bulk generators such as hotels, color-coded polybags (pink, orange, green, white, and black as in Fig. 1) are used, and CCP also has a bag-marking system so that any noncompliance can be identified and fined. This kind of waste, along with the municipal waste received, is taken to the sorting area inside the Material Recovery Facility (MRF). As indicated below, the differently colored polybags indicate the type of waste in them (Fig. 3):

- Orange bag: Mixed plastics like PET bottles and PE foils.
- White bag: Recyclables—Coated foils, Tetra Pak.
- HDPE, Shoes, cloth, thermocol, etc.
- Pink bag: Paper, newspaper, cartridge, Tetra Pak.

Fig. 2 Waste being delivered to a single point within a colony by domestic help employed by a society

Fig. 3 Innovative way of collecting waste in Panaji using orange, green, pink, white, and black bags

- Green bag: wet biodegradable organic waste, garden cuts, coconuts.
- Black bag: Metal (Fe, non-Fe), glass (colored, white).

1.1.2 Ambikapur, Chhattisgarh

Ambikapur shows one of the best examples of plastic waste collection. Ambikapur town in Chhattisgarh state was the fourth cleanest city in India as per the *Swachh Survekshan* rankings 2020 (India Today, 2020). *Swachh Survekshan* is an initiative launched by the Ministry of Housing and Urban Affairs and *Swachh Bharat* Urban. Under *Swachh Survekshan*, cities are rated based on online surveys and ranked based on their cleanliness. In 2020, a total of 4242 Indian cities were surveyed under *Swachh Survekshan*, making it the world's largest cleanliness survey (Ministry of Housing and Urban Affairs, 2020). Ambikapur is one of the few cities in India to have achieved 100% segregation of waste at source—courtesy of centers effectively decentralizing the waste management process. This accolade has been achieved by Ambikapur through the close working of women self-help group (SHG) members along with Ambikapur Municipal Corporation (AMC). The AMC has a contractual agreement with the SHGs for carrying out waste collection services. The agreement entitles the SHG to provide the SHG women workers with their basic salary, a uniform and gear, including a cap, mask, shoes, and gloves. Every day, 447 SHG women of 48 wards (each major Indian town classified into administrative units or wards) collect household waste door-to-door and store it in solid and liquid resource management (SLRM) centers where organic, recyclable, and nonrecyclable wastes are packed separately after vigorous segregation. There are 17 SLRM facilities in Ambikapur.

Fig. 4 SLRM Facility in Ambikapur

In this SLRM facility, the inorganic waste is segregated into 20 categories. An SLRM facility in Ambikapur is shown in Fig. 4. The recyclables then get transferred to the city's only tertiary segregation center, where the waste is cleaned and further classified into 156 categories (Purohit, 2018). The various recyclables, especially plastics, recovered from SLRM centers and tertiary segregation centers are then deposited in makeshift warehouses, which are called treasury, where they get sold off through a transparent process (Ministry of Housing and Urban Affairs, 2019). The maintenance of waste infrastructure, along with the handling of finances and the salaries of the personnel employed in Ambikapur's waste management, is the responsibility of AMC (Kausr, 2019).

1.1.3 Indore, Madhya Pradesh

The city of Indore, Madhya Pradesh, is another success story of managing waste exceptionally well. The city is deemed the cleanest city in India as per the *Swachh Sarvekshan* 2020 ranking (The Hindu, 2020a). Indore has achieved 100% source segregation of waste. The residential households, as a part of the behavioral change drive, were trained and updated on the need to segregate waste into three fractions, namely wet, dry, and hazardous waste.

The door-to-door waste collection is done through partitioned tipper auto vehicles having three separate collection bins (for wet, dry and domestic hazardous waste), with Indore Municipal Corporation (IMC) being the facilitator (Fig. 5). To

Fig. 5 Door-to-door collection in Indore

improve the city's collection of waste and make it a success, IMC started closely monitoring the routes of each waste tipper through GPS trackers and understanding the waste-handling behavior of the local people. Waste collection timings were rescheduled as per these observations (Scroll, 2019). The collected dry and wet waste is taken to one of the 10 transfer stations of the city, where the staff makes sure that the waste is properly inspected for thorough segregation. From these transfer stations, the segregated wet waste is taken to the central wet waste processing facility and processed to compost. The bulk generators process all their wet waste on their premises, bypassing the central processing plant. As per Solid Waste Management Rules 2016, "Bulk Waste Generator" refers to buildings occupied by the Central Government Departments or Undertakings, State Government Departments or Undertakings, Local Bodies, Public Sector Undertakings or Private Companies, Hospitals, Nursing Homes, Schools, Colleges, Universities, other Educational Institutions, Hostels, Hotels, Commercial Establishments, Markets, Places of Worship, Stadia and Sports Complexes, etc., having an average waste generation rate exceeding 100 kg per day of all waste streams put together (Ministry of Housing and Urban Affairs, 2017). The dry waste is taken to a central dry waste processing facility—MRF—where waste is segregated into different components such as metal, rubber, board, and plastic. Valuable recyclables and plastic are sold to waste aggregators, whereas other waste components, such as nonrecyclable plastics, are sent to either cement plants for coprocessing or sold to the Madhya Pradesh Rural Road Development Board for the construction of roads (Indore Municipal Corporation, 2020).

The success of the Indore waste management model is due to the induction of behavioral change components into waste management. Several segregation campaigns were initiated by IMC, aided with tools like flip charts, pamphlets, banners, roadside plays, and demonstrations to encourage proper waste management by the citizens. Another reason for the success of the model is the collaboration of IMC with several NGOs and students for awareness building and campaigns (Ministry of Housing and Urban Affairs, 2019).

1.1.4 Madukkarai, Tamil Nadu

Another notable example of streamlined waste management is that of a small town called Madukkarai in the Coimbatore district of Tamil Nadu. This town holds the World Record for conducting the "Largest Recycling Lesson" for 13,000 participants in a 55-minute awareness session held on August fifth, 2015 (Madhavan, 2015). The credit for Madukkarai's exceptional achievements in waste management goes to the initiative "Clean & Green Madukkarai."

Organizations like the ACC Madukkarai Cement Works, Hand in Hand (NGO) along with Madukkarai Special Grade Panchayat (Madukkarai village council) have evolved and launched this program, which is implemented by an NGO, Thidakazhivu Melanmai Thittam—SHG Group consisting of 55 women. These women, called "green friends," initiate the collection process (Fig. 6). A total of two "green friends" carry one pushcart collecting door-to-door waste from 200 households daily and segregating it into different bins—one for wet waste, one for hazardous waste and eight for dry waste (plastic above 40 μm, plastics below 40 μm, papers, cardboards, glass, rubber and leather items, oil/milk pouches, and e-waste).

To facilitate a smooth collection process, the citizens of Maddukarai were educated about the bell ringing (by waste collectors) as a signal for bringing out and handling over their waste (outside their doors) to the collectors. The collected garbage is transferred to the community bins, which are strategically placed across all 18 wards. The community bins are shifted to the dump yard, where the wet garbage is converted into organic manure in vermicompost pits. The plastic and recyclable

Fig. 6 "Green Friends" collecting waste in Madukkarai. (*Source*: ACC)

wastes are segregated. The recyclables are sold to recycling industries, while other nonrecyclable plastic wastes (plastic below 40 microns) are downcycled and sent to ACC Cement Works for energy recovery and coprocessing. The plastic wastes are used as fuel (India Sanitation Coalition, 2017). The project does not only promote effective waste management practices but also aims at bringing positive behavioral change in the community. The success of this program is owed to community participation; there has been a drastic change in the behavior of people that was achieved through several community engagement and awareness generation measures on effective waste management. The continuous education and awareness programs have resulted in 100% participation of the community in effective waste management (India Sanitation Coalition, 2017).

1.2 Global Context

Implementing source segregation has been challenging for cities across the world. A few of the best practices from the global perspective have been presented below.

1.2.1 Hernani, Spain

Hernani is a city in the Basque Country of Spain. Together with nine other municipalities, it is part of the San Marko *mancomunidad* (a free association of municipalities), created to manage solid waste jointly. In 2010, Hernani started the process of door-to-door waste collection, owing to the public backlash caused by the establishment of two new incinerators within the administrative limits of the town. The municipality distributed two small bins per household, placed hooks to hang the bins and bags at the front of houses and buildings, removed the large containers from the streets, established waste segregation as mandatory, and launched door-to-door collection (Fig. 7).

Citizens began to segregate organics, light packaging, paper and cardboard in front of their houses. Each stream has a designated pick-up day: organics on Wednesdays, Fridays, and Sundays; light packaging on Mondays and Thursdays; paper and cardboard on Tuesdays; and residuals on Saturdays. Light packaging is placed in bags, and the government sells reusable bags for this purpose. Paper and cardboard are tied in bundles or placed in boxes or bags. Organics are placed in the bins provided by the government, and the residuals are disposed of in bags. The collection is done by a public company called Garbitania, created by the government of Hernani. The collection is done at night, with a complementary shift during the morning. Each bin and each hook have a code that identifies the household that uses them. This allows the government to monitor separation in each household. If the collector identifies a stream that does not correspond to that collection day, s/he puts a sticker with a red cross on the bin and does not collect that waste. The information is passed on to the administration office, and the

Fig. 7 Door-to-door collection bins installed in Hernani, Spain. (*Source: GAIA*)

household receives a fine for noncompliance. In Spain, the municipalities that have implemented door-to-door collection programs have promoted the creation of citizens´ committees to monitor their implementation. Moreover, local Zero Zabor (zero waste) groups have emerged in these cities. The residents of Hernani have welcomed the local Zero Zabor community and always comply with their guidance.

1.2.2 San Francisco, United States of America

San Francisco, a major metropolitan area in the State of California in the United States, has established itself as a global leader in waste management. In 2002, the city set an ambitious goal to achieve zero waste to disposal. Since then, legislation has pushed the city, residents, and businesses to increase their recycling rates (The United States Environmental Protection Agency, 2017). In 2012, the city achieved 77% waste diversion, the highest in the United States, with a three-part approach: enacting strong waste reduction legislation, partnering with a waste management company to innovate new programs, and working to create a culture of recycling and composting through incentives and outreach (Global Anti-Incenerator Alliance, 2012). Thus far, they have been performing exceptionally well, courtesy of their innovative tri-stream collection system. Dubbed "fantastic three," the model provides three separate bins for all commercial and residential users, namely a black bin for waste for the landfill, a blue bin for recyclable materials, and a green bin for compostable material (Pollans, 2012, Fig. 8). This segregation system was

Fig. 8 "Fantastic Three" bins in San Francisco, California. (*Source: San Francisco Department of Public Works*)

mandated by San Francisco in 2009 under the Mandatory Recycling and Composting Ordinance, which requires all San Francisco residents to separate recyclable materials, compostable materials and landfilled trash (The United States Environmental Protection Agency, 2017).

San Francisco has developed a robust collection and pricing scheme with Recology—a private waste management company. In this city, businesses and residences segregate waste at the source, and double-chambered back-loading trucks pick up the trash and recycling bins. Smaller side-loading trucks pick up the compostable waste. The "fantastic three" program was one of the first in the United States to scale up the collection and composting of biodegradable waste. The recyclables are diverted to MRFs, which are owned and operated by Recology. Recology further inspects and segregates recyclables into 27 categories in these facilities before selling them to the city's registered recyclers. The profits from selling these recyclables are used by Recology to subsidize their collection fee and facilitate curbside garbage collection (Global Anti-Incenerator Alliance, 2012).

The city also has a thriving informal recycling sector, thanks to the state-wide bottle bill that places a 5 or 10 cent value on glass and plastic bottles and over 20 recycling centers in the city where residents or collectors can redeem them (Centre for Public Impact, 2016). Collection of waste is the sole responsibility of a Recology (Brigham, 2018). The unique part of this partnership is that the city has not directly contracted itself with Recology; rather, each resident and business in the city contracts Recology individually. This vigorous segregation, collection and waste treatment has helped San Francisco achieve a huge reduction in waste, as more than 85%

of waste that would have been landfilled is now being recycled and reused. Also, reduction in landfill waste has accounted for a significant reduction in GHG emissions (C40 Cities Climate Leadership Group, 2016).

Another initiative to counter plastic pollution in marine ecosystems was the New Plastics Economy Global Commitment led by the Ellen MacArthur Foundation in collaboration with UN Environment. Other leading institutions that were signatories to this initiative include WWF, the World Economic Forum, the Consumer Goods Forum, Veolia and the International Union for Conservation of Nature (IUCN). Additionally, more than 50 academic universities and other educational or research organizations, including the MIT Environmental Solutions Initiative, Michigan State University, and University College London. The primary stakeholders, with 200 businesses that were part of the plastic packaging value chain, include Nestlé, PepsiCo, Unilever, L'Oréal, WalMart Inc., Carrefour, Tetra Pak and Coca-Cola Company. The initiative urged all the signatories to continue raising their level of ambition and enhance collaboration to move forward from commitments to actions at scale. The project has 6 primary motives (Ellen MacArthur Foundation and UN Environment, 2019) (Table 1):

- Eliminate problematic or unnecessary plastic packaging through redesign, innovation, and new delivery models as priorities.
- Reuse models are applied where relevant, reducing the need for single-use packaging.
- All plastic packaging is 100% reusable, recyclable, or compostable.
- All plastic packaging is reused, recycled, or composted in practice.
- The use of plastics is fully decoupled from the consumption of finite resources.
- All plastic packaging is free of hazardous chemicals and the health, safety, and rights of all people involved are respected.

Table 1 The Global Concern of Marine Litter

Marine debris coming from land-based sources has created a lot of issues in our ocean ecosystems. Poor waste management practices along the coast, including littering and dumping, have resulted in the generation of debris. This debris from land could be blown, swept, or washed off towards the sea
Littering and dumping in and near the coastal areas is one of the key sources for land-based debris that reaches the marine environment. These debris include street litter like cigarette butts, syringes, food packages, beverage containers, and plastic polybags that contribute to marine litter. Marine litter is responsible for huge economic losses. The communities living in coastal areas must bear the increased expenditure on beach clean-ups, public health, and waste disposal activities. The tourism industry is also heavily impacted due to marine pollution
To allay this growing issue, several alliances and initiatives have been implemented at regional, national, and international levels. One such alliance is "ocean Conservancy's fighting for trash free seas" initiative, comprising several stakeholders (from various sectors), namely, WWF, Nestlé, PEPSICO, Bank of America, Starbucks foundation, P&G, Coca-Cola company, world plastic council and many more. Since its inception in 1986, this initiative has collected close to 340 million pounds of trash, which would have otherwise ended up in the oceans (Ocean Conservancy n.d.)

2 Overview of Relevant Plastic Waste Recycling Schemes and Best Environmental Practices

2.1 Indian Context

Examples of sustainable plastic waste management solutions in India include the use of plastic waste in road laying, coprocessing of plastic waste materials, and conversion of plastic waste to fuel through pyrolysis at a central or state level, as described below:

2.1.1 Plastic Road Making

According to the 2019 Guidelines on the use of Plastic Waste in Road Construction, the percentage of shredded plastic is specified at 6–8% of the weight of bitumen. It also states that plastic waste conforming to Low Density Polyethylene (LDPE), High Density Polyethylene (HDPE), Polyethylene Terephthalate (PET) and Polyurethane could be potentially used in road making (Ministry of Railways, 2019). Common forms of plastic waste used in road making are littered plastics such as carry bags, milk pouches, wrappers, disposable plastic utensils and PET bottles (Ministry of Railways, 2019). There are several advantages to using plastic waste in road making:

- Higher resistance to deformation.
- Higher resistance to water-induced damages.
- Increased durability and improved fatigue life.
- Improved stability and strength.
- Safe disposal of plastic waste.

In India, plastic road making has been in practice for a long time. Jambulingam Street in Chennai was one of India's first plastic roads, built in 2002. In 2003–2004, KK Plastic Waste Management Ltd., Bengaluru, signed an MoU with the BBMP (Bruhat Bengaluru Mahanagara Palike) to lay 250 km of roads in Karnataka. The Central Pollution Control Board (CPCB) reported that polymer roads are more durable with fewer potholes and edge flaws. The Indian Centre for Plastic in the Environment (ICPE) supported the use of waste plastic in making tar roads. In 2015–2016, the National Rural Road Development Agency laid around 7500 km of roads using plastic waste. As of 2020, India has built nearly 100,000 km of roads using plastics (Chaturvedi, 2020).

2.1.2 Coprocessing of Plastic Waste

Waste materials like plastic waste, segregated nonrecyclable solid waste, and a few hazardous waste can be used as an alternate fuel and raw material, thereby substituting the use of coal. Waste materials must be preprocessed (before combustion) to fit the chemical and physical specifications acceptable to kilns, which include calorific value and/or material value, water content, ash content, and concentration of sulphur, chlorine, and heavy metals. Utilization of waste in the cement manufacturing process contributes towards energy efficiency, recycles waste and conserves resources. There are several advantages of coprocessing, which are stated below (CII, Swiss Confederation, World Resources Forum, EMPA and Toxics Link, 2016):

1. Due to the high temperature in the cement kiln (1250–1450 °C), all types of waste, including plastic waste, can be effectively disposed of.
2. Oxygen-rich atmosphere aids complete combustion of waste material and increases the efficiency of incinerators.
3. Reduces the overall greenhouse gas emissions and conserves fossil fuel resources.
4. No ash is left over after coprocessing.
5. Provides an integrated solution to waste management and facilitates a circular economy.

The CPCB has prescribed guidelines on the coprocessing of plastic waste as an alternate fuel. A total of 65 cement plants have been authorized for coprocessing plastic and hazardous waste (CPCB, 2019). A few of the best practices for plastic waste coprocessing witnessed by specific cement plants are:

Vapi Plastic Waste CoProcessing from Paper Mills The industrial town of Vapi in Gujarat hosts several small- and large-scale paper mills where nonrecyclable plastic waste gets generated. Initially, it was being sent to a landfill, which is not the most suitable disposal technique for plastic waste due to its nonbiodegradable and impervious nature. To solve this problem, Ambuja Cements Limited's Kodinar unit agreed to coprocess the plastic waste being generated from Vapi paper mills and use it as an alternative fuel. From 2012 to 2016, a total of 154,018 metric tonnes (MT) of plastic waste from Vapi paper mills had been coprocessed, which would have otherwise been disposed of in landfills (CPCB, 2017). This resulted in a mutually beneficial outcome for both the paper mills and the cement plant, as the Vapi paper mill was able to eliminate their plastic waste, while Ambuja Cement Plant saw a reduction in their coal expenses (Shah, 2018).

Coprocessing of Packaging Waste in Kymore, Madhya Pradesh The ACC Kymore Cement Works has partnered with Unilever India (Hindustan Unilever Limited), wherein the rejected and outdated products from Unilever were coprocessed in the cement kiln. The various types of rejected materials, such as nonhazardous shampoo sachets, shampoo bottles, toothpaste, creams, and lotions, were disposed of successfully at ACC Limited, Kymore Cement Works. Kymore Cement Works also coprocesses packaging waste from other large companies such as Cadbury, Coca-Cola, Nivea and Bournvita. The average quantity coprocessed at this facility is about 300 MT/month (ACC, 2019).

2.1.3 Conversion of Plastic Waste to Fuel

Pyrolysis refers to the breaking down of polymers into smaller molecules by thermal decomposition at temperatures of 300 °C–400 °C in the presence of a catalyst (aluminium oxides, fly ash, red mud, and calcium hydroxide) in an inert atmosphere. Rudra Environmental Solutions, Pune (Maharashtra), has set up a pyrolysis plant where 1 tonne of plastic waste can be converted to 600–650 litres of fuel with an almost 60% conversion rate. M K Aromatics Ltd. has set-up two plants in Goa to convert plastic waste to fuel. Hydroxy Systems Pvt. Ltd., Hyderabad, has adopted a different approach, and its facility converts around 13–15 tonnes of plastic waste per month into approximately 500 litres of fuel. The adopted process is safe, controllable, pollution-free, and has been approved by the State Pollution Control Board.

2.1.4 Other Plastic Waste Recycling Initiatives

Other Indian examples of plastic waste recycling initiatives undertaken by corporates, NGOs or corporations at local/regional levels include:

City Corporation of Panjim (CCP) One innovative plastic waste recycling scheme in India is witnessed in Panjim, Goa, called the "Shop with Your Waste' campaign. The City Corporation of Panjim launched this campaign under the Waste NAMA project implemented by GIZ and The Energy and Resources Institute (TERI) on October 2, 2020, under which citizens could exchange dry recyclables for daily use consumables such as soaps, shampoos, or groceries. This barter system model, commonly known as the "Zero Baht Model," is established in the capital of Goa to inhibit a sense of value for waste among its citizens and to provide a way to incentivize its citizens who have been segregating their waste diligently for years. This campaign encourages the citizens to purchase essentials in exchange for dry waste such as plastic and cardboard. The campaign also strives to improve the socio-economic fabric of the city by providing new opportunities to existing entrepreneurs and waste workers (Herald Goa, 2020). This program has further expanded since its inception, and from December 30, 2020, two more stores joined the corporation to accept recyclable waste in exchange for grocery items. CCP is looking forward to registering more such stores in the future, which include shops in supermarkets, schools, and college canteens (Times of India, 2020).

Hasiru Dala A Bengaluru start-up NGO, works towards achieving plastic recycling through various initiatives. *Hasiru Dala* Innovations aggregates PET bottles at its center, which are sourced from its own network of waste pickers, entrepreneurs, dry waste collection centers, and scrap dealers at a reasonable price. The start-up launched a fair trade plastic recycling initiative with The Body Shop. In its aggregation center, the PET plastic is sorted and baled, followed by the stringent rules of The Body Shop. The World Fair Trade Organization was audited to check compliance with its standards on fair trade. The Plastic for Change mobile application ensured traceability and transparency in the entire supply chain (Hasiru Dala

Table 2 Impact story at Hasiru Dala

Krishna Ayyakannu, a Dry Waste Collection Centre (DWCC) operator, with the help of *Hasiru Dala*, went from waste picking to collecting waste from 24 different locations in the city. Now, he manages a team that provides a monthly waste collection service for 3-star hotels. On average, he collects approximately 10 metric tonnes of post-consumer Tetra Pak cartonnes in a month. The DWCC he operates receives about 15 tonnes of dry recyclable waste every month. Krishna received his first-ever formal loan from Vijaya Bank, using his recyclable waste inventory as collateral (Rang n.d.). S3IDF, *Hasiru Dala* Trust, *Hasiru Dala* Innovations and Plastics for Change came together and collaborated to bring about a change and help in the development of the waste sector. The Alliance aims to engage themselves with the entrepreneurs on a regular basis to better understand the immediate and long-term plans. Just like Krishna, this alliance has touched upon the lives of other entrepreneurs as well and has helped them bag loans to fulfill their requirements (Hasiru Dala, 2017)

Fig. 9 SAAHAS waste segregating kiosk in Bengaluru. (*Source: SAAHAS*)

Innovations, 2020). *Hasiru Dala* Innovations provides end-to-end responsible waste management solutions for the disposal of urban waste. *Hasiru Dala* also empowers the waste pickers by providing them intensive training on responsible waste management and access to waste disposal networks, leadership initiatives and entrepreneurship opportunities, which results in collecting and recycling waste at a fair price (Business India, 2020) (Table 2).

SAAHAS, Bengaluru The entire SAAHAS network in Bengaluru collects nearly 45 tonnes of waste a day from city segregation, processing and transporting it to the recycling companies placed at a different location (Fig. 9). One of SAAHAS MRF's locations, located in Jigani, handles all the dry waste from the nearby IT Parks and multinational companies (Khadilal, 2019). SAHAAS collaborates with communities, administrators, businesses and lawmakers based on the pilot innovative resource management programmes they have carried out in the past. SAAHAS, at present, has been managing decentralized waste facilities that turn biodegradable waste into compost and send the recyclable waste to authorized waste dealers for processing while the waste which cannot be treated is sent to disposal sites (Business call to action n.d.).

Shakti Plastics Industries Shakti Plastics Industries is a business enterprise in India that works in presenting innovative waste management solutions for communities and organizations. They provide end-to-end waste management services that cater to individual collectors, companies, retailers, manufacturers, municipalities, NGOs, and small businesses. They help to recycle tonnes of plastic waste into new value-added products. Shakti Plastics has formalized the unregulated and unorganized waste management sector by sourcing plastic waste from the marginalized waste pickers and ensuring sustainable relationships by providing fair and transparent prices to support their livelihood. They have an Enterprise Resource Planning (ERP) system to track and monitor their activities efficiently.

2.2 Global Context

Project STOP This initiative aims to create sustainable waste management systems in the areas where plastic leakages have occurred from land to oceans. In Indonesia, The Alliance to End Plastic Waste (AEPW) partnered to provide financial support to Project STOP. The project was launched in 2017 by Borealis and SYSTEMIQ. Other strategic partners of Project STOP include Nestlé, the Government of Norway, NOVA Chemicals and Borouge (Project STOP, 2020). It aimed at creating a sustainable, low-cost waste system to capture maximum value from waste in Indonesia. It works towards designing and implementing circular economy solutions to plastic pollution, especially marine plastic (Stuchtey et al., 2019). The AEPW conducted a feasibility study that focused on the Jembrana Island, Bali. This study revealed that the island leaked about 13,200 tonnes of plastic into the ocean every year due to a lack of proper waste management services, open burning and dumping of waste. Working with companies, local governments and community groups, Project STOP supports cities with technical expertise to achieve zero leakage of waste, improve circular economy systems, create new jobs in waste management, and reduce the harmful impact of mismanaged waste on public health, tourism and fisheries. Project STOP adopted a "System Change Scenario" to avoid the leakage of plastics into the marine ecosystems. The scenario was developed based on three key criteria: the impact and relative cost of different system changes; the risk of unwanted consequences for people and the environment; and expert opinions on the feasibility, technology readiness and speed of implementation of different solutions. The system change scenario includes the following five system changes to be achieved by 2025 (World Economic Forum, 2020), which are:

- Reduce or substitute plastic usage to prevent the consumption of more than a million tonnes of plastics per year.
- Redesign plastic products and packaging for reuse or high-value recycling.

- Double plastic waste collection from 39% to 84% by boosting state-funded and informal or private-sector collection systems.
- Double current recycling capacity to process an additional 975,000 tonnes per year of recycled plastic.
- Build or expand controlled waste disposal facilities to manage an additional 3.3 million tonnes of plastic waste per year.

The partnership between AEPW and Project STOP conducts the following activities (Borealis & Alliance to End Plastic Waste, 2019):

- Conduct diagnostic studies to understand how and why plastic waste enters the environment and design a new, tailored system to combat it.
- Build and supply equipment to scale up waste collection, sorting and recycling efforts.
- Hire and train local workers to manage and staff the new waste management system.
- Partner with local organizations to encourage behaviour change at the community level through awareness and educational programs so more citizens engage and fully utilize the systems being created to dispose of waste.

Currently, Project STOP is being implemented in three cities of Indonesia—Persuran, Jembrana and Muncar—and has avoided close to 2000 tonnes of plastics per annum from entering the water bodies (STOP, 2018).

Plastic Waste Management in Tunisia In Tunisia, the National Agency on Waste Management (ANGeD) manages a public system known as "ECO-LEF," which works towards the recovery and diversion of used packaging material, mainly plastics (Fig. 10). Under the ECO-LEF system, recyclable materials are collected by

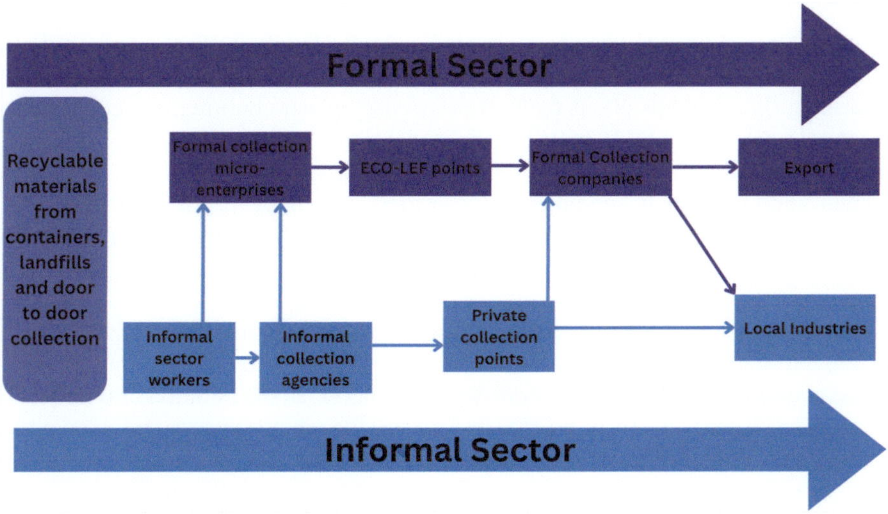

Fig. 10 Working of ECO-LEF. (*Source*: Chaabane et al. (2019))

small companies that are approved and authorized by ANGeD. The authorized collection companies buy the materials from informal collectors and pay them immediately after the recyclables have been weighed. In practice, the collection companies sell what they collect to ECO-LEF; however, they are not under any obligation to do so. The quantities collected are brought to "ECO-LEF points," a network of dedicated storage facilities and prepared for sale to recycling companies (Bünemann et al., 2020). The ECO-LEF system is governed by a decree that specifies how packaging waste should be collected and managed. The initiative is financed primarily from the Central Government's Depollution Fund. The Depollution Fund is a special fund attached to the Tunisian Treasury. There is a scheme within this fund called "eco-tax," which has to be paid by bulk waste generators on the import of and/or production of unfilled plastic packaging materials. Also, hotels are required to contribute 2% of their annual turnover to the Depollution fund to cover solid waste management. ECO-LEF has had a huge impact in controlling plastic pollution in Tunisia. In 2018, 3400 tonnes of plastic waste were collected and recycled under this initiative (Chaabane et al., 2019).

Ocean-Bound Plastics Recycling in Gambia In 2018, WasteAid partnered with UK Aid to run a two-year plastics recycling programme in Gambia. The project aimed at capturing plastic waste from the coastal town of Gunjur and turning it into useful products to stop it from reaching the ocean. With the help of a local partner organization—The Gambian Trust Agency for Rural Development (TARUD)—a recycling center was built wherein WasteAid experts trained local underprivileged residents for plastic waste reprocessing and sustainable waste management (WasteAid, 2018). The first class of 30 trainees completed their course in January 2019. Participants were unemployed youth, marginalized women and people with disabilities, recruited through the TARUD. To enhance the sustainability of the participants, the training program went beyond plastic waste recycling to educate them with business skills training in their local languages. Within 2 months of learning new recycling skills, the team had captured and recycled the equivalent of a million plastic bags. The trained locals also initiated city-wide communication campaigns informing the residents about the harm caused by plastic burning and encouraged them to save their LDPE plastic bags and bring them to the recycling center (Lenkiewicz, 2019). In 2020, the center was named Innovative Plastic Recyclers Gunjur and had already served many customers with floor tiles made by recycled plastics, avoiding a large amount of plastic from being either burnt or dumped in the coastal areas of the city (Gunjur Plastics n.d.).

Kuru-Kuru Shops in Kamikatsu, Japan Situated on Shikoku Island in South Japan, the small town of Kamikatsu holds the honour of being the first municipality in the country to implement the Zero Waste (Nippon, 2018) initiative. Initially, the residents of Kamikatsu were hostile towards the idea of zero waste and were more comfortable dumping their waste conventionally into the landfills. However, in 2005, the Zero Waste Academy, a local nonprofit organization, tied up with the Kamikatsu municipality to provide services to turn waste into something useful and help Kamikatsu work towards its sustainability goals (Henam & Sambyal, 2019). Kamikatsu has set up a waste collection center where the residents are required to drop their clean and dry waste (Global Alliance for Incinerator Alternatives, 2019). From becoming a hub for social interaction among residents, the waste collection center later expanded to include a kuru-kuru shop (kuru-kuru meaning circular in Japanese), where locals can drop off items they no longer need and take away any of the items that were also dropped off there for free. Additionally, to eliminate the use of disposable plastic cups and plates for special occasions, residents can also borrow 8000 tableware items from the Zero Waste Academy for free (Gray, 2019). Next to the waste collection center is a kuru-kuru craft center where local women, particularly older women, make products out of discarded materials. Products sold in these local businesses need to be designed to adhere to the circular economy, where everything made is either reused or recycled (Global Alliance for Incinerator Alternatives, 2019). Effective implementation of the zero-waste model in Kamikatsu not only saves the town close to six million Japanese Yen ($56,000 approx.) every year but also helps the city earn close to three million Japanese Yen ($28,000 approx.) annually from the sale of recyclables (Global Alliance for Incinerator Alternatives, 2019).

3 Approaches to Reduce Plastic Pollution

3.1 National Level (India)

To combat the severe issue of plastic waste in the country, the Government of India's Ministry of Environment, Forest and Climate Change (MoEFCC) notified an updated Plastic Waste Management Rules in 2016. The Rules provided for steps such as mandatory levels of thickness of 50 microns on plastic carry bags and for addressing Extended Producers Responsibility (EPR) as a tool that establishes accountability on producers (manufacturer, importer, and/or brand owner) to handle their post-consumer waste. The EPR mandate would require producers to establish a system for collecting back the plastic packaging waste (milk pouches, plastic bottles, etc.) generated from their products. As per the MoEF&CC's draft implementation guideline on EPR (2020), depending on the regional operations of the producer, the EPR plan has to be submitted to (a) the Concerned State Pollution Control Board (SPCB) if the producer is operating in 2 or fewer states/union territories (UTs); (b) the Central Pollution Control Board (CPCB) if the producer is operating in more than 2 states/UTs (Ministry of Environment, Forest and Climate Change, 2020).

The draft guideline document by MoEF&CC also suggests three EPR models that could be used by the producers to comply, which include the following:

1. *Fee-based model*: It requires the producers to contribute to the EPR corpus fund at a central or state level depending on their regions of operations. The amount of contribution will be determined based on the amount of plastic introduced by the producer into the environment.

2. *Producer Responsibility Organization (PRO) based model*: This model aims to establish a PRO to lead the implementation of the collection of plastics on the ground, while the funding required for the operation would be paid by the producers.
3. *Plastic credit model*: The producer is not required to recycle their own packaging but to ensure that an equivalent amount of packaging waste has been recovered and recycled to meet their obligation (Ministry of Environment, Forest and Climate Change, 2020).

A few examples of Indian corporates practicing EPR efficiently are described below:

Dabur Under their plastic waste management EPR activity, Dabur has collected over 12,000 tonnes of post-consumer plastic waste (both recyclable and nonrecyclable) in over 300 Indian cities, directly from the end-users with the help of around 15,000 local ragpickers, covering 100 collection centers and 30 local NGOs/local partners. Moving forward on its commitment towards becoming a Plastic Waste Neutral Company, Dabur has launched an initiative—My 10 kg Plastic. Along with the Indian Pollution Control Association (IPCA), they have been running a series of programmes involving government agencies, community members, residential societies, schools and offices to build awareness about segregation and the safe disposal of plastic waste. Through this programme, a unique plastic waste collection strategy has been adopted by Dabur, wherein they are inviting individuals to collect and send 10 kg of plastic waste generated in their homes to become Plastic Waste Neutral Citizens. A special waste collection vehicle will reach their homes at a scheduled time to weigh and collect the plastic waste. The depositor is awarded reward points for every kg of plastic waste submitted. They can accumulate these points and redeem them at the microsite for a variety of gifts and products (Dabur India Limited, 2019).

Bisleri "Bottles for Change Initiative" was launched by Bisleri International Pvt. Ltd., an FMCG company, by partnering with three NGOs in Mumbai, India. They collect the plastic bottles, crush and convert them to flakes, which are then used to create products such as cloth fabric, handbags, window blinds, and public benches. Also, they have been active in creating awareness amongst citizens about waste segregation, recycling plastic bottles and best practices towards a clean environment (Bisleri n.d.-a, b). To ease the collection process, in 2019, Bottles for Change introduced a mobile app in Mumbai aimed at bringing plastic waste producers, collectors and recyclers under one platform. The app allows waste generators to locate waste collectors, who in turn sell it to recyclers. By the end of 2019, this initiative had recycled close to 3000 tonnes of plastic with the help of 500,000 individual members, 600 housing societies and 300 corporate hotels in Mumbai (Jacob, 2019).

On a national level, UNDP India worked with Hindustan Coca-Cola Beverages Private Limited (HCCBPL) and Recykal, a waste management-focused digital technology company, to improve the way processes and systems are used to reduce the damage that plastic waste does to the Indian environment. This was done in partnership with the Government of India. The partnership, which is called Project Prithvi (Project Earth), supports sustainable plastic waste management in India by collecting, sorting, and recycling all plastics. This initiative contributes to India's transition to a circular economy, aligning with the Swachh Bharat Mission, Plastic Waste Management Rules, 2016, and the UN's Sustainable Development Goals (SDGs, Fig. 11). Under this project, UNDP and HCCBL have established several MRF facilities called *Swachhta Kendras* (Cleanliness Centres), wherein several formalized workers called *Safai Sathis* (Cleanliness Friends) take up the initiative of sorting, segregating and collecting plastics for recycling. As part of this partnership, Recykal will leverage its cloud and mobile applications to raise consumer awareness on segregation and recycling. It will enable the collection and channelization of plastic waste from schools, colleges, corporates, residents welfare associations (RWAs), and bulk generators to connect with recyclers, closing the loop of the recycling ecosystem (Business Line, 2020). The initiative heavily focuses on empowering the informal sector—which is considered to be the most essential stakeholder in implementing the circular economy—through various inclusion activities like providing soft skills training, the opening of bank accounts, and enabling digital transactions for waste collectors, rag pickers and aggregators (The Hindu, 2020b). By 2020, this project has been successfully implemented in 28 Indian cities and plans on expanding to 50 cities and towns of various states across India by the year 2024.

3.2 State Level (India)

As many as 28 Indian states and UTs have imposed a partial plastic ban, mainly on plastic carry bags (CPCB, 2019). A few case studies on the ban on carry bags are described below:

Sikkim Sikkim was the first Indian state to ban the use of nonbiodegradable materials like plastic carry bags since 1998 (Forest, Environment and Wildlife Management Department, 2016). This ban was issued by the government of Sikkim as a countermeasure to combat the frequent landslides in the state. It was found that plastic bags were choking the drains and streams of the state capital—Gangtok, which hindered and diverted the flow of water, resulting in a spate of landslides (Toxics Link, 2014). Owing to the severity of this issue, the plastic carry bag ban was welcomed and practiced by the people of Sikkim. Furthermore, in 2016, the state took more drastic measures when they introduced bans on disposable cutlery and containers made out of polystyrene materials such as styrofoam.

With penalties and awareness campaigns, the ban has proved successful; most Sikkim locals now opt for plates made of leaf, bagasse and even areca nuts (Cavalletti, 2020). Sikkim has adapted well to the plastic bans. Despite some amount of plastic bags still being sold in the state, Sikkim fares well in its anti-plastic stance, with 66% of shop owners preferring paper bags or newspapers for packaging over

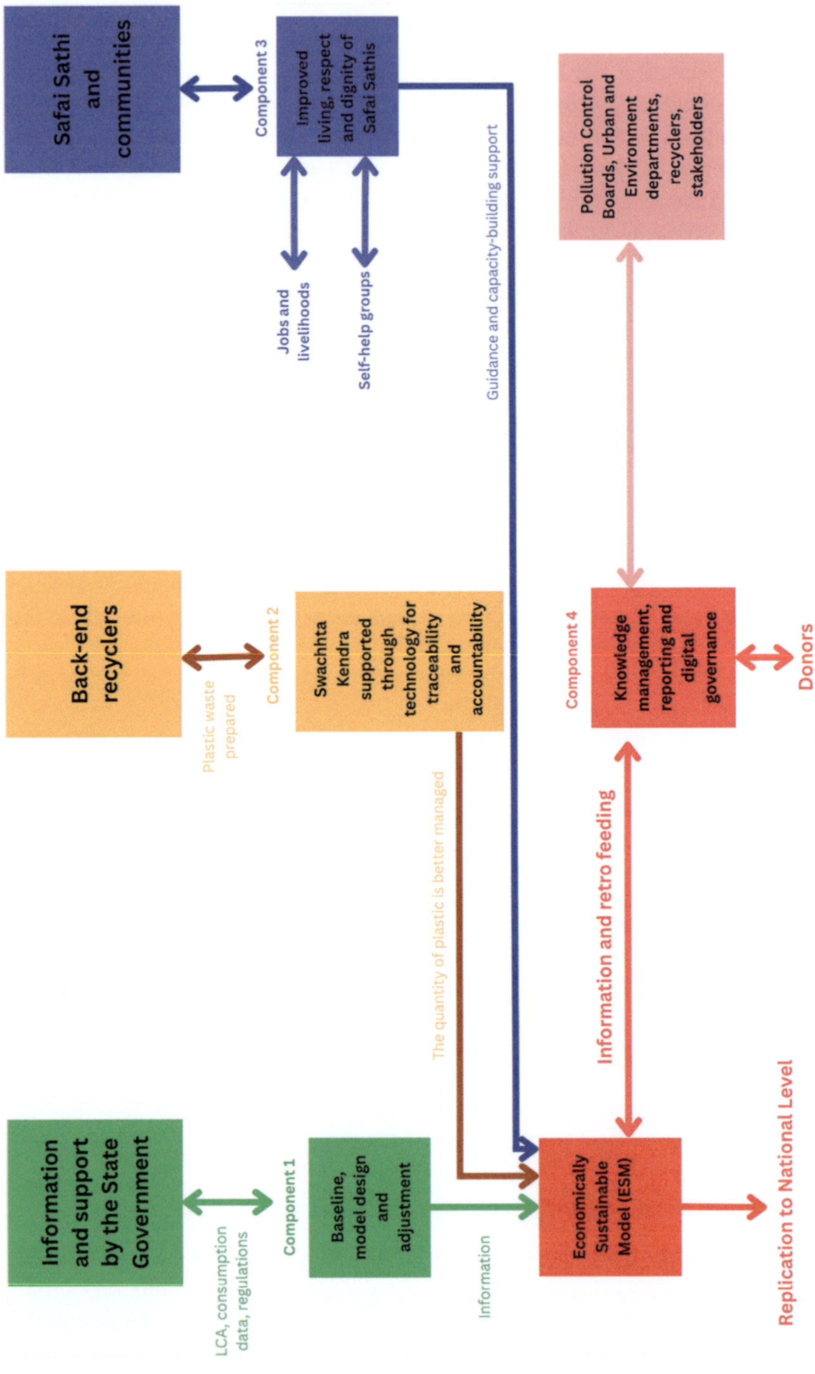

Fig. 11 Schematic representation of the Project Prithvi Model. (*Source:* UNDP and HCCBL websites)

plastic carry bags (Fernandes, 2020). Furthermore, Sikkim is planning to gradually phase out plastic water bottles since a large number of tourists are generating plastic waste in the form of used water bottles (United Nations Environment Programme, 2018).

Maharashtra In March 2018, the state of Maharashtra announced a complete ban on plastics. Enforced by the Maharashtra Pollution Control Board (MPCB), this ruling completely banned the manufacturing, use, sale, distribution and storage of plastic materials such as bags, spoons, plates and other disposable items, along with common packaging materials such as thermocol. Municipal Corporations, along with district and local authorities, monitored the ban and levied rigid fines and penalties, which ranged from INR 5000 for first-time offenders to INR 25,000 for third-time violators, along with imprisonment of up to 3 months (Economic Times, 2018). However, plastics used in medicine packaging, milk packaging, and solid waste handling, along with plastics used in exporting and packaging for horticultural purposes, were exempt from this ban. Despite Maharashtra's sincere effort to reduce plastic pollution in the state, the plastic ban is reported to not have yielded the expected results due to the following reasons:

- The lack of effective communication prior to the introduction of the ban. The state-wide ban on plastics was announced 3 months prior to its implementation. This was not sufficient for the public to fully understand, adapt and figure out alternatives to plastics.
- Secondly, lack of proper monitoring from the municipal corporations and local authorities was another barrier. MPCB asked the local authorities and urban local bodies (ULBs) to prepare squads for enforcing, monitoring and penalizing the non-compliers. This decentralized monitoring was ineffective, leading to improper implementation (Deshpande, 2020).
- Finally, the backlash from the plastic industry is owing to the impact of the ban on the economy. According to an estimate, the plastic industry in Maharashtra is worth INR 15,000 crore ($2.20 billion) a year and employs nearly 300,000 people (Shantha, 2018) (Table 3).

3.3 Local/ULB (India)

At the ULB level, one notable example of Corporate Social Responsibility (CSR) displayed by two multinational companies in India is the Alag Karo Har Din Teen Bin (Segregate daily in three bins) initiative undertaken and supported by multiple stakeholders, including the Municipal Corporation of Gurgaon (MCG), Tetra Pak, Coca-Cola India Pvt. Ltd., Deutsche Gesellschaft für Internationale Zusammenarbeit (GIZ), and SAAHAS (Fig. 12). Launched in September 2017 in Gurugram, India, the objective of the programme is to establish and sustain segregation at the source and develop the capacities of the waste collectors (formal and informal) to ensure high recycling rates in Gurugram. The entire project was supported by multiple stakeholders. SAHAAS was the implementation agency and worked closely with

Table 3 Complete ban on plastic bags—boon or bane

It is notable that the Plastic Waste Rules 2018 amendment rolled back the clause for "Explicit Pricing of the carry Bag.". A study titled "Consumer responses to Incentives to reduce plastic bag use: Evidence from the field experiment in urban India" carried out by the Delhi School of Economics (DSE) highlighted that in developing nations, framing legislation on a complete ban alone is not idealistic. Instead, there is a need for low-cost informative interventions, along with the availability of plastic bag substitutes and subsidies (tax) on the usage of plastic bags. This study included experiments in three stages, before which a two-week baseline study was conducted to assess and record transactions in each shop and for different types of plastic bags used by consumers. The three interventions included dissemination of information, a cashback scheme, and availability of reusable bags. In the first, campaigning and persuasion on the demerits of using plastic bags was explained, which further motivated consumers to shift their behavior from using plastic to cloth bags. Secondly, cashback schemes were introduced, in which a consumer not using a retailer-provided plastic bag would receive 1%–2% cashback on his purchase. Thirdly, alternatives to plastic bags, such as cloth bags, were introduced, wherein consumers were asked to bring their own bags. In the end, shops were visited to understand and gauge if behavioral changes took place during the run of the experiments persisted in the long run (Gupta, 2011). Countries like France, Denmark, Ireland, Italy, Malta, Kenya, Rwanda, South Africa, Tanzania, and Uganda introduced a ban on single-use plastic in their countries too; there are also levies on the usage of plastic bags. In these countries, taxation promoted the reuse of plastic bags or switching to alternatives (Musa et al. 2013). Hence, it is observed that instead of completely banning plastic bags, it is better to levy taxes on the usage of plastic bags (Gupta, 2011)

Fig. 12 SAAHAS waste segregation

the stakeholders to ensure the success of the project. The main idea was to implement source segregation of waste in residential complexes, educational and commercial establishments and further aimed to bring about behavioral change amongst citizens. This enabled the public to segregate waste into three bins and developed capacities of waste workers (both formal and informal) to ensure high recycling rates, diversion of waste from landfills, and reduction in waste burning and dumping. As part of this project, waste audits, Information Education and Awareness (IEC), training, community awareness, door-to-door campaign and monitoring activities (check the level of segregation) were implemented. Waste workers were trained on good practices on waste collection and handling. To further spread the message, schools, colleges, commercial establishments, and housing societies were identified as the major waste generators. Various awareness campaigns on waste management were organized for waste generators and collectors (GIZ n.d.). Awareness was created through organizing public events in malls, as well as through workshops and training sessions for RWAs, government officials, estate management and housekeeping agencies. By the end of 2018, about 24,000 households in 50 RWAs were sensitized on source segregation of waste and about 1 lakh people were covered under the awareness programmes. About 17,500 households in nearly 32 housing complexes achieved 90% segregation and had segregated waste up to 18,400 kg daily. About 14 apartment complexes installed in situ composting plants that helped many households reuse their wet waste through composting (SAAHAS, 2019).

3.4 Global Context

3.4.1 Financially Resilient Palau

Palau is a small island in the Pacific Ocean and is part of Micronesia. Its GDP is majorly dependent on tourism. For years, Palau had been a victim of improper waste management and was in a huge litter crisis (ADB, 2014). To manage this issue, Palau enacted the Beverage Container Recycling Regulation in 2011, under which they launched a deposit fee program. A recycling facility was instituted in Koror, Palau and was named Redemption Centre (Fig. 13). The deposit fee program requires the consumers to deliver their beverage containers to the redemption center and redeem a part of the fee that has been incorporated in the selling price. The redeemed containers are then counted, packaged, and physically removed from the island by exporting them to a bona fide recycling facility in countries like Japan and South Korea (Ministry of Public Infrastructure, Industries and Commerce & Ministry of Finance, 2011). The finances of the deposit fee program included a \$0.10 deposit fee per beverage container, which is levied on the distributors at the time of import (which is reflected in the selling price); \$0.05 is redeemed by the consumers after depositing this beverage carton at the redemption center; \$0.025 is given to the redemption center for their services; and \$0.025 is retained by the government as the recycling fund (Bureau of Public Works, 2013). At the redemption center, plastic bottles are counted mechanically at a rate of 14 PET bottles per second, based on which refund slips are issued that can be redeemed at Koror State Treasury. The cost of exporting used containers is borne by the recycling fund. There are three main bodies governing this scheme in Palau:

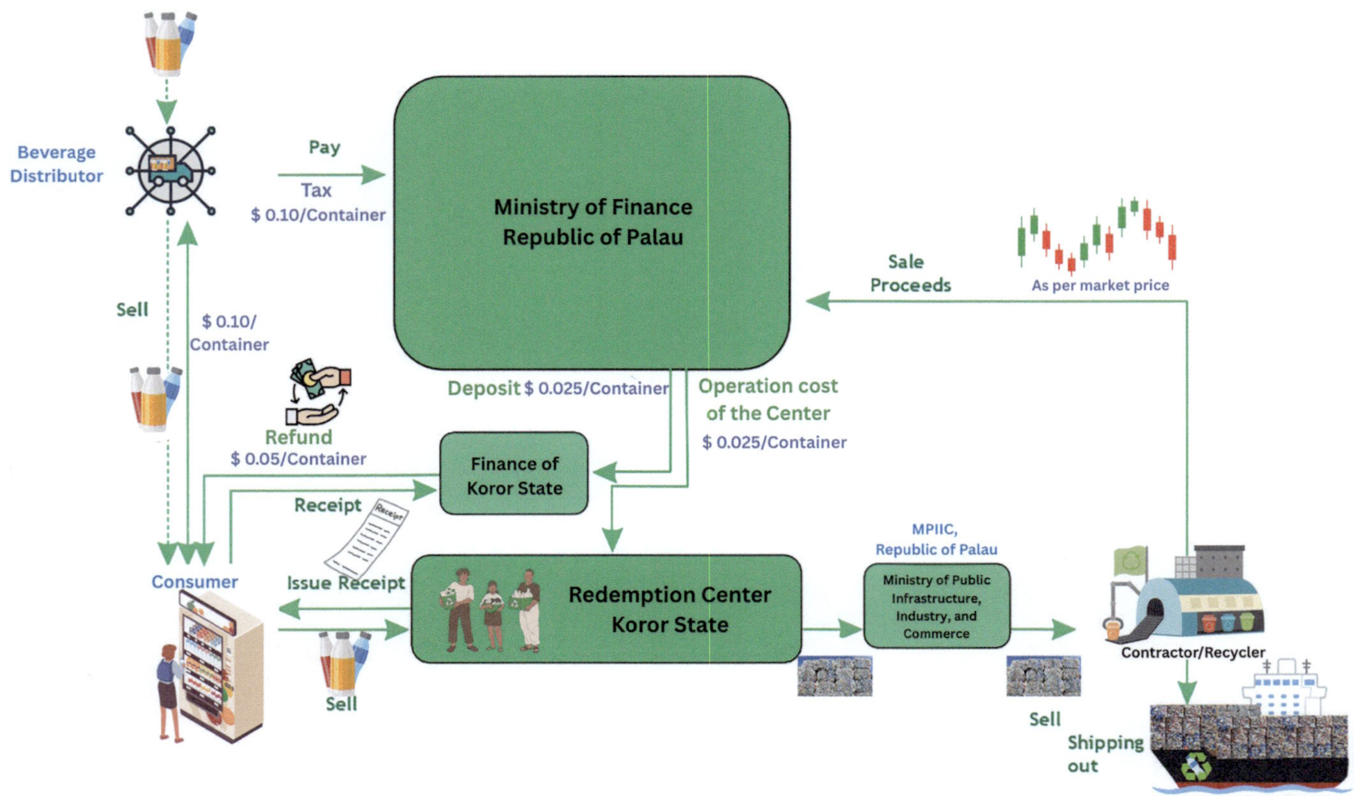

Fig. 13 Overview of the deposit fee program. (*Source*: Ministry of Public Infrastructure, Industries and Commerce, Palau)

- Ministry of Public Infrastructure, Industries and Commerce, who is responsible for the implementation of the recycling program, monitor the redemption center and export the beverage containers out of the island.
- Ministry of Finance is responsible for the monitoring, collection, and management of the funds.
- Koror State Government is responsible for the operation of the redemption center.

This program has been very effective in managing the litter produced in Palau. In 2018, Palau was able to retrieve nearly 90% of the used beverage containers and exported nearly 100 tonnes of PET bottles outside the island (Bureau of Public Works, 2019).

3.4.2 Every Bottle Back Initiative

Initiated by the American Beverage Association (ABA), this initiative brings three of the fiercest beverage competitors—Coca-Cola, Pepsico and Keurig Dr. Pepper—into an alliance to reduce plastic pollution in the United States of America. Cofounded by the World Wildlife Fund (WWF), The Recycling Partnership and Closed Loop Partners, this initiative aims at recovering plastic bottles after use and remaking them into newer bottles so as to reduce the plastic footprint of the three largest plastic-polluting companies in the country. Every Bottle Back includes four core components (American Beverage Association n.d.):

- Reducing the use of new (virgin) plastic in the U.S.: Progress will be measured using WWF's ReSource. Plastic accounting methodology and data collected will be used to drive actionable solutions.
- Investing $100 million to improve sorting, processing, and collection of discarded plastic bottles in the four key U.S. regions. The industry-backed fund, which will be administered by the Recycling Partnership and Closed Loop Partners will attract an additional $300 million in matching grants and investment to increase the amount of recycled plastic available to be remade into beverage bottles.
- Launching a public awareness and community outreach campaign to promote recycling and promote the value of 100% recyclable bottles. According to a poll conducted by Public Opinion Strategies (POS) on behalf of ABA, nearly half of U.S. consumers are unaware that the country's leading beverage companies are making 100% recyclable bottles.
- Working together to use packaging to remind consumers that PET bottles are 100% recyclable and can be remade into new bottles. U.S. beverage companies were supposed to introduce voluntary messaging on packaging labels in 2020.

As a part of this initiative, Coca-Cola has committed to collecting and recycling as many bottles as it sells by 2030 (The Coca-Cola Company, 2020). Keurig Dr. Pepper will have 100% recyclable or compostable packaging and use 30%

post-consumer recycled content across its portfolio by 2025 (Keurig Dr Pepper, 2020). PepsiCo will make 100% of its packaging recyclable, compostable or biodegradable by 2025 (PepsiCo, 2020).

Coca-Cola Zero-Waste Urban Environment In 2018, The Coca-Cola Company launched "Zero Waste Cities," a new program that aims to make Thessaloniki the first city in Greece with the potential to become a zero-waste municipality. Among the program's innovations is "Print Your City." Through an online platform, residents can nominate useful objects that will be 3D printed using plastic waste and installed in various neighborhoods across the city. The first 10 pieces of functional urban furniture demonstrating the concept were placed at the Thessaloniki seaside. Other program elements include a Coca-Cola Foundation-funded waterfront cleanup, outreach and education, a waste collection system, and the first-ever prevention study on how the city can better handle its waste. Coca-Cola further plans to roll the program out to additional cities in Europe, the Middle East and Africa (Coca-Cola, 2019). Earlier, Coca-Cola initiated this program in Brazil.

Acknowledgments This research was supported by the Royal Norwegian Embassy in New Delhi through the INOPOL Project (Project No: IND-22/0007).

References

ACC. (2019). *ACC limited sustainability report 2018–2019*. ACC. Retrieved January 3, 2021.

ADB. (2014). *Solid waste management in the Pacific: Palau country snapshot*. ADB. Retrieved January 10, 2021.

American Beverage Association. (n.d.). *Our New Initiative*. Retrieved January 12, 2021, from Innovation Naturally website: https://www.innovationnaturally.org/plastic/

Bisleri. (n.d.-a). *Plastic is not bad. How you dispose plastic is bad.* Retrieved July 06, 2020, from Plastic Recycling Bisleri: https://www.bisleri.com/plastic-recycling

Bisleri. (n.d.-b). *Plastic Recycling*. Retrieved December 30, 2020, from Bisleri website: https://www.bisleri.com/plastic-recycling

Borealis & Alliance to End Plastic Waste. (2019). *Alliance to end plastic waste joins project STOP to help tackle plastic waste in the environment in Indonesia*. Borealis Group. Retrieved January 5, 2021.

Brigham, K. (2018, July 14). *How San Francisco sends less trash to the landfill than any other major U.S. city*. Retrieved January 12, 2021, from CNBC website: https://www.cnbc.com/2018/07/13/how- san-francisco-became-a-global-leader-in-waste-management.html

Bünemann, A., Brinkmann, J., Löhle, D. S., & Bartnik, S. (2020). *EECO-LEF in Tunisia: A case study*. GIZ. Retrieved January 8, 2021.

Bureau of Public Works. (2013). *Manual for beverage container deposit fee program*. Ministry of Public Infrastructure, Industries and Commerce. Retrieved January 10, 2021.

Bureau of Public Works. (2019). *Solid waste management in Palau*. Ministry of Public Infrastructure, Industries and Commerce. Retrieved January 10, 2021.

Business call to action. (n.d.). *Sahaas zero waste*. Retrieved December 24, 2020, from Business call to action: https://www.businesscalltoaction.org/member/saahas-zero-waste

Business India. (2020). *The company's business model empowers waste pickers through intensive training on responsible waste management and access to waste disposal networks, leadership initiatives and entrepreneurship opportunities*. Retrieved December 23, 2020, from business India: https://businessindia.co/climate-change/waste-management/hasiru-dala-innovations-provides-responsible-solutions-for-urban-waste-disposal

Business Line. (2020, June 8). *Recykal in pact with UNDP, HCCB for plastic waste management*. Retrieved January 3, 2021, from the Hindu business Line website: https://www.thehindubusinessline.com/news/recykal-in-pact-with-UNDP-hccb-for-plastic-waste-management/article31778159.Ece

C40 Cities Climate Leadership Group. (2016). *Good practice guide: Waste to resources*. C40 Cities. Retrieved January 12, 2021.

Cavalletti, I. (2020, July 15). *Sikkim the Indian state that is 100% organic and plastic free*. Retrieved December 30, 2020, from eco-Connect website: https://eco-nnect.com/articles/2020/6/30/sikkim-the-indian-state-that-is-100-organic-and-plastic-free

Centre for Public Impact. (2016, April 18). *San Francisco's Zero Waste 2020 target*. Retrieved January 12, 2021, from Centre for Public Impact website: https://www.centreforpublicimpct.org/case-study/zero-waste-san-francisco/

Chaabane, W., Nassour, A., Bartnik, S., Bunemann, A., & Nelles, M. (2019). Shifting towards sustainable tourism: Organizational and financial scenarios for solid waste management in tourism destinations in Tunisia. *Sustainability, 11*(13), 3591. Retrieved January 8, 2021.

Chaturvedi, A. (2020, July 10). *1 lakh km of road built using plastic waste, govt aims to double it*. Retrieved January 6, 2021, from Hindustan Times website: https://www.hindustantimes.com/India-news/1-lakh-km-of-road-built-using-plastic-waste-govt-aims-to-double-it/story-iwmkiKjlG86BYIDlg2aLtM.Html

CII, Swiss Confederation, World Resources Forum, EMPA and Toxics Link. (2016). *Co-processing of non-recyclable hazardous plastic waste in cement kilns*. Sustainable Recycling Industries. Retrieved January 3, 2021, from https://www.sustainable-recycling.org/wp-content/uploads/2018/03/Co-Processing-Non-Recyclable- plastics-in-cement-kiln-25-11-2016.pdf

Coca-Cola. (2019). *2018 Business & Sustainability Report*. The Coca-Cola Company. Retrieved January 4, 2021.

CPCB. (2017). *Guidelines for co-processing of plastic waste in*. CPCB. Retrieved January 3, 2021, from https://cpcb.nic.in/uploads/plasticwaste/Co-processing_Guidelines_Final_23.05.17.pdf

CPCB. (2019). National plans to make river stretches pollution free. Retrieved from https://cpcb.nic.in/NGTMC/National_Plans_24.07.2019.pdf

Dabur India Limited. (2019). *Integrated annual report 2019–20*. Dabur India Limited. Retrieved December 30, 2020, from https://dabur.com/digital-annual-report/pdf/Dabur-IR-2019-20.pdf

Deshpande, T. (2020, February 20). *Maharashtra takes steps to end single-use plastic by May 1*. Retrieved December 30, 2020, from The Hindu website: https://www.thehindu.com/news/national/other-states/maharashtra-takes-steps-to-end- single-use-plastic-by-may-1/article30865146

Economic Times. (2018, June 23). *Plastic ban in Maharashtra: 7 key things to know*. Retrieved December 30, 2020, from the economic time website: https://economictimes.indiatimes.com/news/politics-and-nation/plastic-ban-in-Maharashtra-7-key-things-to-know/whats-plastic-ban-when-was-it-implemented/slideshow/64708773.Cms

Ellen MacArthur Foundation and UN Environment. (2019). *New Plastics economy global commitment June 2019 report*. Ellen MacArthur Foundation. Retrieved January 6, 2021.

Fernandes, K. (2020, July 3). *International plastic bag free day 2020: Role model Sikkim*. Retrieved December 30, 2020, from the CSR journal website: https://thecsrjournal.in/international-plastic-bag-free-day-2020-alternatives-Sikkim/

Forest, Environment and Wildlife Management Department. (2016). *Environmental initiatives of the state government (1994–2016)*. Government of Sikkim. Retrieved December 30, 2020, from http://sikenvis.nic.in/WriteReadData/Publication/Environmental%20Initiatives%20of%20Sikkim%2015%20Aug%202016.pdf

GIZ. (n.d.). *Scavengers to managers: a 3-bin revolution for Delhi's twin city*. Retrieved December 24, 2020, from GIZ website: https://www.giz.de/en/worldwide/80897.html
Global Alliance for Incinerator Alternatives. (2019). *Small town, big steps: The story of Kamikatsu, Japan*. GAIA. Retrieved January 6, 2021.
Global Anti-Incenerator Alliance. (2012). *On the road to zero waste: Successes and lessons from around the world*. GAIA. Retrieved January 12, 2021.
Gray, A. (2019, January 21). *The inspiring thing that happened when a Japanese village went almost waste-free*. Retrieved January 6, 2021, from world economic forum website: https://www.weforum.Org/agenda/2019/01/the-inspiring-thing-that-happened-when-a-japanese-village-went-almost-waste-free/
Gunjur Plastics. (n.d.). *Innovative plastic recyclers Gunjur*. Retrieved January 7, 2021, from Gunjur Plastics website: https://gunjurplastics.com/
Gupta, K. (2011, November). Consumer responses to incentives to reduce plastic bag use: Evidence from a field experiment in urban India. *Economics & Environment*, 65-11.
Hasiru Dala. (2017). *S3IDF - Hasiru Dala trust—Hasiru Dala innovations—Plastics for change our beneficiary feedback*.
Hasiru Dala Innovations. (2020). *Bengaluru startup launches plastics recycling initiative with The Body Shop*. Retrieved December 23, 2020, from Hasiru Dala Innovations: https://hasirudalainnovations.com/news-detail/2/Bengaluru-startup-launches-plastics-recycling-initiative-with-The-Body-Shop
Henam, S., & Sambyal, S. S. (2019, December 26). *Ten zero-waste cities: Kamikatsu, Japan's zero-waste miracle town*. Retrieved January 6, 2021, from down to earth website: https://www.downtoearth.org.in/news/waste/ten-zero-waste-cities-kamikatsu-Japan-s-zero-waste-miracle-town-68577
Herald Goa. (2020, October 2). *CCP to launch 'Shop With Your Waste' campaign*. Retrieved December 21, 2020, from Herald Goa website: https://www.heraldgoa.in/Goa/CCP-to-launch-%E2%80%98Shop-With-Your-Waste%E2%80%99-campaign/165937#:~:text=Citizens%20will%20be%20provided%20essentials%20in%20exchange%20of%20dry%20waste&text=PANJIM%3A%20In%20an%20innovative%20scheme,in%20exchange%20of%20d
India Sanitation Coalition. (2017). Clean and green Madukkarai: A solid waste management initiative. India Sanitation Coalition. Retrieved January 1, 2021, from http://www.indiasanitationcoalition.org/resources/Case-Study-ACC-Coimbatore.pdf
India Today. (2020, August 20). *Swachh Survekshan 2020: Full rankings*. Retrieved December 21, 2020, from India Today website:https://www.Indiatoday.in/india/story/swachh-survekshan-2020-full-list-rankings-results-cleanest-city-India-1713178-2020-08-20
Indore Municipal Corporation. (2020). *Solid waste management*. Retrieved December 21, 2020, from Smart City Indore website: https://www.smartcityindore.org/solid-waste/
Jacob, U. (2019, December 19). *How Bisleri is using its 'bottles for change' initiative to encourage people to recycle plastic*. Retrieved December 30, 2020, from your story website: https://yourstory.com/socialstory/2019/12/bisleri-backed-initiative-plastic-awareness-recycling
Kaur, C. (2019, November 23). *Chhatisgarh's small town Ambikapur scores big in waste management*. Retrieved January 4, 2021, from the Times of India website: https://timesofindia.indiatimes.com/city/raipur/chhatisgarhs-small-town-ambikapur-scores-big-in-waste-management/articleshow/72197965.cms
Keurig Dr Pepper. (2020). *2019 corporate responsibility report*. Keurig Dr Pepper. Retrieved January 12, 2021.
Khadilal, A. (2019, March 28). *Trash is sustainable at Saahas, a city enterprise*. Retrieved December 24, 2020, from Deccan Herald: https://www.deccanherald.com/city/bengaluru-infrastructure/trash-is-sustainable-at-saahas-a-city-enterprise-725588.html
Lenkiewicz, Z. (2019, May 26). *The Gambia: Plastics recycling training update*. Retrieved January 7, 2021, from WasteAid website: https://wasteaid.org/the-Gambia-plastics-recycling-training-update/

Madhavan, K. (2015, August 6). *Coimbatore sets Guinness world record*. Retrieved January 1, 2021, from the Hindu website: https://www.thehindu.com/news/national/tamil-nadu/coimbatore-sets-guinness-world-record/article7505480.ece

Ministry of Environment, Forest and Climate Change. (2020). *Guideline document of uniform framework for extended producers responsibility*. MoEFCC. Retrieved December 28, 2020, from http://moef.gov.in/wp-content/uploads/2020/06/Final-Uniform-Framework-on-EPR-June2020-for-comments.pdf

Ministry of Housing and Urban Affairs. (2017). *Bulk Solid Waste Generators*. Ministry of Housing and Urban Affairs. Retrieved January 4, 2021.

Ministry of Housing and Urban Affairs. (2019). *Plastic waste management—Issues, solutions & case studies*. MoHUA. Retrieved December 21, 2020, from http://164.100.228.143:8080/sbm/content/writereaddata/SBM%20Plastic%20Waste%20Book.pdf

Ministry of Housing and Urban Affairs. (2020). *Swachh Survekshan 2020*. Retrieved January 4, 2021, from Swachh Survekshan 2020 website: http://www.swachhsurvekshan2020.org/

Ministry of Public Infrastructure, Industries and Commerce & Ministry of Finance. (2011). In Retrieved January 10, 2021 (Ed.), *Beverage container recycling regulations*. Republic of Palau.

Ministry of Railways. (2019). *Guidelines on use of plastic waste in road construction*. Government of India. Retrieved January 6, 2021.

Musa, H. M., Hayes, C., Bradley, M. J., Clayson, A., & Gillibrand, G. (2013). Measures aimed at reducing plastic carrier bag use: A consumer behaviour focused study. *Natural Environment, 1*(1), 17–23.

Nippon. (2018, July 13). *The Kamikatsu Zero waste campaign: How a little town achieved a top recycling rate*. Retrieved January 6, 2021, from Nippon website: https://www.nippon.com/en/guide-to-japan/gu900038/

Ocean Conservancy. (n.d.). *Fighting for trash free seas: Ending the flow of trash at source*. Retrieved January 6, 2021, from ocean Conservancy website: https://oceanconservancy.org/trash-free-seas/

PepsiCo. (2020). *2019 sustainability report*. Pepsi Company. Retrieved January 12, 2021.

Pollans, L. B. (2012). *Greening infrastructural services: The case of waste Management in san Francisco*. Massachusetts Institute of Technology. Retrieved January 12, 2021.

Project STOP. (2020, February 20). *Pasuran overview*. Retrieved January 5, 2021, from the Stop Ocean Plastics website: https://www.stopoceanplastics.com/en_gb/pasuruan/

Purohit, M. (2018, January 19). *Waste away, Ambikapur shows way*. Retrieved December 28, 2020, from India water portal website: https://www.indiawaterportal.org/articles/waste-away-Ambikapur-shows-way

Rang D. (n.d.). *Krishna Ayyakannu*. Retrieved December 24, 2020, from Rang De: https://www.rangde.org/borrower/microcredit/India/karnataka/krishna/33384

SAAHAS. (2019). *Implementing source segregation of waste in residential. communities: an initiative under the Alag Karo program in Gurugram*.

Scroll. (2019, October 3). *How Indore became India's cleanest city (and how others can follow)*. Retrieved January 2, 2021, from Scroll India website: https://scroll.in/article/939210/how-indore-became-indias-cleanest-city-and-how-others-can-follow

Shah, H. (2018, September 3). *How Gujarat succeeded in co-processing plastic waste in cement kilns and promoting circular economy*. Retrieved January 3, 2021, from down to earth website:https://www.downtoearth.org.in/blog/environment/how-Gujarat-succeeded-in-co-processing-plastic-waste-in-cement-kilns-and-promoting-circular-economy-61518

Shantha, S. (2018, April 8). *Insufficient planning may be the undoing of Maharashtra's plastic ban*. Retrieved December 30, 2020, from the wire website: https://thewire.in/environment/insufficient-planning-may-be-the-undoing-of-maharashtras-plastic-ban

STOP. (2018, October 15). *Project STOP overview*. Retrieved January 5, 2021, from Stop Ocean Plastics website: https://www.stopoceanplastics.com/en_gb/muncar/

Stuchtey, M. R., Dixon, B., Danielson, J., Hale, J., Wiplinger, D., & Bai, P. (2019). Project STOP: City partnerships to prevent oceans Plastics in Indonesia. *The Journal of Field Actions, 19*, 86–91.

The Coca-Cola Company. (2020). *World without waste*. Coca-Cola Company. Retrieved January 12, 2021.

The Hindu. (2020a, January 5). *'Cleanest city' Indore earns 4 crore annually through proper waste management*. Retrieved January 2, 2021, from the Hindu website: https://www.thehindu.com/news/national/other-states/cleanest-city-Indore-earns-4-crore-annually-through-proper-waste-management/article30485625.Ece

The Hindu. (2020b, June 8). *Pact for better plastic waste management inked*. Retrieved January 3, 2021, from the Hindu website: https://www.thehindu.com/news/cities/Hyderabad/pact-for-better-plastic-waste-management-inked/article31782366.ece

Times of India. (2020, December 30). *Two more stores join CCP's Shop-With-Your-Waste' plan*. Retrieved December 31, 2020, from Times of India: https://timesofindia.indiatimes.com/city/goa/two-more-stores-join-ccps-shop-with-your-waste-plan/articleshow/80016134.cms#:~:text=Panaji%3A%20The%20Corporation%20of%20the,in%20exchange%20for%20grocery%20items.&text=The%20waste%20has%20been%20sent%20for%20

Toxics Link. (2014). *Plastics and the environment: Assessing the impact of the complete ban on plastic carry bag*. Toxic Link. Retrieved December 30, 2020, from https://toxicslink.org/docs/Full-Report-Plastic-and-the-Environment.pdf.

United Nations Environment Programme. (2018, April 26). *How the Indian state of Sikkim is working to end plastic pollution*. Retrieved December 30, 2020, from United Nations Environment Programme website: https://www.unenvironment.org/news-and-stories/story/how-indian-state-Sikkim-working-end-plastic-pollution

United States Environmental Protection Agency. (2017, January 19). *Managing and transforming waste streams—A tool for communities*. Retrieved January 12, 2021, from USEPA website: https://www.epa.gov/transforming-waste-tool/zero-waste-case-study-san-francisco

WasteAid. (2018, November 11). *Press release*. Retrieved January 7, 2021, from WasteAid website: https://wasteaid.org/wp-content/uploads/2018/11/Press-Release_UK-Aid-funds-WasteAid-plastics-recycling-in-the-Gambia.pdf

World Economic Forum. (2020). *Radically reducing plastic pollution in Indonesia: A multistakeholder action plan National Plastic Action Partnership*. World Economic Forum. Retrieved January 5, 2021.

Regulatory Framework, Policies, and Programs in Plastic Waste Management: Global and Indian

Girija K. Bharat, Avanti Roy Basu, Abdullah Atiq, Emmy Nøklebye, Ingvild Skumlien Furuseth, Twinkle Dev, and Hans N. Adam

1 Introduction

Plastic is an essential material owing to the host of benefits it offers, including convenience, flexibility, lightness, durability, and water-resistance. The first-ever plastic Bakelite, a thermosetting phenol formaldehyde resin, was produced in 1907 in New York. Much later, only in the 1950s, the rapid growth in global annual plastic production started with two million tons, which increased manifold to 381 million tons by 2015 (Ritchie & Roser, 2018) (Fig. 1). If current production and waste management trends continue, it is estimated that 12,000 million metric tons of plastic waste will end up either in landfills or in the open environment by 2050 (Geyer et al., 2017). Primary plastic production is not directly related to plastic waste generation because it is influenced by the polymer type and lifetime of the end product (Fig. 2). Mismanaged plastic waste has a high chance of entering the seas and oceans. Globally, about 80% is carried from land-based sources (through winds and rivers) and the remaining 20% from marine sources (fishing, commercial ships) (Li et al., 2016). The World Economic Forum has predicted that by 2050, the quantity of plastic waste in the oceans will be greater than the number of fish. This prediction is with evident uncertainties, but is an appropriate warning signal for how grim the plastic pollution issue has become. The issue of plastic waste management (PWM) is also very pertinent in the current scenario of the COVID-19 pandemic. As an effort to prioritize, many local, state, and national governments have rolled back

G. K. Bharat · A. R. Basu
Mu Gamma Consultants, Gurugram, Haryana, India

A. Atiq · T. Dev
Environment and Waste Management Division, TERI, New Delhi, Delhi, India

E. Nøklebye · I. S. Furuseth · H. N. Adam (✉)
Norwegian Institute for Water Research (NIVA), Oslo, Norway
e-mail: hans.adam@niva.no

P. Chakraborty et al. (eds.), *Efficient and Enhanced Management of Plastic Waste in India*, https://doi.org/10.1007/978-3-031-89398-8_7

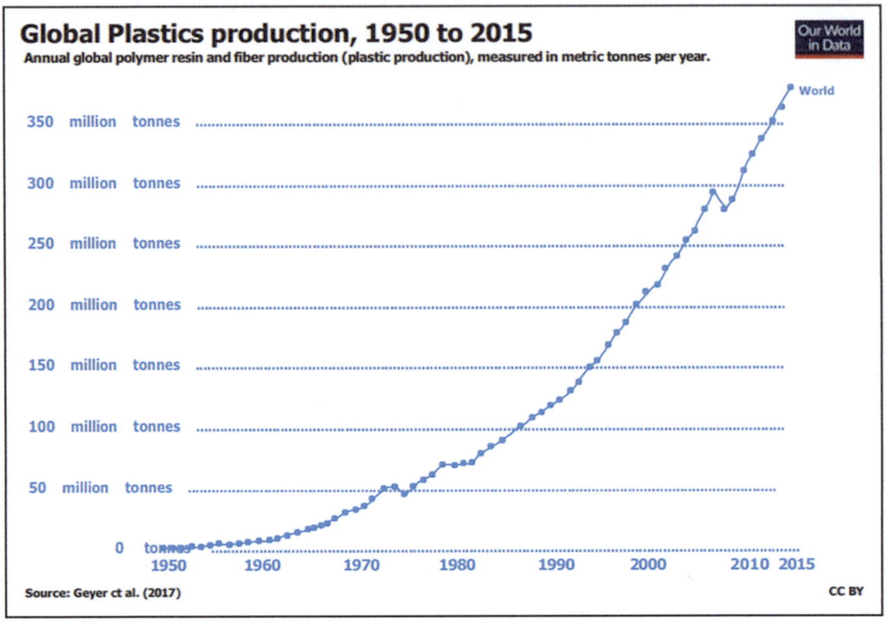

Fig. 1 Global plastics production, 1950 to 2015 (Source: Geyer et al., 2017)

certain progressive steps that they were taking toward PWM. The US state of California has lifted its ban on single-use plastic bags, while other US states (of New York and New Jersey) as well as the United Kingdom have postponed bans that were to be initiated in 2020 (Hughes, 2020). However, many global organizations including the World Economic Forum (WEF) have made efforts toward PWM in this crisis. WEF's Global Plastic Action Partnership (GPAP), an alliance between businesses, national and local governments, community groups, and experts taking significant actions to manage plastic waste, aims to transform the global "take-make-dispose" economy into a circular one, in which "products and materials are redesigned, recovered and reused to reduce environmental impacts." WEF has also laid down a set of guidelines for consumers to help deal with PWM in the COVID-19 crisis period. The advent of plastic pollution, and an accompanying body of literature that testifies to the potentially harmful consequences for human health and ecology, coincided and motivated a renewed impetus to improve plastic waste management, and reduce its unmitigated release into environmental systems. Environmental governance is therefore key. A growing body of literature in this field has addressed similar issues in relation to other environmental concerns, such as climate change, air pollution, and biodiversity loss. While specific approaches and instruments often differ depending on the problem at hand, policy approaches are generally categorized as economic (e.g., taxes, subsidies); legal-regulatory

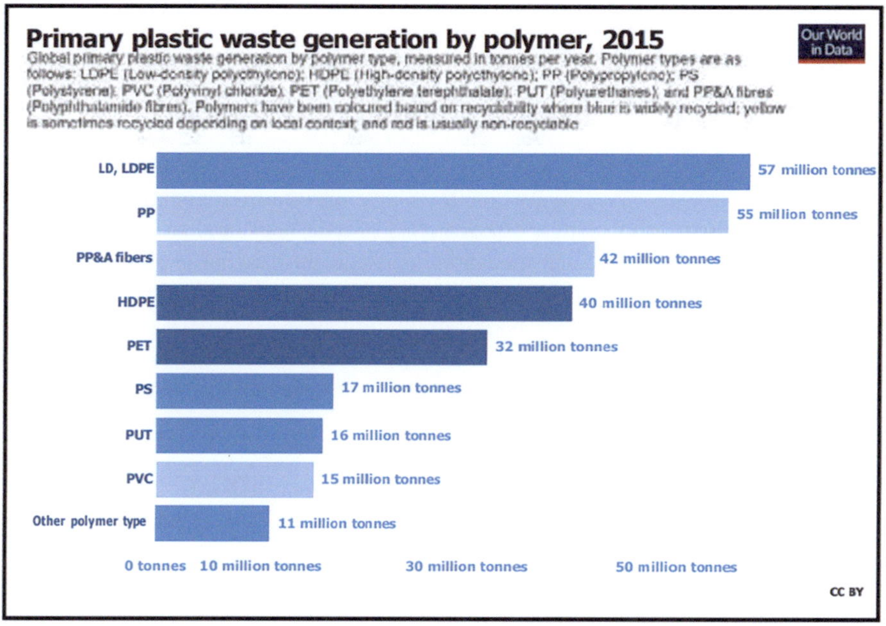

Fig. 2 Primary plastic waste generation by polymer, 2015 (Source: Geyer et al., 2017)

(e.g., bans, mandates), informational (e.g., campaigns, awareness raising), and infrastructural (e.g., waste transport, organizational systems) (Vatn, 2015). In practice, global environmental challenges such as plastic waste management tend to be managed using a mix of different policy types (ISWA UNEP, 2015), including market-based economic tools and regulatory approaches (European Commission, 2011). A similar pattern can be observed in the governance of plastic pollution, where both national and international regulations, alongside circular economy thinking, tend to be dominant. Policy approaches to PWM typically involve a mix of instruments, including informational and infrastructural measures. For example, awareness raising campaigns aimed at influencing individual behaviors are a common feature of PWM strategies. In parallel, there has been a notable rise in calls for better coordination between state and non-state actors, and for the development of contextually appropriate and effective global and local governance solutions at both global and local levels (Vince & Hardesty, 2017). In this chapter, we will focus on policies and regulations aimed at reducing plastic pollution in India and explore how these relate to circular economic thinking. Coupled with scientific disposal and efficient recycling of plastic waste, policy approaches to reduce waste management and pollution play a key role in preventing plastic leakage into terrestrial and aquatic environments. In addition, we will also discuss the largely unregulated—yet important—informal sector, its challenges, opportunities, and its relationship to more formal systems of governance.

2 Relevant Policies and Programs in PWM (International, National, State, ULBs)

2.1 Global Governance of Plastic Waste and Pollution

Of unmanaged plastic waste, many countries have introduced legislation to regulate the plastic products, most commonly targeting plastic bags, and to a lesser extent, microbeads (Anon, 2018). Such measures, particularly bans have played a role in raising global awareness about the environmental impacts of single-use plastics. At the same time, international discussions are underway to establish a legally binding global agreement to end plastic pollution. The following sections present an overview of selected international policies and programs relevant to PWM.

2.1.1 Achieving the Sustainable Development Goals (SDGs) Through PWM

The 2030 Agenda for Sustainable Development, adopted by the United Nations in 2015, sets out a "shared blueprint for peace and prosperity for people and the environment" through 17 sustainable development goals (SDGs). Effective PWM is closely linked to several of these goals, particularly related to health, sustainable cities, responsible consumption, and environmental protection, and can contribute meaningfully to their achievement (Table 1) (Zoë, 2016).

2.1.2 Inclusion of PWM in Basel Convention

The Basel Convention is an international treaty to reduce the movements of hazardous waste between nations, and specifically from developed to less developed countries. During April–May 2019 a total of 187 countries of the Convention agreed to amend it to include plastic waste in a legally binding framework, by making global plastic waste trade more transparent and better regulated. A new partnership on plastic waste was established to organize businesses, governments, academic, and civil society organizations in implementing newer ways to provide resources, tools, best practices, technical and financial assistance, as necessary.

2.1.3 Other Initiatives Under the United Nations

- The Global Partnership on Marine Litter (GPML) was launched at the United Nations Conference on Sustainable Development (Rio + 20) in June 2012. GPML is a multi-stakeholder partnership that brings together all countries working to prevent marine litter and microplastics, thereby making contribution to the achievement of SDG 14.1.

Table 1 Relevance of plastic waste management to the UN SDGs

SDGs	Relevance to PWM
SDG 1: No poverty	Jobs in plastic waste collection and recycling; workers to generate income and provide for their families
SDG 2: Zero hunger	Prevention of food waste from effective reuse/recycling and use of biodegradable plastic packaging
SDG 3: Good health and well-being	Prevention of health risks from plastic recycling and stopping open burning of plastic waste
SDG 4: Quality education	Training and awareness on plastic waste management
SDG 5: Gender equality	Women often bear the brunt of unscientific plastic waste disposal
SDG 6: Clean water and sanitation	Prevention of health risks from minimizing pollution caused by plastic waste disposal on water bodies
SDG 7: Affordable and clean energy	Energy recovery from plastics
SDG 8: Decent work and economic growth	Waste management is one of the largest industries of the world; informal plastic waste sector adds to the economy by converting waste into other goods
SDG 9: Industry, innovation, and infrastructure	Recycling innovation is growing and scalable; forces transition toward circular economy
SDG 10: Reduced inequalities	The poorest are harmed by inefficient plastic waste management
SDG 11: Sustainable cities and communities	Better plastic waste management is crucial for healthy and resilient communities; reduced cost related to plastic waste management for municipalities
SDG 12: Responsible consumption and production	Recycling rates: recycled plastic reduces the need for raw materials; emphasizes the need to shift from waste to resource management
SDG 13: Climate action	Enables carbon savings by preventing emissions from production of virgin plastic and burning/dumping of plastic waste
SDG 14: Life below water	Decreased plastic pollution helps to mitigate ocean pollution and save ocean and sea life
SDG 15: Life on land	Less plastic pollution on land and healthier environments
SDG 16: Peace, justice, and strong institutions	Producer responsibility and governance
SDG 17: Partnerships for the goals	Working together: formal and informal, rich and poor

- UN Environment launched the Clean Seas Campaign in February 2017 that contributes to the goals of the GPML and aims to engage governments, general public, and the private sector to manage marine plastic pollution.
- At the United Nations Environment Assembly (UNEA 2) of the United Nations Environment Program held in Nairobi, Kenya (May 2016), a draft resolution on Marine Plastic Debris and Microplastics was drawn up.
- At the UNEA-3 held in Nairobi, Kenya (December 2017), efforts to address global plastic pollution took a step forward as countries agreed to establish a

specialist group aimed at assessing different options to combat marine plastic pollution.

- At UNEA-4 held in Nairobi, Kenya (March 2019), resolution addressing marine plastics received considerable attention. UNEA-4 considered 29 resolutions and adopted 23, both record numbers. The resolutions focused on a variety of issues related to the circular economy, including chemicals, waste, sustainable mobility, sustainable infrastructure, and food loss and waste, as well as environmental assessment.

2.1.4 "Closing the Loop" in ASEAN Cities

In May 2020, the United Nations Economic and Social Commission for Asia and the Pacific (UNESCAP) and the Government of Japan initiated the "Closing the Loop" project to minimize the impacts of plastic waste pollution on the riverine and marine environment of South East Asia. The project will be piloted in four ASEAN (Association of Southeast Asian Nations) cities, namely Kuala Lumpur (Malaysia), Surabaya (Indonesia), Nakhon Si Thammarat (Thailand), and Da Nang (Vietnam). The project will make use of innovative technologies such as remote sensing, to understand and monitor the sources and pathways of plastic waste entering rivers. The project supports the implementation of the existing "ASEAN Framework of Action on Marine Debris" and the "G20 Osaka Blue Vision" to manage plastic waste, both of which work toward achieving SDGs 11, 12, and 14.

2.1.5 Circular Economy and Plastics Strategy of the European Union

The circular economy (CE) derives from different epistemological fields but has, in recent years, evolved to become the dominant model for sustainable growth and development. The CE entails a shift from the current linear take-make-waste extractive industrial model of consumption, where value is increasingly kept in the economy in a "closed loop" system. The basic principles of CE include utilizing technology to increase resource efficiency, reducing the usage of primary resources, and eliminating waste and negative externalities. Strategies to retain and add value in a CE are based around rethinking progress and redesigning the way our economy works through innovative thinking.

2.1.6 Aims within a Circular Plastic Economy

With relation to plastic pollution, CE solutions aim to restore and regenerate the value of already extracted plastic materials and prevent plastic waste from polluting the environment. Review studies highlight that the scope of current CE implementation primarily focus on selected products, materials, and sectors, while desired system changes to the economy are rarely suggested (Homrich et al., 2018; Kalmykova et al., 2018). The Ellen MacArthur Foundation has played an important role in mobilizing a systemic shift toward a circular plastic economy. In a 2016 report, they highlighted opportunities to deliver better system-wide economic and environmental outcomes through a plastic value chain approach while continuing to harness the benefits of plastic packaging. In 2017, a second report to catalyze action provided a global action plan to move toward 70% reuse and recycling of plastic packaging, while highlighting the need for fundamental redesign and innovation for the remaining 30%.

2.1.7 Challenges for the Circular Economy to Limit Plastic Pollution

While the CE concept has rapidly expanded to become the dominant narrative and political imperative in Europe and North America, the drive for resource recovery in the Global South is often mandated by local resource scarcity and socio-economic necessities rather than ideology (Velis, 2017). Informal waste picking activities are currently dominantly or exclusively responsible for resource recovery of recyclables in the Global South (da Silva et al., 2019). In this regard, and particularly in relation to plastics, a key challenge is to develop highly technological focused CE solutions to become applicable to the dominant labor-intensive informal recycling environments of less financially developed countries. In addition, the market price of fossil fuels for virgin plastic production continues to be lower than for recycled plastics. Stimulating the market demand for recycled plastics would be a potential way of making domestic recycling streams viable and increasingly circular (Ledsham, 2018). The European Union (EU) has become a very active proponent of the CE approach, and a European Strategy for Plastics in a Circular Economy (European Commission, n.d.) was adopted in January 2018. The Plastics Strategy is the first-of-its-kind in the EU. Plastic products are aimed to be better designed, have higher plastic waste recycling rates, and boost a better market for recycled plastics. The Strategy targets ten single-use plastic products found on Europe's beaches and seas, which comprise 70% of all marine litter items including lost and abandoned items. The strategy will contribute to achieving the SDGs (Fig. 3) (Kerscher, 2019).

2.1.8 Joint Efforts of India and Norway

At the Fourth UN Environment Assembly (held from 11 to 15 March 2019 in Nairobi, Kenya), India made a resolution for a ban on single-use plastics, and Norway emphasized the need for a global agreement for manage marine plastic litter. In 2019, the governments of India and Norway signed a Memorandum of Understanding (MoU) on India–Norway Ocean Dialogue and established a Joint Task Force on Blue Economy for Sustainable Development. In 2020, the two governments reviewed the progress made under this MoU, including the establishment of the Marine Pollution Initiative (Royal Norwegian Embassy, 2020a). As part of

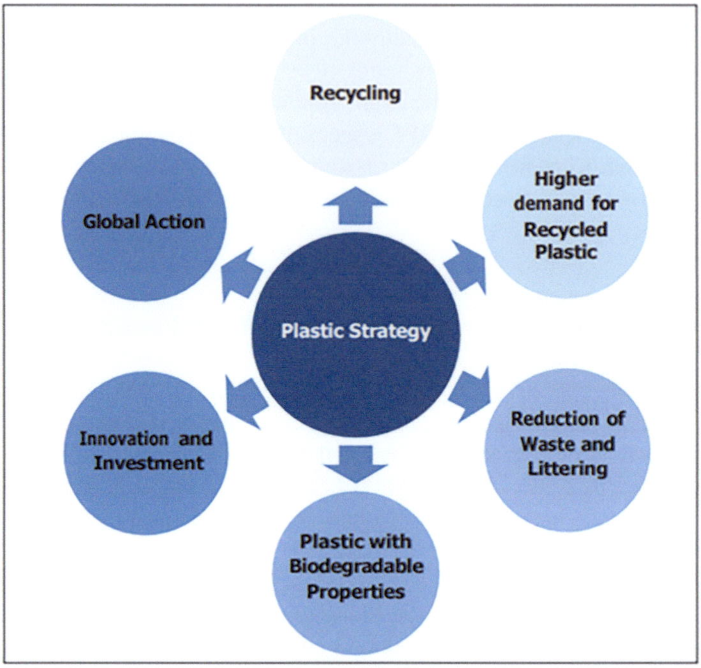

Fig. 3 Six key areas of the European Strategy for Plastics in a Circular Economy, 2018 (Kerscher, 2019)

this collaboration, the two countries signed a Letter of Intent on integrated ocean management to manage marine plastic pollution.

2.1.9 Policies and Programs of Top Plastic Waste Producing Countries

In 2015, China reportedly produced the largest quantity of plastic waste, at nearly 60 million tons followed by the United States at 38 million tons, Germany at 14.5 million tons, and Brazil at 12 million tons (Ritchie & Roser, 2018). The recycling policies and strategies of these top plastic waste producing countries are discussed below.

- *China*: In 2008, China banned retailers from giving out free plastic bags, asked them to charge customers (and keep the profit made) for plastic bags. The production of ultra-thin plastic bags was also banned. In 2013, China adopted the Operation Green Fence Policy to prohibit the import of contaminated recyclable materials followed by the National Sword program in 2017 to curb the illegal import of foreign waste including plastics. In 2020, China presented a major plan to reduce single-use plastics across the country; this included a ban on non-degradable bags in major cities by the end of 2020 and in all cities and towns by 2022. The plan was also to ban restaurant industry from using single-use straws by the end of 2020 (BBC News, 2020b).

- *USA:* In February 2020, USA introduced a Bill for a legislation to reduce single-use plastic nationwide. The Break Free from Plastic Pollution Act of 2020 is the only federal legislation that addresses the issue through source reduction and extended producer responsibility. Beginning on January 1, 2022, the bill will phase out single-use plastic products (US Congress, 2020). Nine US states—California, Connecticut, Delaware, Hawaii, Maine, New York, Washington, Oregon, and Vermont, have imposed ban on single-use plastic bags, followed by five other states, namely Connecticut, Delaware, Maine, Oregon, and Vermont, in 2019 (NCSL, 2020c). The Keep America Beautiful (KAB) initiative aims to promote shared American responsibility to build and maintain clean, green, and beautiful spaces. The KAB Litter Index and Community Appearance Index are step-by-step methods of assessing current litter conditions and other indicators which is used in thousands of communities and by municipalities nationwide. The KAB's key goals include preventing litter, improving recycling, and beautifying communities in the country (Keep America Beautiful, 2021).
- *Germany*: In 2015, Germany introduced a 5-point plan aimed at managing the dependency of single-use plastic and taking steps in adopting a sustainable system (Deutsche Welle, 2018) to become less of a "throwaway society." The Ministry introduced five steps to drastically reduce plastic waste, which included: (a) Avoiding unnecessary products and packaging; (b) Making packaging and other products more environmentally friendly; (c) More recycling, more recycling stations; (d) Preventing plastics from getting into organic waste; and (e) International efforts to limit plastic waste at sea.
- *Brazil*: Plastic recycling in Brazil is primarily related to polyethylene terephthalate (PET) bottles. A few years ago, about 51% of PET bottles was recycled in Brazil—314,000 tons had been recycled in 2014 that reduced to 274,000 tons in the following year, presumably due to low economic activity and reduction in oil prices (Global Recycling, 2020). Key economic activities related to sectors such as textiles and chemicals dropped almost 50%, reflecting negatively on recycling, as these industries were large consumers of raw recycled materials. In 2019, a total of 1055 Brazilian cities were reported to have a recycling program. The country recycles 1% of its waste (as per 2018 reports). At present, GIZ (Deutsche Gesellschaft für Internationale Zusammenarbeit GmbH) supports Brazil with a project called ProteGEEr (support to the implementation of the National Policy for Solid Waste toward climate protection), which ends in 2021. In 2010, Brazil adopted a waste management policy (Law 12.305), which has been partially implemented.

2.2 *The National Perspective: India*

Plastics production underwent rapid growth between 1970s and 1990s, which is at the rate of 2.5 times the GDP growth in India. The plastic processing industry is estimated to grow to 22 million tons (MT) a year by 2020 attributed mainly to rapid consumerism. To facilitate plastic recycling, the Society of the Plastics Industry, Inc. introduced its resin identification coding system in 1988. The seven types of plastic include: (1) polyethylene terephthalate (PETE or PET), (2) high density polyethylene (HDPE), (3) polyvinyl chloride (PVC), (4) low-density polyethylene (LDPE), (5) polypropylene (PP), (6) polystyrene or styrofoam (PS), (7) miscellaneous plastics (polycarbonate, polylactide, acrylic, acrylonitrile butadiene, styrene, fiberglass, and nylon).

2.2.1 Plastic Waste Management Rules, India

To manage and control plastic waste, the Government of India notified Plastic Waste Management (PWM) Rules, 2016 on March 18, 2016, superseding Plastic Waste (Management & Handling) Rules, 2011. In 2018, the 2016 rules were further amended. The rules require carry bags made of virgin or recycled plastic to not be less than 50 μm in thickness (carry bags made up of compostable plastic, complying IS/ ISO: 17088 are exempted). The Amendment Rules phased out manufacture and use of multilayered plastic (MLP), which are "non-recyclable, or non-energy recoverable, or with no alternate use." Further, in March 2021, Ministry of Environment, Forests and Climate Change (MoEF&CC) notified the draft Plastic Waste Management Rules 2021. The amendment documents the inclusion of a few key definitions such as "Single Use Plastic," "Plastic waste processing," and "Non-woven plastic bag." The amendment also proposed to increase the thickness of carry bags made of virgin plastic to 120 μm. Another notable addition is the proposed ban on the manufacture, import, stocking, distribution, sale, and use of specific single-use plastic from January 1, 2022. The amendment is currently in the draft stage and is open for public comments.

The roles and responsibilities of different stakeholders as captured by the rules are described below:

- Institutional waste generators and event organizers are required to manage, segregate, and handover plastic waste to appropriate agency, pay user fee, and spot fine in case of violation.
- Urban local bodies are required to encourage use of plastic waste for road construction or energy recovery or waste to oil or co-processing in cement kilns, and facilitate the development of infrastructure for managing plastic waste.
- Gram Panchayat, formalized local, grass-root level self-governance systems at village/small-town level in India, are required to manage PWM in the rural areas

and channelize recyclable plastic waste to recyclers having valid registration; they are responsible for creating awareness and ensuring to prevent open burning of plastic waste.

- Producers, importers, and brand owners are needed to work out modalities for collecting back the plastic waste within 6 months in consultation with the authorities.
- State Pollution Control Boards (SPCB) or Pollution Control Committees (PCC) are needed to enforce the provisions of PWM Rules, 2016/2018, related to registration, manufacture of plastic products and multilayered packaging, processing, and disposal of plastic wastes.
- The secretary-in-charge of urban development in urban areas and Gram Panchayat in rural area of the state or union territory are responsible to enforce the provisions of PWM Rules, related to waste management by waste generator, use of plastic carry bags, plastic sheets or covers, and multilayered packaging.
- District Magistrate or Deputy Commissioner will help SPCBs/PCCs, Secretary-in-Charge, Urban Development, and Gram Panchayat, as required.
- *Enforcement of the Plastic Rules at the state level*: There are 28 states and 8 union territories in India. The State/UTs Governments have through their own notifications imposed ban on the use of plastic bags and other single-use items within their respective jurisdiction. While 22 States/UTs have issued notifications/orders imposing complete ban on plastic carry bags, seven States/UTs have not imposed any ban on plastic carry bags except fulfilling the thickness criteria of 50 μm (as per the Rules) (MoEF&CC, 2019) (Fig. 4).
- *Import of plastic waste*: The Hazardous and Other Wastes (Management and Transboundary Movement) Rules, 2016 was amended in March 2019 to prohibit the import of solid plastic waste into the country including SEZ and EOUs (MoEF&CC Annual Report, 2019–2020).
- *Actions against violators*: As per provision "12" of PWM Rules, 2016, a total of 21 States/UTs have imposed fines, issued notices and closure directions to the defaulters, and seized materials in their respective States/UTs. Further, there are no violations reported by rest of the States/UTs (MoEF&CC, 2019).
- *Use of properly marked and labeled plastic carry bags*: Rule "14 (1)" of PWM Rules require shopkeepers/retailers to use properly marked and labeled plastic carry bags. Most of the States/UTs (except a few like Gujarat and Nagaland) are not following the proper practice of plastic carry bag labeling, especially the street vendors and small retailers.
- *Enforcement in Gangetic towns and cities:* As an effort to implement the Plastic Rules, the Central Pollution Control Board has issued directions to the Commissioners of 46-million plus cities, 20 state capital cities, and 118 towns located on the bank of the river Ganga to prohibit manufacturing and use of sub-standard plastic bags (less than 50 μm thickness, with or without handle) (MoEF&CC, 2019).

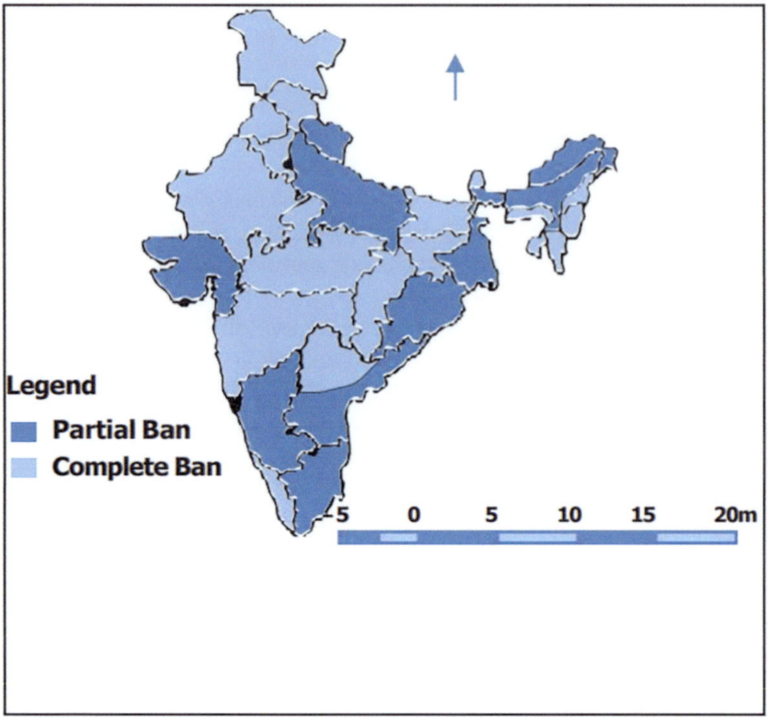

Fig. 4 Status of plastic ban across various states in India (Source: Prasher, 2018)

2.2.2 Plastic Waste Management (Second Amendment) Rules, 2023

On October 16, 2023, the Government of India released a draft notification suggesting revisions to the Plastic Waste Management Rules, 2016, to strengthen the effective implementation of the rules. These amendments seek to enhance the regulation of plastic waste management, in terms of compostable and biodegradable plastics, waste processing, labeling, and extended producer responsibility. The following are the main provisions of this draft notification:

- Manufacturers and importers of plastic raw materials to be added to the list of entities covered by Extended Producer Responsibility (EPR).
- Regulations for biodegradable and compostable plastic management to be introduced.
- Information technology solutions for online reporting and data collection to be incorporated at various levels of waste management.
- Provision of online electronic trading platforms for certificates by registered plastic waste processors for price discovery.
- Mandatory labeling for biodegradable and compostable plastics detailing its environmental conditions under which they can be used.
- The use of plastic raw materials in the manufacture of banned single use plastic (SUP) products is prohibited.

- Putting strategies in place to manage pre-consumer plastic waste generated from products composed of biodegradable plastics.
- Manufacturing facilities are required to process and report on pre-consumer plastic waste in the production of goods made of plastic or containing plastics.
- Strengthen the role of Gram Panchayat, district-level Panchayati Raj Institutions, and local bodies toward waste management.
- Making quarterly reporting and labeling for plastic raw ingredients mandatory.
- In cases where applications for one-time registration (for selling plastic raw/intermediate material) are waiting for longer than 30 days with SPCBs, the Central Pollution Control Board (CPCB) will be the authority for registration.
- Mandating that organizations involved in the processing or recycling of plastic waste prepare and submit annual reports.
- Mandating manufacturers and importers of plastic raw materials to incorporate specific clauses in sale contracts on the use of the materials in compliance with PWM Rules.
- Adding new categories for products made from compostable and biodegradable plastics and revising the categories of plastic packaging and commodities.
- Amending the guidelines on Extended Producer Responsibility (EPR) for plastic packaging and commodities to include the latest regulations.
- Establishing distinct color and labeling standards for plastic packaging and commodities composed of compostable and biodegradable plastics.
- Mandating guidelines for the prevention of open burning of plastic waste.
- Mandating yearly assessment and reporting by organizations on their infrastructure for managing plastic waste and anticipated waste production.
- Empowering the CPCB to revise forms and rules under these regulations.

2.2.3 Extended Producer Responsibility (EPR) Guidelines for Sustainable Plastic Packaging

The Plastic Waste Management Amendment Rules, 2021 notified on August 12, 2021, across the country prohibits the manufacture, import, stocking, distribution, sale, and use of plastic carry bags of thickness less than 75 μm and thickness less than 120 μm with effect from September 30, 2021, and December 31, 2022, respectively. The ban on identified single use plastic items and plastic waste management in the country are monitored effectively through the following online platforms: (a) National Dashboard for monitoring of implementation of comprehensive action plan, (b) CPCB Monitoring Module for Compliance on Elimination of Single Use Plastic, and (c) CPCB Grievance Redressal App. The Guidelines on Extended Producer Responsibility (EPR) for plastic packaging vide Plastic Waste Management (Amendment) Rules, 2022 stipulates mandatory targets on EPR, recycling of plastic packaging waste, reuse of rigid plastic packaging, and use of recycled plastic content. The guidelines also mandate sustainable plastic packaging and reducing the plastic foot print of plastic packaging. Under the Swachh Bharat Mission for solid waste management including plastic waste management, in urban and rural areas of the country, additional central assistance is provided by the Government of India as per the scheme guidelines.

2.2.4 India's Move on Single Use Plastic Products

Two consecutive amendments to PWM Rules (GoI, 2021a, 2021b) were notified subsequently by the Ministry of Environment, Forest and Climate Change (MOEF&CC, 2023). The amendment has brought in a stepwise approach such as increase in the thickness of single-use and non-woven plastic carry bags and ban on 17 problematic SUPP categories with effect from July 1, 2022. The ban exempted SUPPs made from compostable plastics, but it is required for the manufacturers to register with the Central Pollution Control Board (CPCB) before manufacture or sale of compostable products. The amendment also pre-emptively proposes that any notification prohibiting the manufacture, import, stocking, distribution, sale and use of carry bags, plastic sheets, MLP packaging, and SUP commodities issued post the 2021 amendments will face a 10-year delay to enter into force. The GOI had issued a draft notification on January 18, 2022, with the definition of biodegradable plastics and also introducing a new protocol for compostable and biodegradable materials (GoI, 2021a) (Fig. 5).

2.2.5 Plastic-Free Protected Areas

In 2018, the Wildlife Division of the Ministry of Environment Forest and Climate Change (MoEF&CC) issued directions to all States and UT Governments to organize awareness generation programs to sensitize people to avoid the usage of plastic in any form in all protected areas (500 national parks, zoos, and wildlife sanctuaries) and declare them as "plastic-free zones." Protected areas in states like West Bengal, Odisha, Delhi, and Uttarakhand have been declared as "plastic-free zones" (MOEF&CC Annual Report, 2018–2019).

2.2.6 Plastic-Free Schools and Colleges

The MoEF&CC advised that Eco-club of schools and colleges should make their institutions plastic-free. Under the 100-day action plan, the MoEF&CC intended to create single-use plastic-free Eco-clubs in at least 1000 schools. In 2019–2020, a total of 1367 schools across 13 states and UTs attained the 'Single use plastic free' status on their campuses (MoEF&CC Annual Report, 2019–2020).

Fig. 5 Stepwise approach implemented by the GOI to restrict or phase out selected SUPPs by 2022. (Source: CPCB, 2021). *Not provided: Data not available in the source report

2.2.7 Swachh Bharat 2.0 to Tackle Plastic Waste

In October 2019, the Government of India announced that its flagship program, Swachh Bharat Mission, will aim to manage plastic waste by eliminating single use plastic by 2022 as part of "Swachh Bharat Part 2." The key objective of SBM Part 1 was to achieve an open-defecation free (ODF) country by 2019. SBM Part 2 has deployed a set of strategies toward recycling and managing waste, including single-use plastic, within a span of 3 years.

2.2.8 Swachh Bharat Mission: Urban 2.0 'Managing Plastic Waste'

Swachh Bharat Mission (SBM-U) 2.0 has been launched on October 1, 2021 for a period of 5 years with a vision of achieving Garbage-Free Status for all cities through 100% source segregation, door to door collection, and scientific management of all fractions of waste including safe disposal in scientific landfills. To ensure that no untreated fecal sludge or used water is discharged into the environment, and all used water (including sewerage and septage, graywater, and black water) is safely contained, transported, and treated, along with maximum reuse of treated used water, a new component of Used Water Management (UWM) for Cities with less than 100,000 population has been included. It also aims at remediation of all legacy dumpsites and converting them into green zones.

3 Current Waste Management Scenario in India

3.1 Waste Management Practices

Assessment and quantification of plastic waste generated in 60 major cities in India showed that the demand for plastic raw materials has doubled mainly due to rapid urbanization and growing consumerism (CPCB, 2015). Some of the most common problematic practices of managing plastic waste in India include:

- Dumping of plastics along with other solid waste, mostly due to the lack of awareness and the need for land to dispose of the large amount of wastes generated from households and other areas.
- Inefficient or lack of source separation of waste and segregated collection and transportation in the waste supply chain.
- Burning of plastic waste leading to many environmental and health effects is a perennial problem faced by municipalities and authorities.

The practice is due to illegal practices, ignorance, lack of awareness, and inaccessible plastic waste collection infrastructure. Some of the successful waste management practices in Indian cities are:

Hyderabad: The city has a separate primary collection system for dry and wet solid wastes followed by its transportation to designated collection points, from where they are taken to different transfer stations. The dry recyclables including plastics are then sorted out through mechanized technology into various materials.

Bengaluru: The city has set up 164 dry waste collection centers (DWCCs) to facilitate the dry waste management system. These centers have appropriate infrastructure capable of purchasing, collecting, aggregating, and processing both high value and low value dry waste including plastics. The dry waste from domestic generators is collected (at doorstep) twice/thrice a week by auto tippers, properly segregated, and the recyclable non-biodegradable waste (including plastic waste) is sent to DWCCs, material recovery facilities (for sorting and baling of dry waste), and then to authorized recycling units to be turned into raw materials (for new products). Non-recyclable, non-biodegradable waste having calorific value more than 1500 kcal/kg is used as RDF (refuse-derived fuel) combustible fraction of waste or supplied as feedstock to solid waste-based power plants or cement kilns (BBMP, 2020). During 2018–2019, the estimated plastic waste generation in India was 3,360,043 tons/annum. Plastic waste is utilized for various purposes: used in cement plants for co-processing (14 states), used for polymer bitumen road construction (15 states), recycled (16 states), sent to "waste to energy" plants (two states), utilized in waste to oil plants (one state), and used as RDF (three states). It is observed that there is a gap of information on the current implementation status and practices of plastic waste management because the SPCBs/PCCs were unable to obtain relevant data from all ULBs of the states (CPCB, 2018–2019). The estimated amount of plastic waste generated, the status of implementation of PWM Rules, and the overall plastic waste management practices in each Indian state/UT during 2018–2019 are provided in Table 2.

Box 1 PWM Practices in Urban Gujarat

Based on a study conducted by the Central Pollution Control Board (CPCB) on the assessment and quantification of plastics waste generation in major Indian cities, an account of the plastic waste management practices (and quantity of plastic waste generated) in a few cities in Gujarat is given below (CPCB, 2015).

Surat: The average PW was generated about 124.68 kg/MT (12.46% of total solid waste). About 94.44% of PW was of HDPE/LDPE material, high value plastic material that is suitable for recycling. The Surat Municipal Corporation (SMC) of Surat has six different transportation sites, where plastic waste is segregated before going into the dumping ground.

Vadodara: An average of about 45.69 kg/tons of plastic waste is generated, out of which 36.65 kg/MT (79%) are HDPE/LDPE plastic waste comprising carry bags, milk pouches, packaging items, etc.

Ahmadabad: An average of about 105.07 kg/tons of plastic waste is generated, out of which 92.416 kg/MT (88%) are HDPE/LDPE plastics.

Dwarka: The PW generated was about 80.79 kg/MT (8.07%), out of which 45.8 kg/MT (56.7%) are HDPE/LDPE plastics. Plastic waste consists of waste plastics films and carry bags littered on main roads and streets.

Gandhinagar: The average PW generated in Gandhinagar was about 48.06 kg/MT. The minimum PW generation was about 28.14 kg/MT and the maximum was about 65.20 kg/MT. The MSW along with plastic waste is directly dumped in the dumping yard without any post treatment operations.

Rajkot: The average PW generated was about 69.28 kg/day. About 62.06% of PW was made of HDPE/LDPE material. At the dumpsite of Rajkot, the MSW collected from various locations of the city are transported to three different stations where PW is segregated.

Source: CPCB Annual Report PWM 2017–18.

Table 2 Status of implementation of Plastic Waste Management Rules in Indian States/UTs (CPCB, 2018–2019)

S. no	State/UT	Estimated plastic waste generation (tons for annum/TPA)	Implementation of PWM rules	Plastic waste management practices
1	Andaman & Nicobar Islands	1850	Complete ban on plastic bags	Door to door waste collection from households by authorized unorganized waste collectors; segregated plastic waste is used in recycling and road construction
2	Andhra Pradesh	66,314	Complied with the Rule	ULBs are collecting and segregating plastic waste and sent to recyclers; recycled waste used in cement plants and partly for road constructions
3	Arunachal Pradesh	3787.37	Partial ban on use of plastic carry bags	Not provided
4	Assam	32,277.87	Ban on single use plastic items including plastic carry bags irrespective of thickness	Non-recyclable plastic waste utilized in road construction on a pilot project. Efforts being made to use it in cement kilns
5	Bihar	68,903.328	Complete ban on plastic bags irrespective of thickness in urban and rural areas	Door to door waste collection
6	Chandigarh	11,715.4	Complete ban on plastic carry bags	Door to door waste collection from households and used at RDF processing plant
7	Chhattisgarh	6000	Complete ban on plastic carry bags	Collected and segregated plastic waste used in recycling through the registered recyclers and non-recyclable plastics stored separately and sold as alternate fuel and raw material to authorized cement plants for co-processing
8	Daman, Diu, & Dadra Nagar Haveli	1947.7	Complete ban on all kinds of plastic bags	Not provided

(continued)

Table 2 (continued)

S. no	State/UT	Estimated plastic waste generation (tons for annum/TPA)	Implementation of PWM rules	Plastic waste management practices
9	Delhi	224,810	Not provided	Door to door waste collection from households; collected waste sent to waste energy plant
10	Goa	32,580.52	Rules implemented	1. 26279.38 T baled RDF sent for co-processing in cement kilns 2. 6057.62 T waste recycled 3. 243.52 tons of inert waste is disposed in sanitary landfill
11	Gujarat	356,873	Partial ban; ban in Gandhinagar city	86,553.62 MT of plastic waste has been used for co-processing in cement plants
12	Haryana	68,735.26	Complete ban on usage of plastic carry bags	Door to door waste collection and segregation through private bodies. Municipal Corporation Sonepat framing a proposal for use of plastic waste in road construction
13	Himachal Pradesh	3672	Complete ban on usage of plastic carry bags	Door to door waste collection from households and the collected waste utilized in waste to energy plant, road construction, and co-processing units
14	Jammu & Kashmir	34,367.37	Rule 4(c) is implemented and ban imposed on <50 μm plastic products	Rule 5 is being followed. Rule 6 to be implemented yet, action plan is being submitted
15	Jharkhand	51,454.53	Total ban on carry bags	Segregation of waste has been initiated in almost all ULBs. Reverse vending machine has been installed at Ranchi, Dhanbad, and Jamshedpur cities for recycling of plastic bottles as pilot project. In Chas and Jamshedpur plastic has been used for road construction and co-processing in cement plant

(continued)

Table 2 (continued)

S. no	State/UT	Estimated plastic waste generation (tons for annum/TPA)	Implementation of PWM rules	Plastic waste management practices
16	Karnataka	272,776	Complete ban on manufacture, use, and sale of carry bags	Informal waste pickers are engaged in door-to-door waste collection system and collected plastic waste is utilized as: 25, 500 T for co-processing units in cement industries, 47,500 T in recycling
17	Kerala	133,316	Complied	Resource recovery facilities: 117; material collection facilities: 513
18	Lakshadweep	148	Complete ban on import and use of polythene and plastic material for packing carrying of consumer goods	Door to door waste collection and the segregated waste sent to Kochi for recycling
19	Madhya Pradesh	72,327.39	Complete ban on the plastic carry bags in the entire state	Collected plastic waste is utilized as: (i) 52,559 MT, recycled; (ii) 18,283 MT in co-processing; (iii) 1222.5 in road construction and RDFs
20	Maharashtra	409,630	Complete ban on plastic carry bags irrespective of size and thickness and other single-use plastic products	Plastic waste collection efficiency = 67%
21	Manipur	12453.8	Implemented in the state as per rules; complete ban on usage of plastic carry bags, plastic sheets, wrapping sheets made by recycled material	Not provided
22	Meghalaya	1263	Use and sale of plastic bags <50 μm have been prohibited	Waste is collected and segregated into dry and wet waste. Plastic waste is segregated from the dry waste at the dumpsite and collected for compaction and bailing. 1 km road was constructed by using plastic waste

(continued)

Table 2 (continued)

S. no	State/UT	Estimated plastic waste generation (tons for annum/TPA)	Implementation of PWM rules	Plastic waste management practices
23	Mizoram	13.306	Action plan for implementation of the Rules, 2016—prepared by the Municipal Corporation	Plans for utilization of plastic waste for road construction Collection centers for plastic waste have been established in urban towns and rural areas NGOs (YMA) and village councils (panchayats) are mobilized thus taking active part in PWM
24	Nagaland	268.18	Complete ban on plastic carry bags irrespective of thickness	Collected by rag pickers and sold to scrap dealers/material recovery center who in term sell to recyclers Polymer bitumen road is constructed using plastic waste
25	Odisha	90138.98	Prohibited use polythene sheets <50 μm thickness; use, manufacturing, stocking, and recycling of carry bags in the state. All plastic carry bags are banned in five municipal corporations and Puri	Major ULBs have been instructed to send segregated plastic waste to ACC Ltd Bargarh for co-processing in cement kiln
26	Punjab	119414.64	Complete ban on sale, recycling, use, manufacturing, stocking, and recycling of carry bags	None of the ULB have submitted action plan for plastic waste management. Three ULBs have started door to door segregation of solid waste and segregated recyclable plastic waste channelized through rag pickers
27	Puducherry	8433	Complete ban on single use plastics	100 MT. road constructed utilizing plastic waste; 272 MT MLP waste collected and sent for co-processing in Dalmia cements
28	Rajasthan	104,704.383	Complete ban on manufacturing, storage, import, sell, or transport of plastic carry bags	Door to door waste collection by the local bodies and efforts made to segregate the waste
29	Sikkim	5.66	Complete ban on plastic carry bags	Not provided

(continued)

Table 2 (continued)

S. no	State/UT	Estimated plastic waste generation (tons for annum/TPA)	Implementation of PWM rules	Plastic waste management practices
30	Tamil Nadu	401,091	Ban on manufacture, store, supply, transport, sale, or distribute of use and throwaway plastics irrespective of thickness including plastic carry bags	Plastic waste collected is segregated and low value plastic and MLP used for road laying, recyclable plastic waste sold to recyclers, and non-recyclable plastic waste disposed to cement industries for co-processing. Action plan submitted
31	Telangana	183014.65	Ban on <50 μm thickness plastic carry bags	Door to door waste collection by ULBs. Segregated plastic waste utilized as 6.34 MT for road construction; 494 tons recycled; 4-5TPD of PW being sent to cement industries for co-processing
32	Tripura	26.2	Complete ban on all kinds of plastic carry bags	20 ULBs collected and disposed as per PWM rules; action plan submitted
33	Uttarakhand	31,093	Complete ban on sale, use, and storage of plastic carry bags, plates, glasses, cups, and packaging	The collected and segregated plastic is recycled through the 25 recyclers and got the profit of 75.5 lakhs
34	Uttar Pradesh	254,401.8	Complete ban on manufacture, sale, use, storage, transport, import, and export of single-use plastic carry bags and other products in Nagar panchayat, Nagar Palika, Nagar Nigam, or industrial township	Plastic waste is utilized as:(i) 21.37 MT in cement industries for co-processing (ii) Non-recyclable plastic waste for the production of fuel, carbon black, and gas (iii) Waste to oil generation plant (3TPD) in Jhansi, etc. the use of plastic waste for road construction has been initiated.
35	West Bengal	300,236.12	Partial ban on several orders restricting use and sale of plastic carry bags in ecologically fragile areas and in certain heritage and tourist spots	Polymer bitumen roads have been constructed by some ULBs like Kalyani, Chandernagore, Ashokenagar, Kalyangarh, and new Kolkata development authority

3.2 Existing Plastic Manufacturing/Recycling Units

Rule 13 (1) of Plastic Waste Management Rules, 2016, requires all the plastic manufacturing/ recycling units to be registered with the concerned SPCBs/PCCs. A new category is the multilayered plastic (MLP) units (preferred due to properties like waterproof and lightweight, as it reduces shipping volume and contributes to increased shelf life of products) that have been included in the 2018 amended Rules. During 2018–2019, there were 4773 registered plastic manufacturing/recycling units (including 7 compostable units) in India. Also, there were 1084 unregistered plastic manufacturing/recycling units in 12 states/UTs. There are no plastic bag manufacturing units in five states (Table 3) (CPCB, 2018–2019). In the state of Gujarat, the State Pollution Control Board has made it mandatory for manufacturers to implement an action plan for disposal and recycling of plastic waste. The state has a total of 1016 registered units (including 1002 plastic manufacturing/recycling units, 4 compostable units, and 10 multilayered plastic units). There are no unregistered plastic manufacturer/recycler present in Gujarat (CPCB, 2018–2019).

Table 3 Registered and unregistered plastic units in Indian States/UTs (CPCB, 2018–2019)

S. no	State	No. of registered manufacturing/recycling units (Rule 13)			No. of unregistered manufacturing/ recycling units
		Plastic	Compostable	Multilayer	
1	Andaman & Nicobar Islands	No plastic manufacturing and recycling units			Nil
2	Andhra Pradesh	86 (80 producers and 6 recyclers)	Nil	Nil	
3	Arunachal Pradesh	Nil	Nil	Nil	Nil
4	Assam	13	0	4	Nil
5	Bihar	Nil	Nil	Nil	1
6	Chandigarh	7 recycling units	Nil	Nil	225 producers, 203 brand owners, and 36 recyclers
7	Chhattisgarh	16 (8 producers, 8 recyclers)	Nil	Nil	Nil
8	Daman, Diu, and Dadra Nagar Haveli	261	NP	NP	Not provided
9	Delhi	18 plastic/multilayer units	Nil	Nil	Not provided
10	Goa	34	Nil	1	Not provided

(continued)

Table 3 (continued)

S. no	State	No. of registered manufacturing/recycling units (Rule 13)			No. of unregistered manufacturing/ recycling units
		Plastic	Compostable	Multilayer	
11	Gujarat	1002	4	10	Nil
12	Haryana	Nil	Nil	59	Nil
13	Himachal Pradesh	Not provided	Not provided	8	Nil
14	Jammu & Kashmir	275	Nil	28 in Jammu	Not provided
15	Jharkhand	59	Not provided	Not provided	72
16	Karnataka	95	Nil	Nil	Nil
17	Kerala	1403	Nil	Nil	43
18	Lakshadweep	Nil	Nil	Nil	Nil
19	Madhya Pradesh	108 (45 manufacturers; 63 recyclers)	0	31	Nil
20	Maharashtra	276	1	11	162 (143 producers and 19 recyclers)
21	Manipur	4 (1 for Mfg. and 3—Partial recycling units)	Not provided	Not provided	Nil
22	Meghalaya	4	Nil	Nil	Nil
23	Mizoram	Nil	Nil	2	Nil
24	Nagaland	6	Nil	Nil	Nil
25	Odisha	8	Nil	Nil	Nil
26	Punjab	129	0	3	16
27	Puducherry	49	Nil	4	42
28	Rajasthan	2	Nil	7	Nil
29	Sikkim	Nil	Nil	Nil	Nil
30	Tamil Nadu	189 (91 plastic mfg. units (producers) and 98 recyclers)	1	8	280 plastic recycling units
31	Telangana	242 (brand owners/ producers—27)	Nil	2	Nil
32	Tripura	23 plastic items/bags manufacturing units	Nil	Nil	Nil
33	Uttarakhand	30	Not provided	27	Not provided
34	Uttar Pradesh	56	1	77	23
35	West Bengal	84	0	5	Not provided

4 Indian PWM Capacity and Expertise: From National to Local Level

4.1 Utilization of Plastic Waste in Road Construction

Utilization of plastic waste for road construction is being implemented in various cities in India. In November 2015, the Ministry of Road Transport & Highways, Government of India has made it mandatory for road developers to use waste plastic along with bituminous mixes for road construction to manage plastic waste in urban India. India has built 100,000 km of roads in at least 11 states (Swachh India, Karelia, 2018) using discarded plastic. Roads made of plastic are durable against extreme weather conditions, are cost-effective, and pothole resistant. However, concerns remain about microplastics shedding, potential leaching toxic additives, and long-term environmental impacts, particularly during wear and degradation. Some of the city-based initiatives on road construction made of plastic waste are given below.

- Chennai city used 160,000 kg of plastic waste to build 1.035 km of roads. Certain areas in the city like the N.S.C. Bose road, Halls road, Ethiraj Silai street, and Sardar Patel street have plastic roads.
- In Bangalore, 3000 km of road length has been laid out by using 12,000 tons of the plastic waste collected from the city.
- In Hyderabad, a 4000 sq. ft. pavement outside the city's first dog park was constructed with 1500 recycled plastic tiles. Each plastic tile weighs up to 300 g and is made from 600 plastic bags. It should be noted here that although the use of plastic materials for road construction is often touted as a solution, it is not free from several technological, material, and environmental challenges. For example, health impacts from heating plastics and leakage of chemicals and microplastics into the environment remain under research and are seen as challenges when plastic materials are integrated into road construction projects (Chin & Damen, 2019).

4.2 Building Material, Roads, and Other Items Made of Plastic Waste

CSIR-NPL of India has developed a sustainable and patented technology to manufacture tiles from plastic waste on a commercial scale. They plan to set up a recycling unit, which will have the capacity to produce 5 lakh tiles per month. The product has been tested for flammability, water absorption, and mechanical strength, as per the standards of the American Society for Testing Materials. Plastic waste has been used for constructing over 1 lakh km of roads in 11 Indian states. These road stretches are expected to last for more than 10 years, as plastic materials make the roads sturdy and heat resistant. The melting point of plastic roads is around 66 °C, compared to 50 °C of conventional roads, and the former is 8–9% more economical and costs less compared to a conventional road (Khanna, 2019). In May 2011, an NGO facilitated the construction of a children's school in New Delhi, using used PET bottles instead of conventional bricks (Pandey et al., 2017). There are several such case studies of construction of school buildings and furniture from plastic waste. In Hyderabad, a parking shelter at Miyapur Metro Station was replaced by a house made of recycled plastic waste. The plastic house was priced at Rs 1.5 lakhs, which is very economical compared to a regular steel shelter. It was made of used tetra paks, bottles, and plastic bags as raw materials, making the structure heat-proof, water-proof, and fire-proof. In Hyderabad, 775 recycle bins were installed using 23,500 kg of plastic waste, each bin made of 30 kg of plastic.

4.3 *Conversion of PET Bottle Waste into Textile Products*

A petro-chemical company is collecting PET bottle waste from across India and converting it into textile products. It has tied up with 150 vendors in India to provide PET bottle bales for processing into textile products. The company is installing reverse vending machines at prominent locations for collection of PET bottle waste and creating consumer awareness. The process involves using PET bottle waste as raw materials and elimination of the wet coloring process in the product of polyester staple fiber making, thereby helping to reduce pollution. Every bag or T-shirt made from PET bottles reduces the water usage by 1400 L, redeems eight waste PET bottles from the landfill, and reduces pesticide usage and carbon footprint by 50% and 32%, respectively (Ministry of Housing and Urban Affairs, 2019). In another case study (Nayak, 2013), a textile manufacturing company that specializes in converting PET bottles into polyester yarn had an annual capacity to process 18,000 tons in 2013 and planned to increase to 48,000 tons by next year. They believe that by recycling 10 billion PET bottles, one million square yards of landfill space can be saved and the release of 0.25 million tons of carbon dioxide can be eliminated. There are about 20 companies who annually convert 300,000 tons of PET bottles into polyester fiber. The industry often depends on waste pickers for its raw material supply and is vulnerable to fluctuations in input and output prices, such as rising costs of recovered plastics or falling market prices for recycled products.

4.4 *Recycling of Used Beverage Cartons*

Tetra Pak (carton packaging company) has set up collection centers for managing used beverage cartons (UBCs) in India (Manuja, 2019). UBCs can be handled and recycled like multilayered plastics and is made up of 21% polymers, 75% paperboard, and 4% aluminum. Through raising mass awareness programs, Tetra Pak runs this program in 20 selected cities of India, in addition to three cities in Kathmandu, Sri Lanka, and Bangladesh. It was observed that recycling rates increased from 29% in 2011 to 43% in 2015 and about 54% in 2019. The UBCs are collected and recycled or upcycled through different products and interventions like school desks, sheets used for making mobile toilets, and material recovery facilities.

4.5 Co-Processing of Plastic Waste in Cement Plants

Rule "6" of Plastic Waste Management Rules requires municipal authorities to set up, operationalize, and coordinate municipal waste management systems. Co-processing means using waste materials as an alternate fuel or raw material in industrial processes such as cement plants or power plants. CPCB has prescribed guidelines on the co-processing of plastic waste as an alternate fuel. A few states and UTs, namely Gujarat, Madhya Pradesh, and Meghalaya, transport their plastic waste to cement plants for co-processing. In Gujarat, the co-processing of plastic waste in cement industries was considered due to the availability of cement plants in the state. The advantages of co-processing of plastic waste in cement industries include high temperature and residence time in cement kilns, which can contribute to the reduction of greenhouse gas emissions. Historically, industries producing recycled paper generate plastic waste as a by-product. The recycled paper production in the Vapi industrial area in South Gujarat generated about 400 MT of plastic waste, which were initially recycled into products like roofs and benches, but the process was discontinued due to technical and financial constraints. This led to the dumping of plastic waste in landfills and open areas, including in remote locations. After several trials, the co-processing of waste at cement plants was found to be technically feasible (Shah, 2018). In general, non-recyclable plastic waste is used in the co-processing of plastic waste in cement kilns (Deepthi et al., 2018).

4.6 Refuse Derived Fuel from Plastic Waste

RDF is produced from domestic and commercial waste including plastics. Contaminated plastics, multilayer and other plastic packaging materials mostly end up in landfills. These wastes can be processed and converted to RDF, which carries significant calorific value, and can be used as alternative fuel in various industries. In October 2017 (CPHEEO, 2018), an Expert Committee was formed by the Ministry of Housing and Urban Affairs (MoHUA) to prepare "norms for RDF from MSW for its utilization in cement kilns, waste to energy plants and similar other installations" for enhancing the use of MSW-based RDF in various industries. The main objective of the Committee was to present guidelines and recommendations and propose regulatory and fiscal incentives for the use of RDF in various industries as per the requirements of *Swachh Bharat* Mission. The Committee deliberated the factors influencing the sustainability of co-processing of waste in cement kilns as a business model, keeping in mind the issues and challenges in the supply chain framework in India.

4.7 Bio-Based Polymers

Bio-based products can be developed using different techniques and raw materials. One approach involves using recycled polymeric materials and blending them with biopolymers, whereas the other approach is to develop the composites from only biopolymers without the incorporation of any kind of synthetic polymer. Lab samples of these bio-based plastics are developed by The Energy and Resources Institute (TERI) (Deepthi et al., 2018). New formulations of natural fiber reinforced completely bio-based thermoplastic composites have been developed by the MoEF&CC in collaboration with industrial partners for injection molded consumer goods. These composites are expected to be completely compostable under specific conditions and are therefore a good environment-friendly alternative (MOEF Annual Report, 2018–2019), mostly due to its biodegradability and enhanced performance after reinforcement. However, the existing challenges for the fabrication and use of bio-based polymer composites include the lack of consistency in characterization of materials (among the proposed systems), which makes it difficult to draw a direct comparison with different (other) materials (Murawski et al., 2019).

5 Existing Infrastructure (India, State, ULB)

5.1 Budget of PWM

The investment potential for managing India's plastic waste (financial capital, operations, and maintenance) over the 20-year period from 2011–2012 to 2031–2032 (8% of MSW) has been estimated at USD 5.1 billion (Rabo Research, 2018). The Government of India had allocated a budget of Rs. 122,940 million for the financial year of 2020–2021 for Swachh Bharat Abhiyan (SBM or Clean India Mission, a country-wide campaign initiated by the Government of India in 2014) that is now aimed to achieve a Swasth and Swachh India (Clean and Healthy India). SBM covers all rural and urban areas. The urban component, SBM Urban [SBM(U)], is implemented by the Ministry of Urban Development, and the rural component, SBM Gramin [SBM(G)] by the Ministry of Drinking Water and Sanitation. The allocation for SBM(G) was unchanged from the previous budget, at Rs 9994 crores. Funds for SBM(U), however, were reduced by Rs 350 crores to Rs 2300 crores. Solid waste collection, source segregation, and processing will be in focus. The target for 2020–21 was to include 1000 new wards to door-to-door collection and

7000 new wards with source segregation (Bandela, 2020). This means that the major share of the allocated budget—rural and urban—is used for solid (and liquid) waste management including plastic waste management. Information on the exact allocation for plastic waste management is not clear. In 2019, the Indian government initiated a plan for one-time funding of Rs 6 crores to develop plastic waste management centers to each of the following four cities: Surat (Gujarat), Patna (Bihar), Bengaluru (Karnataka), and Varanasi (Uttar Pradesh) (The Indian Express, 2019). For the year 2020–2021, the Government of Gujarat has made a provision of Rs. 200 crores under Swachh Bharat Mission (Urban). Under the urban development sector, a new scheme 'Gujarat Clean and Plastic free' was announced wherein rag pickers collecting dry waste and plastic waste will be given financial assistance, for which Rs. 56 crores has been allocated. Also, a provision of Rs. 803 crores has been made under Swachh Bharat (Gramin). Under Panchayat & Rural Development sector, Rs. 88 crores will be provided to all Gram Panchayats for monthly grant of four rupees per person for door-to-door collection of solid waste (including plastic waste).

5.2 *Infrastructure of PWM*

Lack of adequate infrastructure for segregation and collection is the key reason for inefficient plastic waste disposal. Many municipal corporations in India lack adequate systems for collection and segregation of plastic waste. In many Indian states, the technology and infrastructure needed to dispose of plastic waste in a cost- and resource-efficient way are not yet available. Some states and cities have taken the lead in empowering municipalities and developing the required infrastructure. The Government of India has committed support to developing plastic waste management infrastructure through Swachh Bharat Abhiyan initiative.

5.2.1 *Swachhta Kendras* in Urban India

India generates approximately 9.4 million tons plastic waste per annum or 26,000 tons of waste per day. Out of this, approximately 5.6 million tons per annum (58%) or 15,600 tons of waste per day plastic waste is recycled by the informal sector, and 3.8 million tons per annum plastic waste (9400 tons of waste per day) is left uncollected or littered (Ministry of Housing and Urban Affairs, 2019). The United Nations Development Programme (UNDP) along with Hindustan Coca-Cola Beverages Private Limited and Hindustan Unilever Limited are implementing Plastic Waste Management Programme (2018–2024) by building on the existing WM systems. The project promotes collection, segregation, and recycling of all kinds of plastics to move toward a circular economy. The project will help to institutionalize Material Recovery Centers or Swachhta Kendras within governance framework structures and improved socio-economic conditions of waste pickers. About 66,000 metric tons of plastic have been collected and processed. It aims to develop technology-supported knowledge management through promotion of cloud-based traceability, accountability, and digital governance along waste value chain. The project is currently operational in 36 cities, with 22 Swachhta Kendras established for sustainable PWM practices and has reached out to 5500 Safai Sathis to institutionalize workers from the informal sector (UNDP, 2018–2024). Chapter 7 of this report has further details.

5.2.2 PWM Units in Rural India

A plastic waste management unit for collection and recycling of plastic waste has been planned to be set up in each of the 8000 blocks, as per Swachh Bharat 2.0 under the Jal Shakti Ministry (Vishnoi &, Bureau 2019). With a focus on recycling, every village will be required to segregate waste and send its plastic waste to the block level PWM unit. The unit will work at converting the aggregated waste into bales, shred, and transport them to aggregators for recycling. Currently, waste pickers and aggregators are uninterested in collecting and recycling single-use plastic as there is no value attached to it. The Government will ensure that these block-level units are sustainable and emerge as successful business models. A deadline of 2021–2022 has been set for completing targets of making necessary arrangements for collection and transportation of plastic waste with one storage facility per village.

6 The Role, Practice, and Organization of the Informal Sector: What Is the Informal Sector and Why Does It Matter?

Existing systems of waste recovery in India feature an array of actors linked to each other by an internal logic dictated by economic, social, and cultural relationships (Gill, 2010; Doron & Jeffrey, 2018). The processes involved with picking, trading, and recovering waste constitutes what is referred to as the "waste economy" (Harriss-White, 2020). The relationships underlying the Indian waste management economy are oftentimes informal, they comprise formally unregulated activities which are not officially registered nor recognized by official waste management (WM) structures (Deepthi et al., 2018). The informal waste sector (IWS) is estimated to be responsible for recycling between 50% and 80% of India's waste (Annepu, 2012; Bhaskar & Chikarmane, 2012; Chandran et al., 2018), but figures can vary widely due to limitations of data availability and methodologies. The pivotal role of informal structures in providing essential WM services across Indian cities has been emphasized in numerous studies and country-specific reports (Deepthi et al., 2018; ESCAP, 2018; UNEP et al., 2019). However, national plastic WM strategies since the 1990s have so far not sufficiently tapped into the sector's potential (SBM Urban, 2019). This illustrates the state's complicated relationship with the IWS economy, where the state on the one hand relies on its services, yet does not fully recognize its contribution (Harriss-White, 2020). However, this aspect contradicts with the fact that at times, such waste management-related services by waste pickers in India tend to be paid as waged or performance-based labor like segregation of value-added plastics and other materials. Waste pickers are part of the IWS and typically collect different types of plastics and other materials that have been discarded at the (Gill, 2010; Doron & Jeffrey, 2018). IWS work in developing countries has conventionally been portrayed as small-scale, unskilled labor providing a source of livelihood for poor individuals (Haan et al., 1998; Moreno-Sánchez & Maldonado, 2006; Wilson et al., 2006). However, an increasing body of literature challenges the narrative of the marginalized and unorganized view of IWS by highlighting the interconnected socio-economic relations and value networks surrounding the trade and distribution of PW (Chen, 2012; Gill, 2010; Harriss-White, 2020). These narratives allude the need to rethink categorical views of informal and formal labor. Structural challenges that affect IWS workers include the absence of labor regulations, lack of access to social welfare and government insurance schemes, and minimal protection against health and unemployment shocks. The COVID-19 pandemic illuminates these vulnerabilities, as informal workers are not reached by government relief efforts (ICTD, 2020). It is thus relevant to engage with Chu and Michael's (2019) argument that IWS workers are particularly exposed to emerging risks and economic precariousness, as a result of being invisible to, or discriminated by, development policies under the official state system. This highlights that IWS workers' social networks and political voice is often unheard and de-legitimized within legal and political structures, from which the need to empower IWS workers' voice, visibility, and validity has become a reoccurring theme (Pillay, 2017; Goelnitz, 2018). Beyond the recognition of delegitimized and devalued IWS labor, informality has stereotypically been criminalized as a homogenous entity (Haan et al., 1998; Wilson et al., 2006). However, the informal economy as a whole is more heterogeneous and complex than these narratives suggest (Chen, 2012). Bhan and Gajera (2018) criticize literature across the board of conventional disciplines for failing to distinguish between illicit informal activities and legal but unrecorded informality. Accordingly, informal laborers are entrepreneurs and micro-businesses that pay informal fees locally and are highly skilled within their occupations. This is important, because beyond exposing the misinterpretations of informal labor as illegitimate, it also highlights the skills and knowledge that can be harnessed through increasingly inclusive approaches to WM. In the context of informal PW value networks in Delhi, Gill (2010) argues that entrepreneurial acumen is essential for survival within the competitive capitalist structures of the IWS economy. Growing body of literature supports this entrepreneurial framing of IWS workers (Gunsilius, 2010; Chandran et al., 2018; ESCAP, 2018; Harriss-White, 2020). While these narratives emphasise opportunities for developing entrepreneurial skillsets, other studies underscore the persistence of structural inequalities, socio-economic insecurity, and vulnerability (Gidwani & Maringanti, 2016). Such optimistic portrayals of entrepeneurship can thus be interpreted as survival strategies within systems of constraints and exclusion (Gill 2010; Chen 2012). These studies challenge stereotypical representations of IWS workers as merely marginalized or passive. This calls for a reconceptualization of what is meant by informality and positions IWS actors as central knowledge holders in a circular plastics economy.

6.1 Plastics and the Informal Sector

In India, the large informal workforce has been involved in collecting, dismantling, sorting, shredding, cleaning, compounding, pelleting, and manufacturing plastic waste for remanufacturing into new products. The IWS fills the void that is often left by municipalities due to lack of infrastructure, resources, and capacity to implement a formal and efficient waste collection and treatment system. The IWS also contributes to waste and cost reduction, and recovers valuable materials which otherwise would be lost. Estimates suggests the IWS contributes to divert about 4.7 million tons of plastic per year from the public waste collection system (WBCSD, 2017). It is also suggested that 15–25% of plastic waste handled by the IWS is unsuitable for further recycling (Nizzetto & Sinha, 2020). This includes the composite plastics (with different polymers that cannot be segregated), and plastics that have deteriorated in quality over time (after much recycling). These materials are not bought by the pelleting industries because they are considered valueless and are therefore a loss for the shredding and sorting units. In 2019, the estimated annual fluxes into the environment were from recyclable plastics [shredding and sorting (3.3–4.8 Mt), grinding and pelleting (2.4–3.6 Mt), remanufacturing (2.4–3.6 Mt)], and from non-recyclable plastics including single-use plastics (3.3 to 4.8 Mt., due to littering, landfilling, or incineration), end-of-life recyclables (0.8 to 1.2 Mt), and losses during production and transportation (0.2 to 0.4 Mt) (Nizzetto & Sinha, 2020). The Indian capital city of Delhi is one of the main recycling hotspots in India. It generates 5% of plastic waste but produces roughly 25% of the recycled plastics in India. In Tikri Kalan, a trading and sorting cluster, around 1900 tons of materials are traded every day. Within Tikri Kalan, three distinct groups exist—one formal PVC market and two informal satellites (Badi Tikri and Choti Tikri) complement the function of this main market. In the city of Surat, Gujarat, the Surat Municipal Corporation (SMC) segregates plastic waste before going into the dumping ground. SMC employs waste pickers on each transportation site for the segregation of value-added plastics and other materials. The primary segregation of plastic waste is carried out at the waste collection site and secondary segregation is done at the transportation site (CPCB, 2015). The Government of India has made concrete efforts to institutionalize the informal sector. As mentioned earlier, the Swachhta Kendras has reached out to 5500 Safai Sathis (workers). With a strong approach toward providing social security to the Safai Sathis, Self-Help Groups have been formed. Regular workshops organized for them at Swachhta Kendras, where they are given assistance with creating bank accounts, obtaining Aadhar cards, availing health check-up, and providing personal training sessions (UNDP, 2018–2024). Such efforts need to be scaled up across the country, as it will be a win–win for both the people involved and the environment. There is a need to better understand both the material dimension of the informal sector's role within PW management as well as how socio-economic impacts involved with informal PW management in India are perceived and experienced at the lowest levels of the recycling chain. In line with national and international goals and regulations to reduce plastic pollution at its source and adopt more circular economy initiatives, it is key to consider how existing IWS structures can be best included in the development of future WM strategies. However, as evidence has shown, top-down policies often overlook the interests and voices of IWS at a local level and fail to account for socio-economic and structural differences across scales. As a consequence, vulnerabilities faced by livelihoods in the IWS are exacerbated, which potentially reproduce adverse socio-economic and environmental impacts.

Acknowledgments This research was supported by the Royal Norwegian Embassy in New Delhi through the INOPOL Project (Project No: IND-22/0007).

References

Annepu, R. K. (2012). Sustainable solid waste management in India. *Columbia University, 2*(01), 1–89. https://www.nswai.org/docs/Sustainable%20Solid%20Waste%20Management%20in%20India_Final.pdf

Annual Report. (2018–2019). Ministry of Environment, Forests & Climate Change (MoEF&CC), Government of India. https://moef.gov.in/uploads/2019/08/Annual-Report-2018-19-English.pdf

Annual Report. (2019–2020). Ministry of Environment, Forests & Climate Change (MoEF&CC), Government of India. https://moef.gov.in/uploads/2017/06/ENVIRONMENT-AR-ENGLISH-2020.pdf

Anon. (2018). *Legal limits on single-use plastics and microplastics: A global review of national laws and regulations*. United Nations Environment Programme (UNEP). https://www.unep.org/resources/publication/legal-limits-single-use-plastics-and-microplastics-global-review-national

Anon. (2020a). *Norway in India, Indian and Norwegian Ministers of Environment commit to explore a Global Agreement to stop Plastic Pollution*. Royal Norwegian Embassy in India. Retrieved February 17, 2020, from https://www.norway.no/en/india/norway-india/news-and-events/new-delhi/news/indian-and-norwegian-ministersof-environment-commit-to-explore-a-globalagreement-to-stop-plastic-pollution. Last accessed May 2, 2020.

Anon. (2020b). Single-use plastic: China to ban bags and other items. *BBC News*. Retrieved January 20, 2020, from https://www.bbc.com/news/worldasia-china-51171491. Last accessed May 6, 2020.

Anon. (2020c). *State plastic bag legislation*. National Conference of State Legislatures (NCSL). Retrieved February 8, 2021, from https://www.ncsl.org/research/environment-and-natural-resources/plastic-bag-legislation.aspx. Last Accessed March 5, 2021.

Anon. (2021). *End Littering. Keep America Beautiful (KAB)*. Retrieved March 2, 2021, from https://kab.org/goals/end-littering

Bandela, D. R. (2020). Down to Earth. *Union Budget 2020–21: Constant decline in Swachh Bharat Mission allocation*. Retrieved November 2, 2020, from https://www.downtoearth.org.in/news/economy/union-budget-2020-21-constant-decline-in-swachhbharat-mission-allocation-69156

Bhan, N., & Gajera, R. (2018). Identifying the user in an informal trade ecosystem. In *Design research society international conference* (pp. 1010–1022). Design Research Society.

Bhaskar, A., & Chikarmane, P. (2012). The story of waste and its reclaimers: Organising waste collectors for better lives and livelihoods. *The Indian Journal of Labour Economics, 55*, 595–620.

Bruhat Bengaluru Mahanagara Palike (BBMP). (2020). *City Statistics, 2020*. Retrieved March 9, 2021, from https://bbmp.gov.in/departmentwebsites/swm/#

Central Pollution Control Board (CPCB). (2018–2019). Annual Report 2018–2019 on the Implementation of Plastic Waste Management Rules (as per Rule '17(4)' of PWM Rules, 2016, as amended 2018). Retrieved March 2, 2021. https://cpcb.nic.in/openpdffile.php?id=UmVwb3J0RmlsZXMvMTExOV8xNTk3MDM3NTM0X21lZGlhcGhvdG8xOTY1Ni5wZGY=

Chandran, P., Arora, K., Abubaker, M., & Shekar, N. (2018). *Valuing urban waste: The need for comprehensive material recovery and recycling policy*. June. Hasiru Dala. https://hasirudala.in/wp-content/uploads/2020/09/Valuing-Urban-Waste-2019.pdf

Chen, M. A. (2012). *The Informal Economy: Definitions, theories and policies* (WIEGO Working Paper No. 1). WIEGO. https://www.wiego.org/wp-content/uploads/2019/09/Chen_WIEGO_WP1.pdf

Chin, C., & Damen, P. (2019). *Viability of using recycled plastics in asphalt and sprayed sealing applications* (No. AP-T351-19). https://trid.trb.org/View/1670017

Chu, E., & Michael, K. (2019). Recognition in urban climate justice: Marginality and exclusion of migrants in Indian cities. *Environment and Urbanization, 31*(1), 139–156. https://doi.org/10.1177/0956247818814449

CPCB. (2015). *Assessment & quantification of plastics waste generation in major cities*. Retrieved May 8, 2020, from http://www.indiaenvironmentportal.org.in/files/file/Assessment%20and%20Quantification%20of%20Plastics%20Waste.pdf

CPHEEO. (2018). *Guidelines on usage of refuse derived fuel in various industries*. Retrieved May 12, 2020, from http://cpheeo.gov.in/upload/5bda791e5afb3SBMRDFBook.pdf

da Silva, C., Weins, N., & Potinkara, M. (2019). Formalizing the informal? A perspective on informal waste management in the BRICS through the lens of institutional economics. *Waste Management, 99*, 79–89. https://doi.org/10.1016/j.wasman.2019.08.023

Deepthi, M. V., Sailaja Bhattacharya R. R. N., Chandrasekhar, K., Roy, P., & Khan, A. (2018). *TERI analysis on challenges and opportunities—Plastic waste management in India*. https://www.teriin.org/sites/default/files/2018-06/plastic-waste-management_0.pdf

Deutsche Welle (DW). (2018). *Germany unveils 5-point plan to reduce plastic waste*. Retrieved November 2, 2020, from https://www.dw.com/en/germany-unveils-5-pointplan-to-reduce-plastic-waste/a-46455503

Doron, A., & Jeffrey, R. (2018). *Waste of a nation: Garbage and growth in India*. Harvard University Press.

ESCAP (2018) 'Closing the loop: Innovative partnerships with informal workers to recover plastic waste in an inclusive circular economy approach'. Workshop report. 13th September. Economic and Social Commission for Asia and the Pacific. Available from: https://www.unescap.org/sites/default/files/Closing%20the%20Loop%20Bangkok%20Workshop_Final%20Report.pdf.

European Commission. (2011). *The role of marketbased instruments in achieving a resource efficient economy*. ECORYS. Retrieved October 29, 2020, from https://ec.europa.eu/environment/enveco/mbi/pdf/studies/role_marketbased.pdf

European Commission. (n.d.) *A European strategy for plastics in a circular economy*. Retrieved May 6, 2020, from https://ec.europa.eu/environment/circulareconomy/pdf/plastics-strategy-brochure.pdf

Geyer, R., Jambeck, J., & Law, K. (2017). Production, use, and fate of all plastics ever made. *Science Advances, 3*, e1700782. https://doi.org/10.1126/sciadv.1700782

Gidwani, V., & Maringanti, A. (2016). The waste-value dialectic: Lumpen urbanization in contemporary India. *Comparative Studies of South Asia, Africa and the Middle East, 36*(1), 112–133. https://doi.org/10.1215/1089201x-3482159

Gill, K. (2010). *Of poverty and plastic: Scavenging and scrap trading entrepreneurs in India's urban informal economy*. Oxford University Press.

Global Recycling. (2020). *Brazil: Challenges or the Future*. Retrieved November 2, 2020, from https://global-recycling.info/archives/4155

Goelnitz, L. (2018). Unheard, unseen, unrecognized: Women in informal employment. Friedrich-Ebert-Stiftung. https://asia.fes.de/news/unheard-unseen-unrecognized-women-ininformal-employment.html

Gunsilius, E. (2010). *Role of the informal sector in solid waste management and enabling conditions for its integration: Experiences from GTZ*. In Transwaste workshop on the Informal Sector.

Haan, H. C., Coad, A., & Lardinois, I. (1998). *Municipal waste management: Involving micro-and-small enterprises*. Guidelines for municipal managers. International Training Centre of the ILO, SKAT, WASTE. Retrieved from http://www.skat-foundation.org/publications/waste.Htm.

Harriss-White, B. (2020). Waste, social order, and physical disorder in small-town India. *The Journal of Development Studies, 56*(2), 239–258. https://doi.org/10.1080/00220388.2019.1577386

Homrich, A., Galvao, G., Abadia, L., & Carvalho, M. (2018). The circular economy umbrella: Trends and gaps on integrating pathways. *Journal of Cleaner Production, 175*, 525–543. https://doi.org/10.1016/j.jclepro.2017.11.064

Hughes, K. (2020, May). *Protector or polluter? The impact of COVID-19 on the movement to end plastic waste*. World Economic Forum. Retrieved May 11, 2020, from https://www.weforum.org/agenda/2020/05/plastic-pollution-waste-pandemiccovid19-coronavirus-recycling-sustainability/

ICTD (2020) 'COVID-19 and the informal economy: Responses, relief, and research'. International Centre for tax and Development. Available at: https://www.youtube.com/watch? time_continue=3&v=LMIPPxP8eRY&feature=emb_title (17th April 2020).

ISWA UNEP. (2015). *Global waste management outlook*. Retrieved October 29, 2020, from https://www.uncclearn.org/wp-content/uploads/library/unep23092015.pdf

Kalmykova, Y., Sadagopan, M., & Rosado, L. (2018). Circular economy—From review of theories and practices to development of implementation tools. *Resources, Conservation and Recycling, 135*, 190–201. https://doi.org/10.1016/j.resconrec.2017.10.034

Karelia, G. (2018, October 26). Swachh India. *One lakh kilometres of roads in india are being made from plastic waste, is this the solution to end plastic crisis?* Retrieved May 8, 2020, from https://swachhindia.ndtv.com/plastic-waste-roads-one-lakh-kilometreindia-20274/

Kerscher, U. (2019). Towards a sustainable future? The EU policies concerning plastics and their didactical potential for primary and secondary teaching. *Discourse and Communication for Sustainable Education, 10*, 47–62. https://doi.org/10.2478/dcse-2019-0005

Khanna, S. K. (2019, August). *Turning plastic waste into eco-friendly building materials & products*. NBM&CW. Retrieved October 29, 2020, from https://www.nbmcw.com/tech-articles/others-article/40473-turning-plastic-waste-into-eco-friendly-building-materials-products.html

Ledsham, N. (2018, November 12). Creating a circular economy for plastics. *Sustainability*. Retrieved from https://sustainability.com/our-work/insights/creating-a-circular-economy-for-plastics/

Li, W. C., Tse, H. F., & Fok, L. (2016). Plastic waste in the marine environment: A review of sources, occurrence, and effects. *The Science of the Total Environment, 566–567*, 333–349.

Manuja, S. (2019). *Used Beverage Carton management study for India*. TERI (22 Nov 2018–21 Mar 2019). Retrieved May 7, 2020, from https://www.teriin.org/project/used-beverage-carton-management-study-indialast

Ministry of Environment Forest and Climate Change (MoEF&CC). (2023). *Extended Producer Responsibility guidelines mandate sustainable plastic packaging to reduce plastic footprint*. https://www.pib.gov.in/Pressreleaseshare.aspx?PRID=1909909

MOEF&CC. (2019, November 22). *Impact of plastic waste on environment*. Lok Sabha (Starred Question 87). Retrieved May 1, 2020, from http://www.indiaenvironmentportal.org.in/files/file/Impact%20of%20Plastic%20Waste%20on%20Environment.pdf

Moreno-Sánchez, R., & Maldonado, J. (2006). Surviving from garbage: The role of informal waste-pickers in a dynamic model of solid-waste management in developing countries. *Environment and Development Economics, 11*, 371–391. https://doi.org/10.1017/S1355770X06002853

Murawski, A., Diaz, R., Inglesby, S., Delabar, K., & Quirino, R. L. (2019). Synthesis of bio-based polymer composites: Fabrication, fillers, properties, and challenges. In K. Sadasivuni, D. Ponnamma, M. Rajan, B. Ahmed, & M. Al-Maadeed (Eds.), *Polymer nanocomposites in biomedical engineering. Lecture notes in bioengineering*. Springer. https://doi.org/10.1007/978-3-030-04741-2_2

Nayak, M. (2013, June 9). New age alchemists. *Business Today*. Retrieved November 2, 2020, from https://www.businesstoday.in/magazine/features/companiesthat-are-making-wealth-from-waste/story/195163.html

Nizzetto, L., & Sinha, S. (2020). Top priority to curb Plastic pollution: Empowering those at the bottom. *One Earth, 2*, 11–15. https://doi.org/10.1016/j.oneear.2020.01.005

Pandey, S. P., Gotmare, S., & Wankhade, S. A. (2017). Waste plastic bottle as construction material. *International Advanced Research Journal in Science Engineering and Technology, 4*(Special Issue 3). https://tinyurl.com/yfy2y65f

Pillay, K. (2017). Power to the waste pickers. *Independent Online*. Retrieved March 12, 2021, from https://www.iol.co.za/mercury/environment/powerto-the-waste-pickers-8838618

Prasher, G. (2018). *Where does the plastic waste in Indian cities go?* Retrieved May 6, 2020, https://citizenmatters.in/india-plastic-waste-world-environment-day-6800

Press Information Bureau (PIB), GoI. (2021a). Retrieved from https://pib.gov.in/Pressreleaseshare.aspx?PRID=1909909

Press Information Bureau (PIB), GoI. (2021b). Retrieved from https://pib.gov.in/PressReleaseIframePage.aspx?PRID=1846448

Rabo Research. (2018, February). *Plastic waste control in India*. Retrieved May 1, 2020, from https://research.rabobank.com/far/en/sectors/fa-supply-chains/Plastic-Waste-Control-in-India.html

Ritchie, H., & Roser, M. (2018). Plastic pollution. *Our World Data*. Retrieved from: https://ourworldindata.org/plastic-pollution

Shah, H. (2018, September 3). *How Gujarat succeeded in co-processing plastic waste in cement kilns and promoting circular economy*. Down to Earth. Retrieved May 6, 2020, from https://www.downtoearth.org.in/blog/environment/how-gujarat-succeeded-inco-processing-plastic-waste-in-cement-kilns-andpromoting-circular-economy-61518

Swachh Bharat Mission (Urban). (2019). *Plastic waste management: Issues, solutions and case studies*. Ministry of Housing and Urban Affairs (MoHUA). Retrieved March 1, 2021, from http://164.100.228.143:8080/sbm/content/writereaddata/SBM%20Plastic%20Waste%20Book.pdf

The Indian Express. (2019, November). *Plastic waste management: Surat among four cities chosen for central project*. Retrieved March 10, 2021, from https://indianexpress.com/article/india/plastic-waste-management-surat-among-four-cities-chosen-forcentral-project-6127752/

UNEP; COBSEA; SEI (2019) Marine plastic litter in East Asian Seas: Gender, human rights and economic dimensions. Bangkok. Available from: https://www.unenvironment.org/cobsea/resources/report/marine-plastic-litter-east-asian-seas-gender-human-rights-and-economic-dimensions

UNDP Plastic Waste Management Programme. (2018–2024). Retrieved April 30, 2020, from https://www.in.undp.org/content/india/en/home/projects/plastic-wastemanagement.html

US Congress. (2020). H.R.5845—Break Free From Plastic Pollution Act of 2020. 116th US Congress (2019–2020). Retrieved May 2, 2020, from https://www.congress.gov/bill/116th-congress/house-bill/5845

Vatn, A. (2015). *Environmental governance: Institutions, policies and actions*. Edward Elgar Publishing.

Velis, C. (2017). Waste pickers in global south: Informal recycling sector in a circular economy era. *Waste Management & Research, 35*(4), 329–331. https://doi.org/10.1177/0734242X17702024
Vince, J., & Hardesty, B. D. (2017). Plastic pollution challenges in marine and coastal environments: From local to global governance. *Restoration Ecology, 25*(1), 123–128.
Vishnoi, A., ET Bureau. (2019). *8,000 block level plastic waste management units on anvil*. Retrieved May 6, 2020, from https://economictimes.indiatimes.com/news/politics-and-nation/8000-block-levelplastic-waste-management-units-on-anvil/articleshow/71805811.cms?from=md
WBCSD. (2017, November 28). *Informal approaches towards a circular economy—Learning from the plastics recycling sector in India*. World Business Council for Sustainable Development (WBCSD), World Resources Forum (WRF), & Swiss Federal Laboratories for Materials Science and Technology (EMPA). https://docs.wbcsd.org/2016/11/wbcsd_informalapproaches.pdf
Wilson, D. C., Velis, C., & Cheeseman, C. (2006). Informal sector recycling in waste management in developing countries. *Habitat International, 30*, 797–808.
Zoë, L. (2016). *Waste and the sustainable development goals*. Retrieved April 30, 2020, from https://wasteaid.org/waste-sustainabledevelopment-goals/

Plastic Waste Generation, Management Issues, and Impact on Marine Ecosystems: Case Study of Tapi and Daman Ganga Rivers, Gujarat, India

Abdullah Atiq, Girija Prasad, Himalaya S. Vardikar, Smita Mohanty, Manisha Jain, Hans N. Adam, Mary Abraham, Emmy Nøklebye, Ingvild Skumlien Furuseth, and Anoop Balachandran

1 Introduction: Plastic Waste Generation in India

India generates 62 million tons of municipal solid waste (MSW), with 19 million tons not collected, and 31 million tons of the collected waste being disposed of in landfills (Agnihotri, 2022). Some of these dumped plastics, particularly high density polyethylene (HDPE), can take hundreds of years to decompose, with estimated half-lives ranging from 58 years (bottles) to 1200 years (pipes) (Chamas et al., 2020). India generates approximately 9.4 million tons of plastic waste annually, equivalent to 26,000 tons per day (MoHUA, 2019). Out of this, around 5.6 million tons per annum is recycled which means around 15,600 tons of waste per day most of which is in the informal sector, and 3.8 million tons per annum of plastic waste is left uncollected or littered (9400 tons of waste per day) (MoHUA, 2019).

Although the average per capita consumption of plastic (11 kg/annum) in India in 2011 was much lower compared to the world's average (28 kg/person) and other Western countries, the impact of plastic waste generation will be high in

A. Atiq
Environment and Waste Management Division, TERI, New Delhi, Delhi, India

G. Prasad · H. S. Vardikar · S. Mohanty
Central Institute of Petrochemicals Engineering Technology, CIPET, Bhubaneswar, Odisha, India

M. Jain (✉) · M. Abraham · A. Balachandran
Mu Gamma Consultants, Gurugram, Haryana, India
e-mail: mjain@mugammaconsultants.com

H. N. Adam · E. Nøklebye · I. S. Furuseth
Norwegian Institute for Water Research (NIVA), Oslo, Norway

P. Chakraborty et al. (eds.), *Efficient and Enhanced Management of Plastic Waste in India*, https://doi.org/10.1007/978-3-031-89398-8_8

India given the high population and improper management of plastic waste. It makes India the 15th biggest plastic waste generator in the world (MoEFCC, 2018). The plastic industry in India has seen steady growth since its establishment in the 1960s and has grown at a compound annual growth rate (CAGR) of 10% in terms of volume from 8.3 Mega Metric Tonnes Per Annum (MMTPA) in 2010 to 13.4 MMTPA in 2015 and was expected to reach 22 MMTPA by 2020 (with a growth rate of CAGR of 10.5%) (FICCI, 2009). The India packaging (predominantly plastics) market was valued at USD 50.5 billion in 2019 and is expected to reach USD 204.81 billion by 2025, with a CAGR of 26.7% during 2020–2025 (Mordor Intelligence, 2020). Comparatively, in 2018, the global production of plastics had increased to nearly 360 million metric tons and is expected to triple and account for a fifth of global oil consumption, by 2050. The global plastic market size was valued at USD 568.9 billion in 2019 and is expected to grow at a CAGR of 3.2% from 2020 to 2027 (Grandview Research, 2020). The plastic industry engulfs the entire spectrum of daily-use items and covers almost every sector of life such as clothing, housing, construction, furniture, automobiles, household items, agriculture, horticulture, irrigation, packaging, medical appliances, electronics, and electrical items. The packaging industry in India has seen a strong penetration of plastics as compared to global standards (FICCI, 2009). It is one of the fastest-growing industries and influences all other industries, either directly or indirectly. India observes a significant regional diversity in the consumption of plastics in the four key geographical regions, with the Western region accounting for 47%, the Northern region for 23%, the Eastern region for 9%, and the Southern region for 21%. The bulk of the consumption in Northern India is from end-use industries of auto-packaging (including bulk packaging), plasticulture applications, and electronic appliances, which are concentrated mostly in UP and Delhi-NCR. However, plastic processing in other parts like Rajasthan, Punjab, Haryana, Uttarakhand, Jammu & Kashmir, and Himachal Pradesh is expected to grow based on increased availability of feedstock and higher focus on the manufacturing sector (FICCI, 2009). As per the Un-Plastic Collective (UPC) Report of the United Nations Environment Programme (UNEP), Confederation of Indian Industries (CII), and WWF India, among the total plastic waste generated, 40% remains uncollected and 43% of the collected part contains single-use plastics such as carry bags, straws, packaging, bottle caps, garbage bags, polyethene terephthalate (PET) bottles, snacks and confectionery packets, etc. (Un-plastic Collective, 2019).

A 2015 report of CPCB on 60 Indian cities showed that a daily average of 3501 tons of plastic waste is generated, out of which 94% plastic waste is composed of thermoplastics or recyclable materials such as polyethene terephthalate (PET) and polyvinyl chloride (PVC) (CPCB, 2015). Although the majority of the plastic wastes are recyclable, they are either left littered and uncollected in the open environment or a significant part of the collected waste is dumped in landfills due to lack

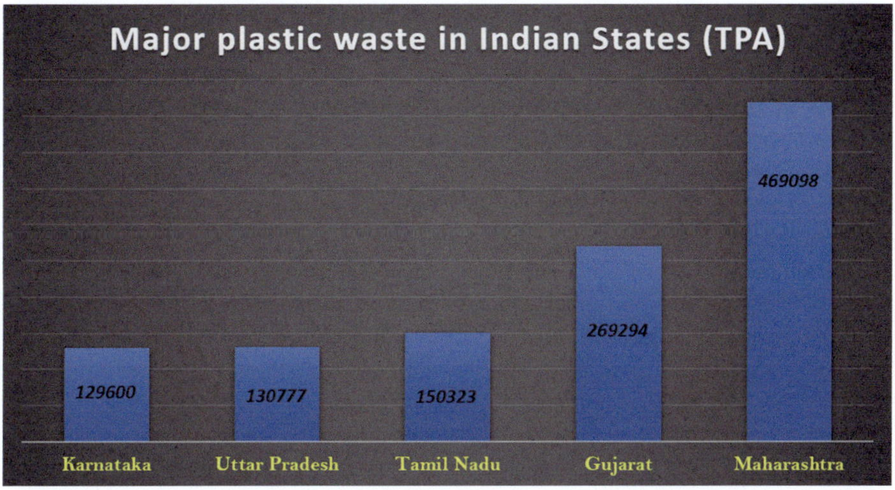

Fig. 1 Major plastic waste-producing Indian states in tons per annum (TPA) (Karnataka 129,600; Uttar Pradesh 130,777; Tamil Nadu 150,323; Gujarat 269,294; Maharashtra 469,098). (*Swachh Bharat Mission (Urban), 2019*)

of knowledge and technology-enabled machinery. Therefore, most plastic wastes are not recycled, reused, or disposed of properly, which points toward the inefficiency of the waste management system (Recykal, 2022). These effortlessly dumped wastes are either landfilled or end up in polluting streams or groundwater sources. Different survey reports have quantified the amount of plastic waste (in tons per annum) generated by different states in India (Fig. 1).

2 Selection of the Case Study

India produces 62 million tons of municipal solid waste annually, including 19 million tons left uncollected and 31 million tons dumped in landfills (Agnihotri, 2022). The country ranks fifteenth in the world in terms of plastic waste generation, and states such as Gujarat, a major industrial center, produce significant quantities of plastic waste, highlighting the need to address the issue comprehensively. The INOPOL project aims to prevent and reduce marine pollution and hazardous waste in India. The State of Gujarat was chosen as a pilot with the potential of scaling up similar efforts at the national level. With an area of 196,244 km^2 and a population projected to be nearly 70 million in 2021, Gujarat has become a state with one of the fastest industrial growth rates in India. The economic landscape is dominated by pharmaceutical, chemical, petrochemical, textile, pesticide, and fertilizer industries. Gujarat, being a major industrial and commercial hub of India, hosts 1002 registered plastic manufacturing/recycling units, four compostable units, and 10 multi-layered plastic units (CPCB Annual Report 2019a). The state comes second in contributing to India's plastic waste generation of ca. 356,873 tons per annum in 2018–2019 (CPCB Annual Report 2019b). Apart from industries, single-use plastic (SUP) waste and littering during religious processions also add to the rise in plastic waste dumped into the state (UNEP, 2018). Domestic wastes such as empty water bottles, garbage bags, carry bags, snacks, and confectionery packets are examples of SUPs that add to the build-up of plastic waste in the state. The two rivers of Tapi (Tapti) and Daman Ganga in Gujarat are highly affected by petrochemical and industrial activities and are major receivers and transport routes for land-based plastic pollution (Dubey & Ujjania, 2013). Tapi River flows through the metropolitan city of Surat, which has experienced significant economic growth and has been estimated to have the highest composition of municipal plastic waste generation (12.47%) among the 60 Indian cities as per a Central Pollution Control Board study (PIB, 2017). Vapi is situated near the banks of Daman Ganga and is the largest industrial area in Gujarat, dominated by chemical and paper industries. Estimates suggest that waste generation in both Surat and Vapi will increase significantly in the next decade. Gujarat is and has been over the past decades one of the fastest-growing states in India. It has a strong industrial and manufacturing base that caters to both the domestic and international markets. Some of the leading industrial activities are related to gems and jewelry, pharmaceutical, chemical, petrochemical, textile, pesticide, and fertilizer industries. For example, 25% of India's plastic is manufactured in Gujarat, largely by micro, small, and medium enterprises (MSMEs). While on the one hand, recent growth has created employment opportunities and prosperity, there are also environmental challenges that accompany this growth, including concerning plastic and chemical pollution. As discussed above, Gujarat state is the second largest generator of plastic waste in India. While a portion of this is captured in the formal and informal waste management and recycling streams, plastic material also escapes from industrial and municipal plastic waste streams. Once this pollution reaches the marine ecosystems through riverine systems, it can have even more significant detrimental effects. Hence, as part of the INOPOL project, plastic pollution was monitored along the catchment areas of the rivers Tapi (Tapti) and Daman Ganga along the cities of Surat and Vapi in Gujarat (Fig. 2).

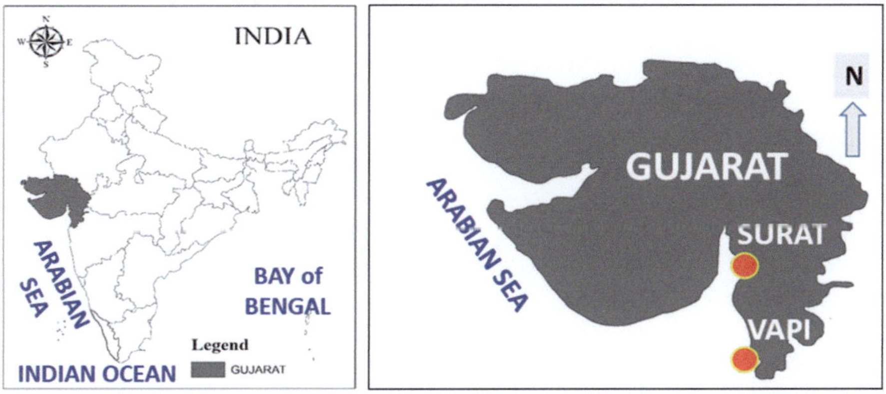

Fig. 2 Study locations in the state of Gujarat

3 Case Study Locations, Surat and Vapi

As mentioned in the previous section, the project focuses on the river catchment regions of Tapi and Daman Ganga in Gujarat, specifically targeting the industrial hubs of Surat and Vapi, where these areas play a substantial role in plastic pollution, and their respective catchments can transport plastics to the oceans (Fig. 3).

The Tapi River flows through the city of Surat, an urban industrial hub, and the Daman Ganga River flows through the city of Vapi, another industrial hub. Based on socioeconomic, geographical, and ecological factors, these catchments are contributing to plastic and chemical waste in the marine ecosystem. The Tapi River is the Peninsula's second-largest westward draining river with a catchment area of 62,225 km^2. At the mouth of River Tapi, Surat is located at 21.1702° N latitude and 72.8311° E longitude. The industrial city is also India's ninth-largest urban agglomeration and one of the world's fastest-urbanizing cities. Surat has been ranked by the Swachh Survekshan ranking (Star Rating for Garbage Free Cities) as the second cleanest city in the country for the years 2020, 2021, and 2022 consecutively (Vikaspedia, n.d.). The Daman Ganga River which originates from the Sahyadri hills in Maharashtra drains into the Arabian Sea is 131.30 km long and has a total catchment area of 2318 km^2. Vapi city is located at 20.3893°N latitude and 72.9106°E longitude and houses Asia's largest industrial area and has small-scale, chemical-based industries as well as plastic recycling units. The Vapi Industrial Estate started by Gujarat Industrial Development Corporation (GDIC) in 1967 stretches over 11.4 km^2 and houses more than 1400 industries, mostly small and medium enterprises largely manufacturing chemicals and paper.

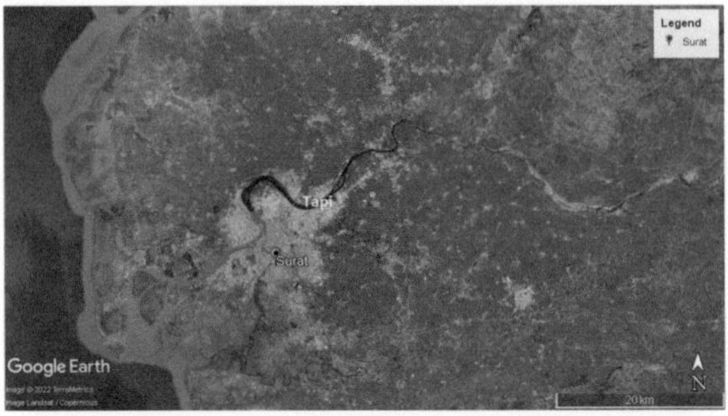

Tapi River
Urban Centre – Surat
Population (Million): 4.4 (2011)
7.1 (2021 estimated)
Urban Area: 474.2 km²
Waste Generation: 2200 TPD

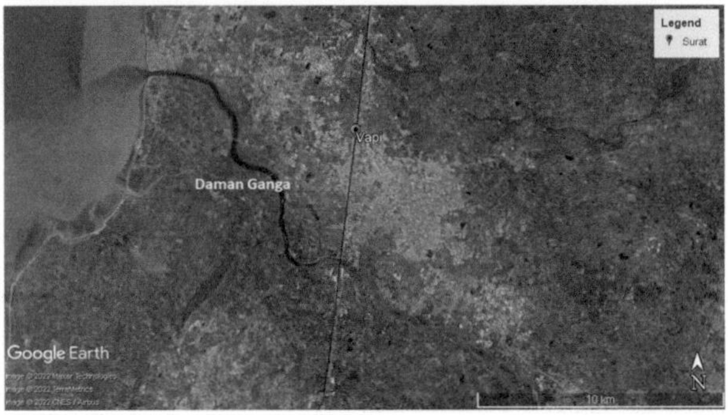

Daman Ganga River
Urban Centre – Vapi
Population (Million): 0.16 (2011)
0.19 (2021 estimated)
Urban Area: 22 km²
Waste Generation: 50 TPD

Fig. 3 Satellite images of river mouth. (*Images taken from Google Earth)

Surat is one of the largest cities in the Indian state of Gujarat, located in the southern part of the state. It is situated at the mouth of the Tapti River (commonly referred to as Tapi River), which flows through the city and flanks the city's southern border. These two rivers drain into the Gulf of Khambhat (Cambay).

Surat has an estimated GDP of $59.8 billion, which is the 9th largest among Indian cities. The city is well known for its diamonds and textile industry, and hence goes by the names "The Diamond City" and "The Silk City." Surat is the largest diamond-cutting and polishing hub in the world and accounts for about 90% of the world's diamond-cutting and polishing exports (World Economic Forum and PWC, 2017). Major companies like GAIL, ONGC, ESSAR, Ultratech Cement, Ambuja Cement, Reliance Industries, Shell, and others have their presence in this city (Business World, 2017). As per GPCB data, Surat has a total of 03 registered plastic waste recyclers that produce granules, fibers, and other materials from plastics (GPCB, n.d.). Surat is expected to further grow economically in the coming years due to its selection in the "Smart City Mission" launched by the Ministry of Urban Development. Under the Smart City Mission, Surat will be developing in various sectors including IT, renewable energy, and solid waste management along with town planning and development (Surat Smart City, n.d.). In the Swachh Survekshan-2020 (annual cleanliness survey of urban areas) Surat ranked the second cleanest city in India. The city has also won recognition for introducing a smart underground waste collection system in 2018. The population of Surat grew from about 2.8 million in 2001 to nearly 4.6 million in 2011, with a population density of 13,680 people per km^2 (Census of India, 2011a). Surat's estimated population in 2019 approached 6.3 million, marking a more than twofold increase since 2001 (Surat Municipal Corporation, n.d.-d). Surat lies between the latitudes 21.112° North and a longitude of 72.814° East (Surat Municipal Corporation, 2018). The city has an approximate area spread of 326.515 km^2 (Surat Urban Development Authority, 2017). Surat witnessed significant post-independence growth, with expanded activities contributing to considerable growth of areas. Surat Urban Development Authority (SUDA), the urban planning authority of Surat, was established under the Gujarat Town Planning and Urban Development Act, 1976 to cover SMC and 195 adjoining villages. SUDA extends over 715 km^2 between the latitudes 21°03′ and 21°19′ North and longitudes 72°41′ and 73°00′ East and lies at 13 m above mean sea level. There are four urban centers or Talukas (zones) in SUDA other than SMC, namely Olpad Taluka, Chorasi Taluka, Palsana Taluka, and Kamrej Taluka (Surat Urban Development Authority, 2017). In the 2035—Development Plan Report, "Extended SUDA" is to be added to include the highly developed industrial pocket of Hazira—one of the major sea ports of India. Extended SUDA will increase existing SUDA's area from 715 km^2 to 985.16 km^2 by the year 2035 (Surat Urban Development Authority, 2017). Surat experiences a tropical climate with temperatures in summer ranging from 37.78 °C to 44.44 °C and 10 °C to 15.5 °C in winter. The average annual rainfall of Surat is approximately 60 in. (i.e. 1525 mm) with about 90% of the rainfall occurring in the monsoon season from June to September. The highest temperature recorded is around 45 °C during May and June. The months of October and November experience retreating monsoon and the winter season runs from November to February. January is the coldest month with the lowest recorded temperature of 4.4 °C (Surat Urban Development Authority, 2017). The maximum reported humidity in Surat is around 80%. The annual average wind speed of Surat is around 18 knots (i.e. 9.26 m/s) in the northwestern direction (Surat Municipal Corporation, 2014). Surat experiences wind flowing from southwest to northeast, and the breeze is flowing from the sea to the main city area. The area commanded by SUDA has a gradual slope at the western and southern sides of the city (Fig. 4, Flow direction map of Surat). The southwestern side of the coastal area slopes toward the Mindhola River. The city is divided into two parts by River Tapi. Urban and industrial development has happened on both sides of the river. The areas of Hazira, Mora, Damka, Limla, Dumas, Bhimpore, Abhawa, Gavier, Sarsana, and Vesu lie at a lower elevation between the river Tapi and Mindhola.

The city is located on a low-lying flood plain and estuarial region having several creeks that lie in the southwestern part. Experts suggest that Surat is prone to climate change-related disasters such as flooding, owing to factors including unplanned rapid urban development, lack of surface runoff, and predictable heavy monsoons (Bouissou, 2014). Having experienced several floods since 1869, the 2006 floods

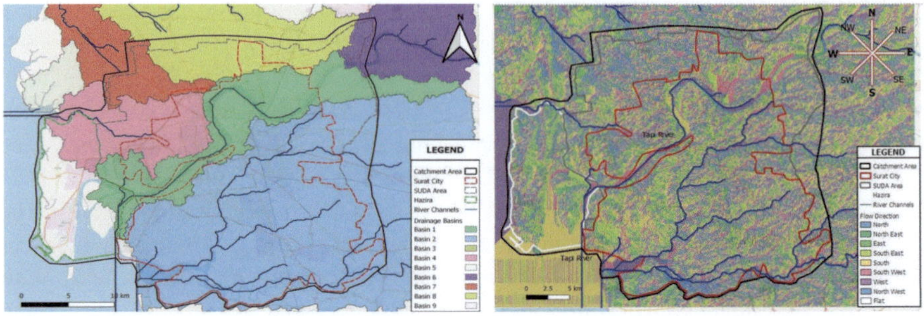

Fig. 4 Drainage and flow direction map of Surat

have been the most devastating affecting almost two-thirds of Surat's population (Waghwala & Agnihotri, 2019). Two expert committees were set up by the state government to examine the cause of the 2006 flood situation. Key causes included inadequate urban development practices like illegal encroachment by industries, housing colonies, and buildings for development purposes. It was also understood that mismanagement of the release of Ukai dam water had made the situation worse (Nidhi Jamwal, 2015). The drainage system continues to be inadequate and being on a lower plane, some areas endure water logging issues. Flood-like situations arise in the Tapi River every year during the rainy season. This results in the closing of the flood gates which further leads to stormwater backflow from drains. This stormwater drainage issue creates problems for the general public. In 2020, Surat experienced a similar flood-like situation, with overflowing stormwater drains submerging close to 6000 houses in low-lying areas (Times of India, 2025).

Vapi, recognized as Gujarat's most industrialized city, has a population of 1.6 million as of the 2011 census, with a density of 7292 persons per km^2. Generating 50 tons of municipal solid waste daily, the city is divided into 14 wards to manage the population and pollution. Known for its chemical-based industries, Vapi's industrial estate covers 11.64 km^2, hosting over 300 medium and large-scale industries. The city has Asia's largest Common Effluent Treatment Plant (CETP) managed by Vapi Nagar Palika. Despite generating 50 tons of waste daily, Vapi lacks waste characterization studies, leading to unprocessed waste dumped beside the Daman Ganga River. Positioned in Southern Gujarat, Vapi spans 22 km^2, with an elevation of 27.33 m above sea level, situated along the banks of the River Daman Ganga.

The river Daman Ganga continues its flow and drains into the Arabian Sea on the coast of Daman (Fig. 5, Flow direction map of Vapi). The city has an industrial dominance and houses one of Asia's largest Common Effluent Treatment Plant (CETP) which is administered by Vapi Municipality. Vapi City experiences tropical weather with three seasons, namely mild winter, moderate summer, and heavy monsoon. The average annual temperature is 26.8 °C with summer ranging from 22 °C to 41.2 °C and 9.9 °C to 36.5 °C in winter. The highest being recorded in April at 42 °C and the lowest reaching below 10 °C in January (Central Ground Water Board,

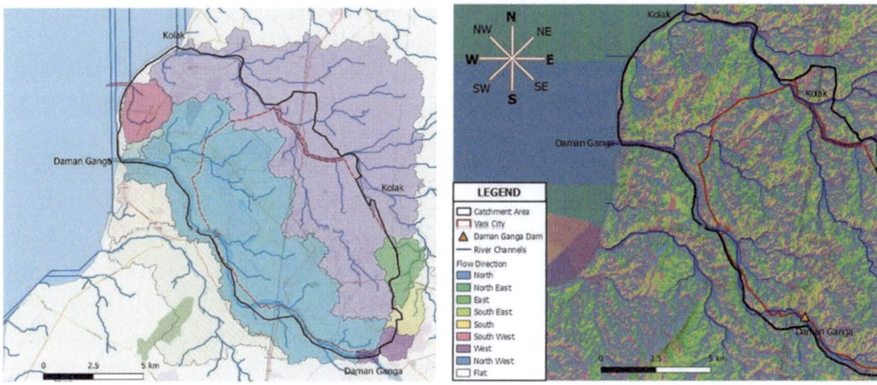

Fig. 5 Drainage and flow direction map of Vapi

2015). The summers are rainier than the winters in Vapi with the average annual rainfall being 90 in. (i.e. 2279 mm) and the average maximum humidity reaching 88.5% while the average minimum being 43%. The maximum and minimum wind velocities of Vapi observed from Valsad observatory are 6.7 knots (i.e. 9.26 m/s) in June to 2.3 knots (i.e. 1.18 m/s) in December (National Water Development Agency, 2014).

Vapi City is acknowledged as Gujarat's foremost industrial hub, surpassing Ahmedabad, Surat, Vadodara, and Rajkot (Industrial Scenario in Gujarat). Originally situated in the west, the town's eastern part is now predominantly industrial and residential. Renowned for its chemical-based industries (MSME, 2013), Vapi experienced significant growth with the establishment of the Gujarat Industrial Development Corporation (GIDC) Industrial Estate. The industrial landscape, dominated by chemical and paper industry plants, houses over 300 medium and large-scale industries, driving the economic growth of the district (MSME-DI). Vapi has an industrial estate area of 1163.77 ha (11.64 km^2) (Central Ground Water Board, 2015). As per GPCB data, Vapi hosts 18 registered plastic waste recyclers that produce a variety of plastic products ranging from plastic bags, granules, pipes, packaging materials, and sheets (GPCB). The most significant step in building the biggest CETP was taken by Vapi Waste and Effluent Management Company. This CETP with treatment capacity of up to 82,000 m^3 wastewater per day is the biggest effluent plant in terms of its capacity, treating the pollutants coming from the industries before they are directly released into the river. The CETP Plant covers an area of 22 acres and treats chemical and biological effluents that are released from Daman Ganga River. About 98% of the chemical, dyeing, and other industries use this CETP, as it can treat up to 54 million tons of waste per day (Sreenivas Janyala, 2006). There are 10,716 units of small and medium enterprises (SMEs) operating in the city in different sectors like chemicals, textiles, engineering, and paper industries (Government of Gujarat, n.d.). The city has been divided into 14 wards. The city's population was 163,630 in 2011.

The population grew from about 71,406 in 2001 to 163,630 in 2011, with a population density of 7292 people living per km^2 (Census of India, 2011b).

4 Plastic Waste Generation and Waste Estimation in the Study Region

As per the Surat Municipal Corporation, the city generated a total of 2200 tons per day (TPD) of MSW at a rate of 0.45 kg per capita per day in 2018–2019, as compared to, the global average was 0.74 kg per capita, reported by World Bank. The highest levels of waste generation were between 2 and 4.5 kg per capita per day with the highest per capita waste generation of 4.54 kg being witnessed in Bermuda (World Bank, 2018). Plastic waste generation in Surat contains 12.47% of MSW in Surat is comprised of plastic (GPCB, n.d.; CPCB, 2015). The plastic waste generation in Surat comprises of 2.12% PET, 1.12% HDPE/LDPE, 1.1% PVC, 0.76% PE, 0.55% PS, whereas the plastic waste pattern for India demonstrates PET as the largest segment at 66.1%, followed by HDPE/LDPE at 9.9%, PVC at 8.66%, PP at 6.43%, PS at 4.77%, and others at 4.14% (as demonstrated in Fig. 6).

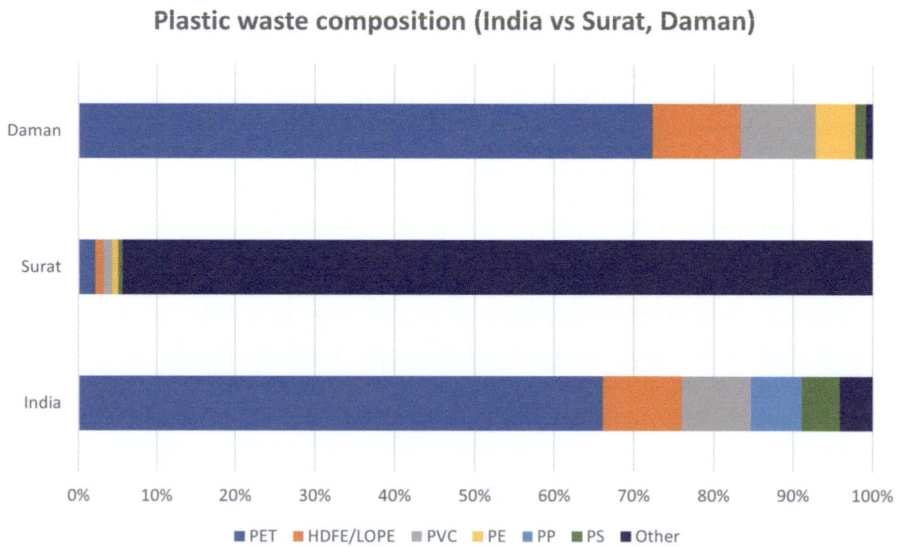

Fig. 6 Plastic waste composition in Surat (MSW%)

Table 1 Waste generation data for Vapi

Year	Waste Generation (TPD)
2011-2012	42[1]
2012–2013	43.06*
2013–2014	44.15*
2014–2015	45.26*
2015–2016	46.4*
2016–2017	47.57*
2017–2018	48.77*
2018–2019	50[2]

[1]Valsad District Committee, 2019, *extrapolation of data, [2]PAS, 2011

As per Surat Municipal Corporation, the share of plastics has now increased to 19%. For estimating waste generation in Surat city different parameters and methodologies have been used, providing an insight into behavioral aspects and standards to estimate waste.

In Vapi, plastic waste estimation is calculated based on the recent District Environmental Action Plan report for Valsad District, which states plastic composition in MSW corresponds to 2% in Vapi (Valsad District Committee, 2019). According to the 2019 Valsad District Environmental Action Plan report, Vapi generated 1 TPD of plastic waste, accounting for 2% of plastics in Vapi MSW (Valsad District Committee, 2019). However, due to a lack of waste characterization studies in the city, determining the properties of the plastic waste generated in Vapi is difficult. Because of its proximity to Vapi (11 km), the features of Daman's plastic waste can be used to anticipate the characteristics of Vapi's plastic waste. Daman was one of sixty cities studied in the CPCB-CIPET assessment in 2011, and the city MSW had 4.64% plastic waste (CPCB, 2015). Vapi Municipality is the authorized body designated to manage municipal solid waste in the city. According to the municipality, the city generated a total of 50 tons of "municipal waste" per day (TPD) (Valsad District Committee, 2019). The MSW spanning 4046 m^2 in Vapi is located beside the river Daman Ganga at Chandor Village where the waste is dumped directly without scientific processing. Table 1 has been developed based on the waste data for Vapi available publicly.

The waste generation data of Vapi is not as widely available and as comprehensive as that of Surat. According to Valsad District Environmental Action Plan report of 2019, Vapi generated a total of 1 TPD of plastic waste (out of 50 TPD of waste generated), which corresponds to 2% of plastics in Vapi MSW (Valsad District Committee, 2019). However, understanding the characteristics of this plastic waste generated in Vapi is a challenge due to the lack of waste characterization studies in the city. In order to predict the characteristics of Vapi's plastic waste, the

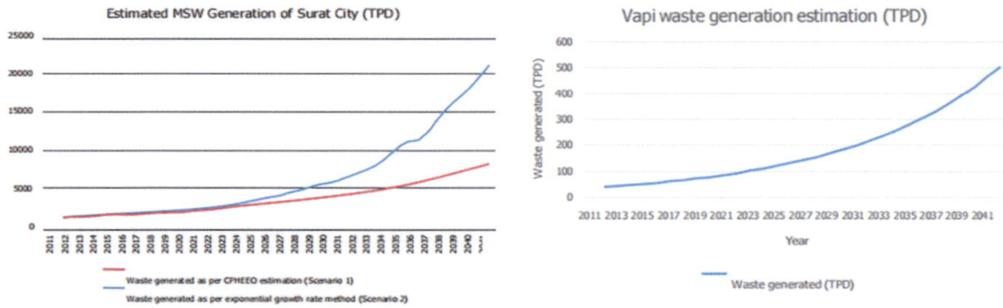

Fig. 7 Waste generation estimation for Surat and Vapi (Source: CPCB, 2015)

characteristics of Daman can be considered owing to its proximity to Vapi (11 km away), along with similar population (Vapi—163,630 and Daman—191,173 as per 2011 census). In the CPCB-CIPET study of 2011, Daman was one of the sixty cities to be assessed and the city MSW consisted of 4.64% plastic waste (CPCB, 2015). It is important to note that Daman's plastic waste generation is only plastic waste (and not quantification). Most authentic and recent data is used to quantify and estimate the plastic waste generated in Vapi. The estimation of waste generation for Surat and Vapi as reported by CPCB in their 2015 report is shown in Fig. 7.

5 Conclusion

The INOPOL project highlights the urgent need for effective plastic waste management solutions in India, particularly in the rapidly industrializing states and cities. It provides a comprehensive analysis of the key challenges related to plastic pollution in the industrial cities of Surat and Vapi in Gujarat. The detailed analysis of the case studies in these two cities presents the implications for future waste management strategies. Gujarat is India's second-largest generator of plastic waste and a major industrial hub; hence, the issues in Surat and Vapi represent the broader national issue. The detailed waste quantification and characterization studies carried out in these two cities demonstrate the contribution of the industrial sector to plastic pollution affecting both terrestrial and marine ecosystems.

The case studies also represent the scale and magnitude of the problem, which is exacerbated by rapid industrial growth. The volume of plastic waste generated, constituting 9–10% of India's 62 million tons of municipal solid waste annually, points out the severity of the situation. The profiles of Surat and Vapi also emphasize both the present shortcomings and the potential avenues for improvement in waste management practices, and the need for integrated waste management. Stringent regulations, advanced waste segregation and recycling processes, and technology must all be combined for effective management of plastic waste. Evidently, integrated plastic waste management will offer several environmental and economic benefits, including job creation in the recycling industry, lowered costs related to better waste management, and reduced environmental impact, especially on marine ecosystems. It may be noted that the success and sustainability of initiatives like INOPOL depend on robust collaboration between government bodies, industry stakeholders, and international partners. Policy frameworks at the national, state, and city levels must also be reinforced to support innovation waste management techniques and sustainable practices. In a nutshell, the case studies in the industrial hubs of Surat and Vapi highlight the urgent need for comprehensive approaches to address plastic pollution in India. Adopting the project's recommendations can have a major positive impact on India's sustainable development, marine ecosystem protection, and plastic pollution control.

Acknowledgments This research was supported by the Royal Norwegian Embassy in New Delhi through the INOPOL Project (Project No: IND-22/0007).

References

Agnihotri, A. K. (2022, September 14). *MSW management: The pitiable situation of municipal solid waste management.* Retrieved from https://www.teriin.org/article/msw-management-pitiable-situation-municipal-solid-waste-management

Bouissou, J. (2014, September 15). *Indian city of Surat anticipates worst effects of climate change.* Retrieved November 12, 2020, from https://www.theguardian.com/cities/2014/sep/15/indian-cities-climate-change-surat

Business World. (2017, June 28). *Richest Cities of India.* Retrieved from Business World Web site: http://www.businessworld.in/article/Richest-Cities- Of-India/28-06-2017-121011/

Census of India. (2011a). District Census handbook Surat. Census of India 2011 - Gujarat - Series 25 - Part XII A. DH_2011_2425_PART_A_DCHB_SURAT. Last Accessed on 14/07/2025 from Census Digital Library at https://censusindia.gov.in/nada/index.php/catalog/422

Census of India. (2011b). District Census Handbook Valsad. Census of India 2011 - Gujarat - Series 25 - Part XII B. DH_2011_2424_PART_B_DCHB_VALSAD. Last accessed on 14/07/2025 from Census digital Library at https://censusindia.gov.in/nada/index.php/catalog/419/study-description

Central Ground Water Board. (2015). *Valsad District.* Ministry of Water Resources.

Chamas, A., Moon, H., Zheng, J., Qiu, Y., Tabassum, T., Jang, J. H., Abu-Omar, M., Scott, S., & Suh, S. (2020). Degradation rates of plastics in the environment. *ACS Sustainable Chemistry & Engineering, 8*, 3494–3511. https://doi.org/10.1021/acssuschemeng.9b06635

CPCB. (2015). *Assessment & quantification of plastics waste generation in major cities.* Ministry of Environment & Forests. https://cpcb.nic.in/uploads/plasticwaste/SOP_PWM_24062024.pdf

CPCB. (2019a). *Annual Report for the year 2018-2019 on Implementation of Solid Waste Management Rules.* Ministry of Environment, Forest and Climate Change. Retrieved November 29, 2020, from https://cpcb.nic.in/uploads/MSW/MSW_AnnualReport_2018-19.pdf

CPCB. (2019b). *Annual Report for the year 2018-2019 on Implementation of Plastic Waste Management Rules.* Ministry of Environment, Forest and Climate Change. Retrieved November 29, 2020, from https://cpcb.nic.in/uploads/plasticwaste/Annual_Report_2018-19_PWM.pdf

Dubey, M., & Ujjania, N. C. (2013). Water quality and pollution status of Tapi river, Gujarat, India. *International Journal of Pure and Applied Zoology, 1*(3), 261–266.

FICCI. (2009). *Survey on the current status of municipal solid waste management in Indian cities and the potential of landfill gas to energy projects in India.* Federation of Indian Chambers od Commerce and Industry. https://www.nswai.org/docs/292_2_ficci_survey_09.pdf

Government of Gujarat. (n.d.). *Economy*. Retrieved November 15, 2020, from Valsad District: https://valsad.nic.in/economy/

Grandview Research. (2020). *India Biodegradable Plastic Market Size & Outlook*. Last accessed on 14/07/2025 from https://www.grandviewresearch.com/horizon/outlook/biodegradable-plasticmarket/india#:~:text=The%20India%20biodegradable%20plastic%20market,12.5%25%20from%202025%20to%202030

GPCB. (n.d.). *Plastic waste recyclers*. GPCB. Retrieved December 2, 2020, from https://gpcb.gujarat.gov.in/uploads/Plastic_Waste_Recyclers.pdf

Ministry of Environment, Forest and Climate Change. (2018). *Beat Plastic Pollution: Good News from India*. MoEFCC, New Delhi. Last accessed on 14/07/2025 from https://moef.gov.in/uploads/2019/10/Beat-Plastic-Pollution.pdf

Ministry of Housing & Urban Affairs, Government of India. (2019). *Smart cities*. Retrieved April 21, 2020, from Ministry of Housing & Urban Affairssmart City website: http://mohua.gov.in/cms/smart-cities.php

Mordor Intelligence. (2020). Packaging industry in India – Growth, trends, forecasts (2020–2025). Retrieved from https://www.mordorintelligence.com/industry-reports/packaging-industry-in-india

MSME. (2013). *Brief Industrial Profile of Valsad District*. MSME-Development Institute.

National Water Development Agency. (2014). *Daman Ganga Pinjal Link Project*. Ministry of Jal Shakti.

Nidhi, J. (2015). *Citizens' reports indict Gujarat government for Surat floods*. Retrieved November 19, 2020, from DownToEarth: https://www.downtoearth.org.in/reviews/citizens-reports-indict-gujarat-government-for-surat-floods-6606

PAS (2011). Annual Performance Assessment Report of Urban Water supply and Sanitation - Gujarat: 2010-11. CEPT University. 2014. Last accessed on 22.07.2025 from https://www.pas.org.in/Portal/document/ResourcesFiles/Databook/Gujarat/Full%20Report/Gujarat_2010-11.pdf.

PIB. (2017). Plastic waste. PIB Delhi. Retrieved from https://pib.gov.in/PressReleseDetail.aspx?PRID=1514709

Recykal. (2022). The rising problem of plastic waste in India. Retrieved May 25,2024, from https://recykal.com/blog/the-rising-problem-of-plastic-waste-in-india/

Sreenivas J. (2006). *Gujarat brings out the scrubbers*. Retrieved November 23, 2020, from The Indian Express: http://archive.indianexpress.com/news/gujarat-brings-out-the- scrubbers/15509/

Surat Municipal Corporation. (2014). *Annual Report*. SMC.

Surat Municipal Corporation. (2018). *Annual Report*. SMC. https://www.suratmunicipal.gov.in/Content/Documents/Departments/VectorBorneDiseasesControl/VBDC_Annual_Report.pdf

Surat Municipal Corporation. (n.d.-d). *Mid-year population estimates*. Retrieved from Surat Municipal Corporation Web site: https://www.suratmunicipal.gov.in/TheCity/City/Stml13

Surat Smart City. (n.d.). *Surat—Smart City Mission*. Retrieved November 25, 2020, from Surat Smart City Web site: https://www.suratsmartcity.com/SmartCityMission/SuratSmartCityMission

Surat Urban Development Authority. (2017). *Development Plan 2035*. Surat Urban Development Authority. https://www.sudaonline.org/wp-content/uploads/2017/03/SUDA-DP-2035-report-final-22.02.2017.pdf

Times of India, (2025). Surat sees pause in rain, but flood woes continue as city remains submerged. Accessed from https://timesofindia.indiatimes.com/city/surat/surat-sees-pause-in-rain-but-flood-woes-continue-as-city-remainssubmerged/articleshow/122054475.cms

Un-plastic Collective. (2019). Last Accessed on 14/07/2025 from https://d2391rlyg4hwoh.cloudfront.net/downloads/unplastic_brochure.pdf.

UNEP. (2018). How the Indian state of Gujarat is taking on plastic pollution. Accessed on 14/07/2025 from https://www.unep.org/news-and-stories/story/how-indian-state-gujarat-taking-plastic-pollution

Valsad District Committee. (2019). Valsad District Environmental Action Plan. Downloaded on 14/07/2025 from https://gpcb.gujarat.gov.in/Home/DistrictEnvironmentPlan

Vikaspedia. (n.d.). Swachh Survekshan. Retrieved May 25, 2024, from https://vikaspedia.in/health/sanitation-and-hygiene/swachh-survekshan

Waghwala, R. K., & Agnihotri, P. G. (2019). Flood risk Assessment and resilience strategies for flood risk management: Case study of Surat ci. *International Journal of Disaster Risk Reduction, 40*, 101155.

World Bank, 2018. What a Waste 2.0: A Global Snapshot of Solid Waste Management to 2050. Accessed on 14/07/2025 from https://documents1.worldbank.org/curated/en/697271544470229584/pdf/What-a-Waste-2-0-A-Global-Snapshot-of-Solid-Waste-Management-to-2050.pdf

World Economic Forum and PWC. (2017). *Migration and Its Impact on Cities*. Accessed on 14/07/2-25 from https://www.weforum.org/publications/migration-and-its-impact-on-cities/

Analysis of Gaps in the Management of Plastic Waste in India and Beyond

Emmy Nøklebye, Manisha Jain, Avanti Roy Basu, and Pallavi Gulati

1 Data Limitations for Waste Generation, Management, and Pollution

Accurate and representative baseline data on waste generation, management (including collection and recovery trends), and leakages is important to inform policy and enable monitoring and enforcement. The Central Pollution Control Board (CPCB) is the prescribed authority for pollution control in India, responsible for the enforcement of regulations, providing technical guidance, and conducting environmental assessments. Data from the CPCB are based on information provided annually by State Pollution Control Boards (SPCBs) (e.g., CPCB, 2019), but monitoring and reporting methodology for data collection vary across SPCBs. The most common mechanisms to measure the amount of plastic waste generated by a city is through the amount received by the urban local bodies' (ULBs) own collection channels which includes street sweeping and door-to-door household collection. This approach does not capture inadequately disposed plastic waste that leaches into the environment, such as illegally burned or openly dumped waste. This makes it challenging to provide reliable estimates for the volume and fate of waste generation and management.

In Gujarat state, there is limited scientific evidence of the mass of plastic waste going into the recycling stream, dumped in landfills, and leakages to the environment (Surat District Profile, 2011). However, preliminary research indicates that significant amount of waste is inadequately managed in landfills, uncontrolled

E. Nøklebye (✉)
Norwegian Institute for Water Research (NIVA), Oslo, Norway
e-mail: emmy.noklebye@niva.no

M. Jain · A. R. Basu · P. Gulati
Mu Gamma, Gurgaon, Haryana, India

P. Chakraborty et al. (eds.), *Efficient and Enhanced Management of Plastic Waste in India*, https://doi.org/10.1007/978-3-031-89398-8_9

recycling facilities, and open burning and dumping (Shah, 2018). Both in Surat and Vapi, industries have been highlighted as key sources of plastic waste generation and pollution. This indicates the need to better understand barriers and enabling mechanisms to achieve increasingly environmentally sound management of waste across different sectors.

Another challenge for aggregating accurate waste management data is the amount of waste managed outside public and private sectors across India, namely the vast amounts of waste which is being handled through the informal sector (Gill, 2010; Doron & Jeffrey, 2018). Recognition and understanding of the role of the informal sector in waste management and broader strategies to reduce pollution form persistent barriers to quantifying key sources and enabling mechanisms toward increasingly sustainable plastic waste management. While some estimates exist (i.e., Chikarmane, 2014; Chandran et al., 2018), there is limited representable data and knowledge of informal recycling structures' role and significance for plastic waste management. Naturally, the waste generation data captured by SPCB does not capture plastic waste collected and recycled through informal recycling channels. For example, the waste handled by recycling units, which have yet to register with the CPCB, would not be captured.

While analyzing existing data, inconsistencies across sources with regard to the interpretation of plastic waste collection and recycling data have been noted. In some instances, official estimates of waste generation per capita are based on unreliable data which does not reflect the rapid growth in plastic consumption and waste generation. Large gray areas need to be accounted for when interpreting numerical data on plastic waste generation, recycling, and management across India. A standardized methodology for measuring plastic waste generation, fate, and leakages could contribute to identifying strategies to increase recycling rates and reduce unmanaged plastic waste and pollution. Relatedly, it is key to acknowledge the role of the informal sector in reducing unmanaged waste and pollution. In this regard, just and sustainable ways of integrating informal sector workers and businesses in formal waste management systems must be identified.

2 Lack of Standardization for Sampling and Monitoring Pollution

Limited knowledge exists on macro- and micro-plastic pollution in India. Standardized methods to record and register stranded macro-plastics exist and are mainly conducted through spectroscopic methods. In contrast to larger, visible types of plastic waste which can be subject to structural identification, sinking and invisible micro-plastics have received limited monitoring. While some studies have sampled macro- and micro-plastics from various beaches and lakes across India (Sarkar et al., 2019; Baroth et al., 2019; Vidyasakar et al., 2018), existing data has largely been collected using different sampling methodologies across different geographical contexts, making it difficult to scale up and compare results.

Research attention has been particularly low in Gujarat state despite its industrial and political importance and its geographical connection to rivers and marine ecosystems. It is key to develop knowledge, sampling, and monitoring capacities in under-researched areas that may be significant sources of waste generation and pollution. Such research would enable a better understanding of the sources and fate of associated pollutants and provide a baseline for developing corresponding and contextually appropriate mitigative measures. In addition to forming an important baseline for continued monitoring of plastic waste, baseline data can drive political momentum and public awareness to reduce plastic pollution.

A shift in the research focus of the scientific community has been seen from macro-plastic to micro-plastic as the emerging contaminants of concern. However, there is a need to study both micro- and macro-levels of contamination simultaneously within a holistic approach to identify and quantify plastic inputs from land to river-systems and marine environments. There are few previous accounts of studies on macro- and micro-plastic contamination in India's lakes, rivers, and marine environments. The need for a coordinated and comprehensive study in the Indian domain that deciphers the highly complex and dynamic riverine system and develops monitoring tools for plastic contamination is imminent. There is also limited research attention to the interlinkages between plastic and persistent organic pollutants (POPs). Macro- and micro-plastic pollution and POPs must be studied in tandem to identify their interlinkages and potential solutions for their management.

3 Policy and Regulatory Gaps

Plastic pollution has largely been framed as a littering issue in the Indian policy context, with policy interventions promoting cleanliness and consumer-oriented measures targeting fast moving consumer goods rather than tackling pollution across the life cycle of plastics (Pathak & Nicher, 2021). Current initiatives and policy priorities branded within the circular economy frame primarily focus on selected material and product categories (e.g., bans on certain single-use plastic products (SUPPs)), rather than setting targets for reducing pollution across the value chain of plastics (including extraction, production, manufacturing, consumption, and end-of-life management). Although India's target to combat plastic pollution has been clear since the Government of India's pledge to eliminate all SUPPs in the country by 2022, the implementation of related regulatory measures continues to face challenges.

The Indian plastic policy environment has rapidly evolved over the last couple of years, with several amendments to the Plastic Waste Management (PWM) Rules (2016) in 2021 and 2022. With the aim to strengthen the circular economy for plastic packaging and waste, key policy developments include bans on SUPPs and the framework for extended producer responsibility (EPR), as mandated by the PWM Rules (GOI, 2021).

The most recent PWM Amendment Rules bans certain categories of SUPPs and incentivize the use of compostable and biodegradable alternatives. Evidence indicates that challenges remain with regard to implementing the SUPP bans in practice (MoEFCC, 2022). The potential for SUPP bans to reduce plastic pollution relies on effective implementation, enforcement, and monitoring, as well as the availability of viable alternatives (Xanthos & Walker, 2017). While banning problematic plastics is generally considered a key step toward reducing unmanaged waste and pollution, SUPP bans can have unforeseen socio-economic and environmental impacts, particularly if these do not account for local realities (Nøklebye et al., 2023). Uncertainties also persist regarding the environmental impacts of biodegradable and compostable plastics (Bolton, 2019). The environmental fates of these alternative materials are uncertain and vary across different environments (EC, 2020). There is a need to consider the potential rate of biodegradation and composability across Indian soil contexts and evaluate whether certain alternative materials may have a higher environmental footprint than SUPPs (Spierling et al., 2018). SUPP bans have also faced backlash from industry players on account of economic concerns (INOPOL, 2022; Nøklebye et al., 2023). Recognizing and counterbalancing these impacts is important to increase public acceptance and effective implementation of bans and other reduction policies.

Industries have highlighted challenges with regard to complying with a rapidly evolving policy environment and the need for adequate infrastructure, technology, and investment for businesses to make the shift toward increasingly circular solutions, particularly considering the limited resources among MSMEs (INOPOL, 2022, p. 58). Some have expressed concerns with regard to the lack of viable alternatives to SUPP and flexible and multi-layer packaging, as well as the costs and additional waste streams caused by biodegradable and compostable SUPP alternatives.

4 Waste Management Capacities, Infrastructure, and Technology Gaps

Existing structures for waste management in India face considerable challenges in managing waste in an environmentally and socially sound manner, a concern underscored by the substantial volumes of plastic waste that remain uncollected and inadequately managed. While reduce, reuse, and recycle are the foundations of the waste hierarchy and 3R initiatives, a significant proportion of waste is disposed of in insanitary landfills or co-processed for energy recovery. Insanitary landfills lacking pollution controls are at high risk of releasing toxic contaminants into surrounding environments, posing direct and indirect health and ecosystem impacts, particularly during heavy rainfalls and uncontrolled landfill fires. While waste-to-energy (WTE) recovery for non-recyclable waste has been promoted as a path toward achieving circular economy objectives, its sustainability implications are increasingly recognized and documented (Zero Waste Europe, 2019; Moon & Morris, 2019).

Plastic waste management challenges can often be linked to limited waste segregation and collection infrastructures, as well as limited awareness among waste generators. Many municipalities lack uniform systems for waste segregation at source, while municipal household collection services have limited reach and are often subsidized by the services of informal workers. Relatedly, the lack of access to technology and infrastructure supporting collection, segregation, processing, and recycling is a barrier to managing waste in an environmentally safe, non-hazardous, cost-efficient, and resource-efficient manner.

Co-incineration of waste in WTE facilities and other industrial facilities, such as paper mills and cement kilns, involves burning collected wastes with coal or biomass to generate electricity. Evidence indicates that plastic waste incineration can be a significant source of fossil fuel emissions, depending on the type of feedstock burned (US EPA, 2018; Moon & Morriss, 2019). The WTE process relies on a steady supply of clean waste of high calorific value. Given the present challenges of waste segregation at source, contaminated waste is often incinerated together with recyclable or compostable waste, which is a less viable alternative considering carbon emissions (Zero Waste Europe, 2019).

While WTE solutions may offer a viable alternative for managing non-recyclable waste of low calorific value, these alternatives are not part of closing the loop within a circular economy perspective. It should also be recognized that there is limited data and evidence available on the wider implications of co-processing of waste, particularly in the Indian context. This research gap will require closer examination before promoting co-processing as a way forward for reducing non-recyclable wastes.

Other sustainability concerns relate to WTE technologies' reliance on a constant flow of feedstock (i.e., recyclables and post-consumer waste), which can maintain or even increase the demand for steady waste streams. This raises the risk of locking societies into a linear economy that justifies a throw-away culture and undermines efforts toward waste reduction (GAIA, 2018). Incinerators may also threaten existing recovery systems, including the informal sector, whose livelihoods may partly depend on the recoverable materials that are used as fuel in incinerators. These associated challenges of WTE should be recognized, while focusing on livelihood empowerment and decent job creation in a just transition toward low emission and non-polluting societies, in India and beyond.

As zero-waste initiatives are gaining ground around the world, holistic and sustainable solutions to the ways we design, produce, dispose, and manage plastic require a system change across all levels of the plastic lifecycle, involving producers, industries, businesses, consumers, civil society actors, and government agencies.

There is a need for investment to support the establishment of organized facilities for environmentally sound plastic waste management and for empowering and building capacities within existing infrastructures for waste collection and handling, including the informal sector. More broadly, capacity building, monitoring tools, training programs, and awareness raising are key elements to bridge infrastructure and capacity gaps.

5 Socio-Economic Implications and Livelihood Insecurities Exacerbated by COVID-19

Socio-economic impacts of plastic pollution and commonly pursued pollution reduction strategies must be better understood to ensure a just and sustainable transition toward tackling the interlinked challenges of pollution, climate change, and biodiversity loss. Plastic waste management is intrinsically linked to broader sustainability goals of eradicating poverty, reducing inequalities, and ensuring societal health and well-being (Kaza et al., 2018). Beyond environmental impacts, unmanaged waste and pollution cause transboundary and irreversible impacts, often impacting people and communities in poor and marginalized situations (Owens & Conlon, 2021; Orellana, 2021). For example, the people working in direct contact with waste and living in the near vicinity of waste and pollution hotspots are particularly vulnerable to health impacts. These impacts must be better understood and accounted for in the process of achieving increasingly safe and environmentally sound management of waste.

Informal waste workers are vulnerable to livelihood insecurities and socio-economic marginalization (ESCAP, 2018; UNEP et al., 2019). Mitigating existing injustices within the waste management system and protecting the livelihoods of those most negatively impacted by the inequitably distributed impacts of plastic pollution are key elements of ensuring the safe and environmentally sound management of plastic waste. Identifying just, inclusive, and participatory models for bridging the gap between the informal and formal recycling sectors could contribute to reducing plastic pollution at the end-of-life stage of plastic products.

Waste reduction strategies may also have unintended socio-economic implications, particularly if these lack sensitivity to local realities, including micro-, small-, and medium-sized enterprises (MSMEs), marginalized communities, and the informal sector (Nøklebye et al., 2023). Engaging with perspectives from industry stakeholders in India, it was noted that the integration of the informal sector can provide social security benefits for formalized workers, contribute to sourcing high-quality recyclable waste, and ensure compliance with ethical standards in international supply chains (INOPOL, 2022, p. 57). A key issue in this regard is to ensure that informal stakeholders are actively participating in the development and implementation of integration initiatives (UN-Habitat & NIVA, 2022).

Understanding the interlinkages between plastics and POPs is of high importance, considering that toxic emissions from plastic production processes and disposal mechanisms may pose significant threats to ecosystems and human health. Limited research attention has been granted to understanding these interlinkages and their broader implications on different environments, human health, and livelihoods. There is an urgent need for understanding and mitigating immediate and long-term livelihood risks among people working in direct contact with waste and those living in areas exposed to potentially hazardous emissions from inadequately managed plastic waste.

The COVID-19 pandemic severely impacted every aspect of the economy, environment, and society in India. The waste management sector has been particularly affected, as cities experienced unprecedented growth in biomedical waste generation and a rise in single-use plastic packaging within e-commerce industries. At the same time, prolonged nationwide lockdowns have had disruptive effects on municipal waste management systems, particularly on its informal components which were largely prevented from operating in certain times and areas of the nationwide lockdowns (INOPOL, 2020). As a result, larger quantities of unsegregated waste were landfilled and incinerated, while livelihoods dependent on waste collection and recovery were severely affected (INOPOL, 2020). Structural challenges that affect the livelihoods of the informal waste sector workers have been particularly highlighted during the COVID-19 pandemic. These results suggest the need for enhanced engagement with the informal sector, assessing their exposure to economic and environmental risks, social recognition, and visibility.

6 Chapter Summary

This chapter has provided an analysis of persistent gaps and challenges in plastic waste management in India, drawing from baseline assessments conducted under the India–Norway cooperation project to reduce plastic and chemical pollution. It has examined these challenges across multiple dimensions, including limitations in data on waste generation and pollution, lack of standardized sampling and monitoring methods, regulatory and policy gaps, and inadequacies in infrastructure, technology and institutional capacity. Beyond technical and regulatory issues, the chapter highlights broader socio-economic dynamics, including the crucial but often overlooked role of the informal sector, and the disruptive impacts of the COVID-19 pandemic on waste systems and marginalized livelihoods. The interlinkages between plastic pollution and persistent organic pollutants (POPs), health risks, and environmental justice are also emphasized, underscoring the need for a more holistic and equity-oriented approach to waste governance.

Addressing these interrelated challenges requires a systemic shift that engages stakeholders across the plastics lifecycle, from production to end-of-life. This includes investing in organized, environmentally sound waste management systems, strengthening monitoring and data collection, supporting technological innovation, and ensuring inclusive participation in policy development and implementation. Crucially, advancing socially just and contextually appropriate pathways for integrating informal actors and ensuring decent work is fundamental to enabling a just transition towards circular, low-emission, and non-polluting societies in India and similar contexts.

Acknowledgments This research was supported by the Royal Norwegian Embassy in New Delhi through the INOPOL Project (Project No: IND-22/0007).

References

Baroth, A., Pant, A., Sah, R., Hussain, S. A., & Chaudhary, P. (2019). Spatiotemporal trends of micro-plastics loading in the sediments of river Ganga: First observation on occurrence, identification and quantification. In *Conference: SETAC Europe 29th Annual Meeting, Helsinki, Finland*. https://tinyurl.com/nhdseaxm

Bolton, D. (2019). Tea bag manufacturers respond to microplastic particle study. *World Tea News*. Retrieved from https://www.worldteanews.com/Insights/tea-bag-manufacturers-respond-microplastic-particle-study

Chandran, P., Arora, K., Abubaker, M., & Shekar, N. (2018). *Valuing urban waste: The need for comprehensive material recovery and recycling policy*. Hasiru Dala. https://hasirudala.in/wp-content/uploads/2020/09/Valuing-Urban-Waste-2019.pdf

Chikarmane, P. (2014). *City report—Informal economy monitoring study: Waste pickers in Pune, India*. Women in Informal Employment Globalising and Organising. Retrieved from https://www.wiego.org/publications/city-report-informal-economy-monitoring-study-waste-pickers-pune-india

CPCB. (2019). *Annual report for the year 2018–2019 on implementation of plastic waste management rules*. Central Pollution Control Board. Ministry of Environment, Forest and Climate Change. Retrieved January 10, 2021, from https://cpcb.nic.in/uploads/plasticwaste/annual_report_2018-19_pwm.pdf

Doron, A., & Jeffrey, R. (2018). *Waste of a nation: Garbage and growth in India*. Harvard University Press.

EC. (2020). *Scientific Opinion No. 10: Biodegradability of plastics in the open environment*. Group of Chief Scientific Advisors, Directorate-General for Research and Innovation. European Commission. Retrieved from https://op.europa.eu/en/web/eu-law-and-publications/publication-detail/-/publication/0c0d6267–433a-11eb-b27b-01aa75ed71a1

ESCAP. (2018). *Closing the loop: Innovative partnerships with informal workers to recover plastic waste, in an inclusive circular economy approach*. Economic and Social Commission for Asia and the Pacific. https://swachcoop.com/pdf/UNESCAP-Closing-The-Loop-Regional-Policy-Guide-Case-Study.pdf

GAIA. (2018). Facts about "waste-to-energ" incinerators. *Global Alliance of Incineration Alternatives*. Retrieved from https://www.no-burn.org/wp-content/uploads/2021/11/GAIA-Facts-about-WTE-incinerators-Jan2018-1-1.pdf

Gill, K. (2010). *Of poverty and plastic: Scavenging and scrap trading entrepreneurs in India's urban informal economy*. Oxford University Press.

GOI. (2021). *Plastic waste management rules, 2021*. Notification. Retrieved August 12, 2021, from https://cpcb.nic.in/uploads/plasticwaste/Notification-12–08-2021.pdf

INOPOL. (2020). *Pilot study for plastic and biomedical waste management during the COVID 19 pandemic in Surat and Delhi*. Norwegian Institute for Water Research (NIVA), Toxics Link & Mu Gamma Consultants (MGC). https://tinyurl.com/2yafkcsj

INOPOL. (2022). *Summary of plastic waste management strategy for Gujarat*. India-Norway Cooperation Project on Capacity Building for Reducing Plastic and Chemical Pollution in India. https://tinyurl.com/3uks8pku

Kaza, S., Yao, L.C., Bhada-Tata, P., Van Woerden, F., (2018) What a waste 2.0: A global snapshot of solid waste management to 2050. *Urban development*. World Bank. Retrieved from https://openknowledge.worldbank.org/handle/10986/30317

MoEFCC. (2022). *Ban on single use plastics*. 12th Ministry of Environment, Forest, and Climate Change. Retrieved December 12, 2022, from https://pib.gov.in/PressReleasePage.aspx?PRID=1882855#:~:text=Awareness%20generation%20and%20capacity%20building,small%20shops%20in%20local%20markets

Moon, D., & Morris, J. (2019). Plastic waste management. In *Plastic & Climate–The hidden costs of a plastic planet* (pp. 55–68) Retrieved from https://www.ciel.org/wp-content/uploads/2019/05/Plastic-and-Climate-FINAL-2019.pdf

Nøklebye, E., Adam, H. N., Roy-Basu, A., Bharat, G. K., & Steindal, E. H. (2023). Plastic bans in India–addressing the socio-economic and environmental complexities. *Environmental Science and Policy, 139*, 2019–2222. https://doi.org/10.1016/j.envsci.2022.11.005

Orellana, M. (2021). *Report of the special rapporteur on the implications for human rights of the environmentally sound management and disposal of hazardous substances and wastes*. A/76/207. United National General Assembly. Retrieved from https://www.ohchr.org/en/documents/thematic-reports/a76207-implications-human-rights-environmentally-sound-management-and

Owens, K. A., & Conlon, K. (2021). Mopping up or turning off the tap? Environmental injustice and the ethics of plastic pollution. *Frontiers in Marine Science, 8*, 713385. https://doi.org/10.3389/fmars.2021.713385

Pathak, G., & Nichter, M. (2021). Ecocommunicability, citizenship, and discourses on plastic control in India. *Geoforum, 125*, 132–139. https://doi.org/10.1016/j.geoforum.2021.04.027

Sarkar, D. J., Sarkar, S. D., Das, B. K., Manna, R. K., Behera, B. K., & Samanta, S. (2019). Spatial distribution of meso and micro-plastics in the sediments of river ganga at eastern India. *Science of the Total Environment, 694*, 133712. https://doi.org/10.1016/j.scitotenv.2019.133712

Shah, S. (2018). *How Gujarat succeeded in co- processing plastic waste in cement kilns and promoting circular economy; down to earth*. Retrieved September 3, 2018, from, https://www.downtoearth.org.in/blog/environment/how-Gujarat-succeeded-in-co-processing-plastic-waste-in-cement-kilns-and-promoting-circular-economy-61518

Spierling, S., Knüpffer, E., Behnsen, H., Mudersbach, M., Krieg, H., Springer, S., Albrecht, S., Herrmann, C., & Endres, H. (2018). Bio-based plastics—A review of environmental, social and economic impact assessments. *Jurnal of Cleaner Production, 185*, 476–491. https://doi.org/10.1016/j.jclepro.2018.03.014

Surat District Profile. (2011). Retrieved March 1, 2021, from https://www.slideshare.net/ourvibrantgujarat/surat-district-profile

UNEP, COBSEA, SEI. (2019). *Marine plastic litter in East Asian Seas: Gender, human rights and economic dimensions*. Retrieved from https://www.unenvironment.org/cobsea/resources/report/marine-plastic-litter-east-asian-seas-gender-human-rights-and-economic-dimensions

UN-Habitat & NIVA. (2022). *Leaving no one behind—How a global agreement to end plastic pollution can enable just transition of the people informally collecting and recovering waste*. UN-Habitat and Norwegian Institute for Water Research (NIVA). Retrieved from https://unhabitat.org/sites/default/files/2022/11/un-habitat_niva_report_leaving_no_one_behind_1.pdf

US EPA. (2018). *Inventory of U.S. Greenhouse Gas Emissions and Sinks: 1990–2016*. United States Environmental Protection Agency. Retrieved from https://www.epa.gov/sites/default/files/2018-01/documents/2018_complete_report.pdf

Vidyasakar, A., Neelavannan, K., Krishnakumar, S., Prabaharan, G., Sathiyabama Alias Priyanka, T., Magesh, N. S., Godson, P. S., & Srinivasalu, S. (2018). Macrodebris and microplastic distribution in the beaches of Rameswaram Coral Island, gulf of Mannar, southeast coast of India: A first report. *Marine Pollution Bulletin, 137*, 610–616.

Xanthos, D., & Walker, T. (2017). International policies to reduce plastic marine pollution from single-use plastics (plastic bags and microbeads): A review. *Marine Pollution Bulletin, 118*, 1–2. https://doi.org/10.1016/j.marpolbul.2017.02.048

Zero Waste Europe. (2019). The impact of Waste-to-Energy incineration on climate. *Policy Briefing*. Retrieved from https://zerowasteeurope.eu/wp-content/uploads/edd/2019/09/ZWE_Policy-briefing_The-impact-of-Waste-to-Energy-incineration-on-Climate.pdf

Future of Bioplastics and Comparison with Conventional Plastics

Smita Mohanty, Girija Prasad, and Manmath Parida

Abbreviations

Bio PE	Bio polyethylene
Bio PET	Bio polyethylene terephthalate
Bio PP	Bio polypropylene
PBAT	Polybutylene adipate terephthalate
PBSA	Polybutylene succinate
PCL	Polycaprolactone
PHA	Polyhydroxyalkanoates
PLA	Polylactic acid
PVC	Poly vinyl chloride
PVOH	Polyvinyl alcohol

1 Introduction

The increasing global concern over the environmental impact of traditional plastics has led to a growing interest in bioplastics. Bioplastics are derived from renewable resources, such as plants and microorganisms, and are biodegradable or compostable, making them a promising alternative to traditional plastics. Bioplastics have a range of applications in various industries, including food packaging, agriculture, and consumer goods. However, their use is not without challenges. This chapter will

S. Mohanty (✉) · G. Prasad · M. Parida
Laboratory for Advanced Research in Polymeric Materials (LARPM), School for Advanced Research in Polymers (SARP), Central Institute of Petrochemicals Engineering Technology (CIPET), Bhubaneswar, Odisha, India

P. Chakraborty et al. (eds.), *Efficient and Enhanced Management of Plastic Waste in India*, https://doi.org/10.1007/978-3-031-89398-8_10

provide a comprehensive overview of bioplastics, including their types, properties, applications, and challenges (Folino et al., 2023).

1.1 Types of Bioplastics

Bioplastics can be divided into two main categories: bio-based and biodegradable. Bio-based plastics are derived from renewable resources, such as corn starch, sugarcane, and potato starch, while biodegradable plastics are those that can be decomposed by microorganisms in the environment. Bio-based plastics can be further classified into three subcategories: bio PE, bio PET, and bio PA. Bio PE is produced from sugarcane ethanol, while bio PET is produced from renewable sources such as corn and sugarcane. Bio PA is produced from renewable sources such as castor oil (Ahsan et al., 2023). Figure 1 presents the examples of bio-based, biodegradable, non-biodegradable, and fossil fuel-based plastics.

Bioplastics have several properties that make them a promising alternative to traditional plastics. They have a lower carbon footprint than traditional plastics, as they are derived from renewable resources, and they have a higher tensile strength

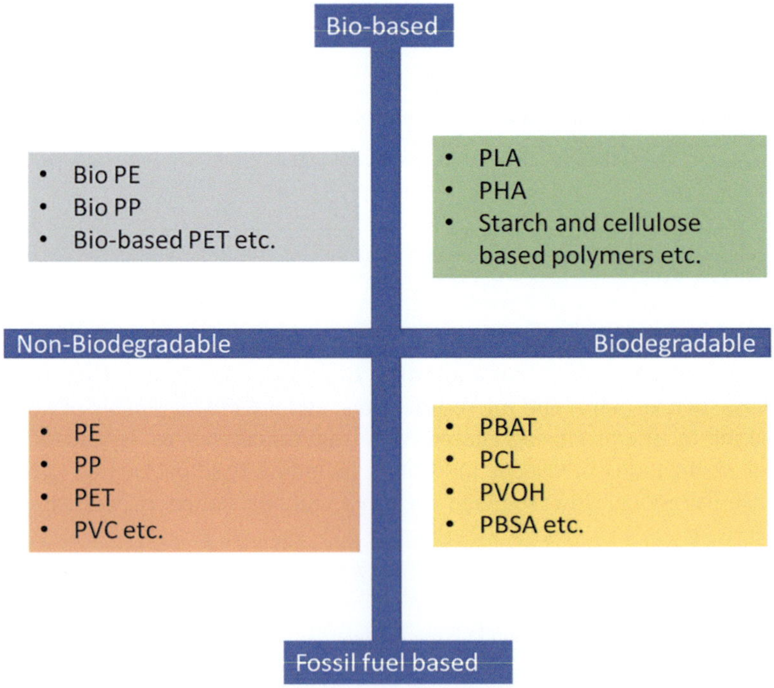

Fig. 1 Bioplastics: Source of materials and biodegradability

than traditional plastics. Bioplastics are also biodegradable or compostable, making them a more sustainable option.

Bioplastics have a range of applications. In the food packaging industry, bioplastics are used to create trays, clamshells, cups, and lids. These products can be composted or recycled, reducing the amount of plastic waste that ends up in landfills. Biodegradable mulch films made from bioplastics are being used as an alternative to traditional polyethylene films in agriculture. Bioplastics are also being used in the production of various consumer goods, such as toys, disposable cutlery, and electronic casings. The automotive industry is also using bioplastics in the production of car interiors, dashboards, and exterior parts.

Bioplastics have the potential to be a more sustainable alternative to traditional plastics, as they are derived from renewable resources and are biodegradable or compostable. However, their use is not without challenges, including cost and biodegradability. Recent research has focused on improving the properties of bioplastics and developing them from waste materials, indicating the potential for continued innovation in this field. Overall, bioplastics have the potential to be a key component in reducing the environmental impact of plastic waste (Gong et al., 2023).

1.2 Challenges Associated with Bioplastics

Despite their advantages, bioplastics face several challenges. One of the main challenges is the cost of production. Bioplastics are often more expensive to produce than traditional plastics, making them less economically viable for some industries. The cost of raw materials, as well as the cost of manufacturing and processing, contributes to the higher cost of bioplastics. Another challenge associated with bioplastics is their biodegradability. Biodegradable plastics require specific conditions to break down, such as high temperatures and humidity, and if not disposed of properly, they can still contribute to plastic pollution. There is also a lack of standardization in the production and labeling of bioplastics, which can lead to confusion among consumers and industries (Jayakumar et al., 2023).

1.3 Recent R&D Efforts

To provide an overview of recent advancements in the field of bioplastics, Table 1 summarizes a selection of successful research findings across different application areas, including packaging, agriculture, disposable products, environmental sustainability, composites, and biomedical uses.

Table 1 List of successful findings in the field of bioplastics with regard to different applications

S. no	Application area	Research titles	Major findings	DOI
1	Packaging	Combination of poly(lactic) acid and starch for biodegradable food packaging	• Potato starch blended with PLA can produce a biodegradable film suitable for food packaging applications • The blend exhibits improved tensile strength and barrier properties	https://doi.org/10.3390/ma10080952
2		Preparation and characterization of biodegradable cassava starch thin films for potential food packaging applications	• Cassava starch-based biodegradable packaging film has good mechanical and barrier properties, as well as biodegradability • Cassava starch-based films show promise as an alternative to traditional petroleum-based plastic films	https://doi.org/10.1007/s10570-021-04199-6
3		Preparation and characterization of edible films based on soy protein isolate-fatty acid blends	• Incorporating glycerol and oleic acid into soy protein isolate-based biodegradable films can improve their mechanical properties and water resistance	https://doi.org/10.1080/03602550801949728
4		Chitosan nanofiber and cellulose nanofiber blended composite applicable for active food packaging	• Biodegradable and antioxidant packaging films based on cellulose nanofibers and chitosan can prevent the oxidation of food products • These films exhibit good mechanical and barrier properties	https://doi.org/10.3390/nano10091752
5		Biodegradable and sustainable packaging materials: a review	• Biodegradable and sustainable packaging materials can reduce environmental impacts associated with traditional petroleum-based plastics	https://doi.org/10.1007/s41742-020-00293-5
6		Bio-based films with improved water resistance derived from soy protein isolate and stearic acid via bioconjugation	• Polyethylene glycol and stearic acid can improve the mechanical and water barrier properties of soy protein isolate-based edible films	https://doi.org/10.1016/j.jclepro.2018.12.277

7	Agriculture	Effect of empty fruit brunch reinforcement in polybutylene-succinate/modified tapioca starch blend for agricultural mulch films	• Tapioca starch-based biodegradable mulch films have potential as sustainable alternatives to conventional plastic mulch films • Films showed good mechanical and thermal properties • Mulch films degraded within 90 days in soil. • Soil nutrient contents were not affected by the mulch films.	https://doi.org/10.1038/s41593-020-58278-y
8		Preparation and characterization of biodegradable cassava starch thin films for potential food packaging applications	• Cassava starch and polyvinyl alcohol (PVA)-based biodegradable mulch films can be a good alternative to petroleum-based plastic mulch films • Films showed good mechanical properties, such as tensile strength and elongation at break • Mulch films degraded within 56 days in soil • Soil nutrient contents were not affected by the mulch films	https://doi.org/10.1007/s10570-021-04199-6
9		Biodegradable films based on rice starch and rice flour	• Rice flour-based biodegradable plastic films have potential for agricultural applications • Films showed good mechanical properties, such as tensile strength and elongation at break • Mulch films degraded within 50 days in soil • Soil nutrient contents were not affected by the mulch films	https://doi.org/10.1016/j.jcs.2009.11.014

(continued)

Table 1 (continued)

S. no	Application area	Research titles	Major findings	DOI
10	Disposable products	Preparation and properties of biodegradable films made of cationic potato-peel starch and loaded with curcumin	• The addition of urea and urea-hydrogen peroxide to potato starch can improve the mechanical and thermal properties of the resulting biodegradable films • The tensile strength, Young's modulus, and thermal stability of the films increased with increasing concentrations of urea and urea-hydrogen peroxide • The addition of urea and urea-hydrogen peroxide can also reduce the water solubility and water vapor permeability of the films, which can improve their durability and shelf-life • The morphology and structure of the films were analyzed by scanning electron microscopy and X-ray diffraction, which showed that the addition of urea and urea-hydrogen peroxide can modify the crystalline structure of potato starch and improve the intermolecular interactions between starch molecules	https://doi.org/10.1016/j.foodhyd.2022.107690
11		Environmental safety of second and third generation bioplastics in the context of the circular economy	• Bioplastics can play a significant role in the transition toward a circular economy by reducing the reliance on fossil-based materials, improving resource efficiency and reducing waste generation and environmental pollution. • The circular economy concept provides a framework for designing and managing bioplastic products and packaging throughout their life cycle, from raw material sourcing to end-of-life disposal or recycling • Bioplastics can be produced from renewable resources, such as starch, cellulose, and lignin, and can be biodegradable or compostable under certain conditions, which can reduce their environmental impact and improve their end-of-life options	https://doi.org/10.1016/j.ecoenv.2023.114835
12		Preparation and characterization of polyvinyl alcohol–starch composites reinforced with pulp	• Bulk density of the composite material decreased when it was reinforced by the addition of pulp in PVA–starch matrix • Water absorption of PVA–starch composite was increased with the increasing addition of pulp/fiber • UTS of the composite decreased with the insertion of starch and after adding pulp it started to increase	https://doi.org/10.1007/s42452-019-1111-2

13	Bio-medical	3D printed poly(lactic acid) scaffolds modified with chitosan and hydroxyapatite for bone repair applications	• PLA/chitosan scaffolds have excellent biocompatibility and biodegradability, making them a promising material for bone tissue engineering • Chitosan and hydroxyapatite were used to functionalize PLA surface • Wetting behavior of PLA surface was improved	https://doi.org/10.1016/j.mtcomm.2020.101515
14		Biodegradable polymeric microspheres as drug carriers for antimicrobial agent	Carrier technology offers a clever method of drug administration by combining the medication with carrier particles, such as microspheres, nanoparticles, liposomes, and niosomes, that control the release and absorption characteristics of the drug	https://doi.org/https://doi.org/10.2174/1574885518666230530095329
15		Chitosan as a wound dressing starting material: Antimicrobial properties and mode of action	• A chitosan-based biopolymer can be used as a potential wound dressing material • The biopolymer exhibited good biocompatibility and enhanced wound healing in animal models	https://doi.org/https://doi.org/10.3390/ijms20235889
		Polyurethane nanocomposite impregnated with chitosan-modified graphene oxide as a potential antibacterial wound dressing	• Polyurethane/chitosan composite films exhibited good antibacterial activity against various bacteria and fungi • The composite films exhibited good biocompatibility and enhanced wound healing in animal models	https://doi.org/10.1016/j.msec.2020.110899
16		Cellulose-based hydrogels as sustained drug-delivery systems	• Cellulose-based hydrogels exhibited good biocompatibility and biodegradability • The hydrogels exhibited sustained drug release over a period, making them potential drug delivery systems	https://doi.org/https://doi.org/10.3390/ma13225270
17		Bioinspired mineral-polymeric hybrid hyaluronic acid/poly (γ-glutamic acid) hydrogels as tunable scaffolds for stem cells differentiation	• A bioinspired hybrid hydrogel composed of chitosan and silk fibroin can be used as a potential scaffold for 3D cell culture and tissue engineering • The hydrogel exhibited good biocompatibility and enhanced cell proliferation and differentiation	https://doi.org/10.1016/j.carbpol.2021.118048
18		Stem-cell and tissue-regeneration analysis in low-dose irradiated planarians treated with cerium oxide nanoparticles	• A bio-based hydrogel composed of gelatin and alginate can be used as a potential scaffold for tissue engineering • The hydrogel exhibited good biocompatibility and enhanced cell proliferation and differentiation	https://doi.org/10.1016/j.msec.2020.111113
19		Evaluation of the in vitro and in vivo properties of a novel biodegradable polymeric scaffold for bone tissue engineering	• A biodegradable polymeric scaffold composed of poly(lactic-co-glycolic acid) (PLGA) and hydroxyapatite (HA) can be used as a potential scaffold for bone tissue engineering • The scaffold exhibited good biocompatibility and enhanced bone regeneration in animal models	https://doi.org/10.1007/978-3-319-09230-0_39-1

2 Petro-Based Conventional Plastics Vis-à-Vis Bioplastics

Bioplastics are gaining attention as a potential alternative to petro-based conventional plastics due to their eco-friendliness and biodegradability. These materials differ from traditional plastics in several ways. First, bioplastics are derived from renewable sources such as plants, bacteria, and fungi, whereas petro-based plastics are made from non-renewable petroleum resources. Second, the production of bioplastics generates fewer greenhouse gas emissions compared to the production of petro-based plastics. Additionally, bioplastics are biodegradable, which means that they can be broken down into natural substances by microorganisms, while petro-based plastics persist in the environment for hundreds of years, contributing to pollution and harming wildlife (Shafqat et al., 2020).

Another significant difference between bioplastics and petro-based plastics is their physical properties. Bioplastics have different mechanical and thermal properties compared to petro-based plastics. For instance, PLA has lower impact strength and thermal stability than conventional plastics, while PHB has lower tensile strength and elongation at break. However, researchers are constantly working to improve the properties of bioplastics through modifications to their chemical structure and processing techniques (Ncube et al., 2020).

Despite the advantages of bioplastics, there are still some challenges to their widespread adoption. One of the main issues is the cost of production, which is higher than that of conventional plastics due to the limited availability of raw materials and the higher energy consumption required for their processing. However, as the demand for sustainable materials increases and technology advances, the cost of bioplastics is expected to decrease (Abe et al., 2021).

The application sectors of bioplastics are diverse and expanding. PLA is commonly used in food packaging, disposable cutlery, and textiles, while PHB is used in the production of medical implants, biodegradable bags, and food packaging. PBS is used in the production of compostable bags, packaging films, and coatings, and PBAT is used in the production of flexible packaging, compostable bags, and disposable gloves (Vega-Baudrit, 2019; Dammak et al., 2020; Yashavanth et al., 2021; Barrino et al., 2023).

Furthermore, bioplastics have the potential to be used in a variety of other applications. For example, researchers are exploring the use of bioplastics in automotive parts, construction materials, and electronic components. Bioplastics can also be blended with traditional plastics to improve their properties and reduce their environmental impact.

From India's point of view, the demand for bioplastics is on the rise due to the country's focus on sustainability and reducing plastic waste. The Indian government has also taken steps to promote the use of bioplastics by setting targets for their use in packaging and reducing the consumption of single-use plastics. Indian companies are also investing in research and development to create innovative bioplastic products and technologies (Bhattacharya et al., 2018).

3 Key Market Drivers and Statistics

The market for bioplastics has been growing steadily in recent years, driven by a combination of factors such as increased consumer awareness and demand for eco-friendly products, government regulations promoting the use of biodegradable and compostable materials, and advancements in technology leading to the development of new and improved bioplastics.

According to a report by Grand View Research, the global bioplastics market size was valued at USD 10.5 billion in 2020 and is expected to grow at a compound annual growth rate (CAGR) of 14.0% from 2021 to 2028 (Research GV, 2021). The report also predicts that the packaging segment will remain the largest application segment for bioplastics, with food and beverage packaging accounting for most of the demand.

The European region has been leading the way in terms of bioplastic production and consumption, with several countries implementing regulations and policies promoting the use of bioplastics. For example, the European Union's Single-Use Plastics Directive, which came into effect in July 2021, bans certain single-use plastic products and encourages the use of alternative materials such as bioplastics (Union, 2021).

In addition to packaging, bioplastics are also finding applications in other industries such as agriculture, automotive, and electronics. For example, bioplastics can be used in agricultural films and mulch films, reducing the use of traditional petroleum-based plastics in agriculture. In the automotive industry, bioplastics can be used in interior and exterior parts, contributing to weight reduction and improved fuel efficiency. In the electronics industry, bioplastics can be used in components such as casings and connectors.

In India, the demand for bioplastics has been increasing due to a combination of factors such as growing awareness about environmental issues, government initiatives promoting the use of eco-friendly products, and a shift in consumer preferences toward sustainable materials. The Indian government has also implemented policies and regulations promoting the use of bioplastics, such as the Plastic Waste Management Rules 2016 which mandates the use of compostable plastic bags and containers for select applications (Bhattacharya et al., 2018).

According to a report by the Indian Bioplastics, the market for bioplastics in India is expected to grow at a CAGR of 30% between 2021 and 2026 (GlobalData, n.d.). The report also highlights the potential for bioplastics in various industries such as packaging, agriculture, textiles, and consumer goods. The packaging industry is expected to remain the largest application segment for bioplastics in India, driven by the growth in e-commerce and the increasing demand for sustainable packaging solutions. A statistics on global production of bioplastic in upcoming years is depicted in Fig. 2.

While the market for bioplastics is growing, there are also challenges that need to be addressed. One of the main challenges is the cost of production, which is currently higher than that of traditional petroleum-based plastics. However, ongoing research and development efforts are focused on reducing the cost of production and improving the properties of bioplastics, making them more competitive in the market (Table 2).

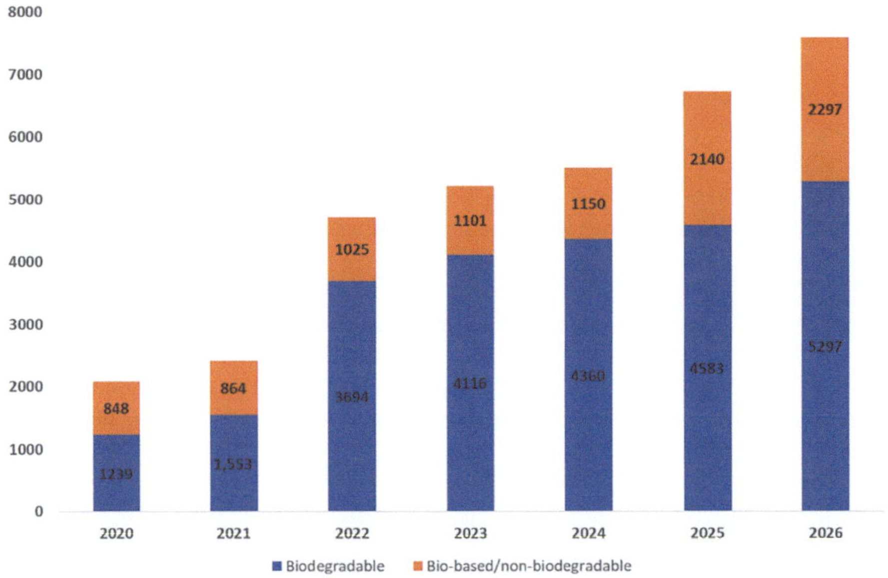

Fig. 2 Statistics on global bioplastic production in upcoming years (in 1000 metric tons)

Table 2 Key players in global bioplastic production

S. no	Company name	Standout products/technology	Proposed applications
1	Synbera Technology (Netherlands) TRL-9	Sugarcane Synterra®PLA	Colored disposable cutlery
2	Biotrem (Poland) TRL-9	Wheat bran based	Tableware and cutlery production
3	CelluComp Ltd. (Scotland) TRL-8	Beet pulp microfibrillated cellulose Curran®	Granules form to use in paints and coatings
4	Paptic Oy (Finland) TRL-9	Wood pulp Paptic	Packaging material, bags and pouches, food packaging
5	Trifilon AB (Sweden) TRL-8	Hemp fibers	Outdoor furniture–bio-based and recycled materials
6	Greengran BV (Germany) TRL-5	Plant fibers	Natural fiber reinforced polymer granules, bio-based matrix compounds (PLA and PHB) biodegradable
7	BASF (GER)	Ecoflex® (PBAT), Ecoflex blends with PLA (Ecovio®) and other materials such as starch	Shrink film, organic waste bags, fruits and vegetable bags, mulch films, molded and thermoformed products, paper coating

(continued)

Table 2 (continued)

S. no	Company name	Standout products/technology	Proposed applications
8	Bewi	PLA-based foam (BioFoam®) which is recyclable and compostable	Filling hollow spaces like beanbags and pillows and for shape molding replacing EPS (as in protective pkg)
9	Biome Technologies (UK)	Potato and Corn starch-based resins	Films for food and industrial packaging, shopper bags, waste bags; coating on paper, flexible films, molded goods, extruded sheets, food wraps, etc.
10	Mitsubishi Chemical Europe (GER) and PTT MCC Biochem (Thailand)	Bio-based PBS (BioPBS™) Petro-based PBS (GS Pla®)	Sealant layer in flexible packaging and coating on paper because of elution resistance to oils, heat sealability, and printability. It can be blended with other biopolymers
11	Biotec (GER)	PLA and Starch blends (BIOPLAST)	Films for carry bags, waste bags, and injection molded articles for food pkg
12	Braskem (Brazil)	Bio-based PE, EVA, and PE wax (I'm greenTM) from sugarcane/ ethanol	For consumer goods packaging and similar applications
13	Danimer Scientific	PHAs (NodaxTM) and their blends	Food and vegetable bags, carry bags, waste bags, cups, lids, straws, plates and diaper linings, etc.
14	DSM (Geleen, NL)	Polyamide-4.10 (EcoPaXX®), Copolyester (Arnitel® Eco)	Automotive, consumer goods, and food contact applications
15	Ecomann (China)	PHA (Ecomann®) & wood powder composites with PHA based bags and 3D printer filament	For food and vegetable storage, waste storage, and 3D printing
16	FKuR (Willich, GER)	PLA blends (BioFlex®), cellulose acetate (Biograde®), Bio-PE (Terralene® = I'm green of Braskem), fibre-filled PLA materials (Fibrolon®)	Bioflex is for household, agricultural, and hygiene films. These are also food contact compliant
17	Futerro	Lactide and PLA from vegetable resources (lactic acid sourced from galactic)	Blown and cast films for food packaging, labels, and laminated films
18	Futumura (UK) (former Innovia)	Cellulose films (Cellophane® & Natureflex®)	Fruit and vegetable bags, bio-waste bags, over wraps, coffee capsules, catering items, and other food packaging. Can replace BOPET and BOPA in barrier laminates
19	Huhtamaki Group	PLA-based food trays, cups, lids, and trays (Bioware®)	For food storage applications

4 Bioplastics: Prominence and Synthesis Processes

4.1 Polylactic Acid (PLA)

Polylactic acid (PLA) is a biodegradable polymer that is becoming increasingly popular as an alternative to traditional petroleum-based plastics. PLA is made from renewable resources such as corn starch or sugarcane and is compostable, making it an attractive option for various applications, including food packaging, disposable cutlery, and textiles (More et al., 2023).

4.1.1 Synthesis Process of PLA

The synthesis process of PLA involves three main steps: fermentation, purification, and polymerization. During the fermentation step, corn starch or sugarcane is fermented with bacteria such as *Lactobacillus* or *Streptococcus* to produce lactic acid. The fermentation yields a solution of about 90–95% lactic acid. The lactic acid solution is then purified to remove impurities and residual water, leaving behind high-purity lactic acid. The purified lactic acid is then polymerized into PLA through either the direct condensation method or the ring-opening polymerization method (More et al., 2023).

In the direct condensation method, the lactic acid is heated with a catalyst, such as tin or zinc, to form a dimer, which then polymerizes into PLA. The reaction is carried out at high temperatures (around 200 °C) and requires careful control of the reaction conditions to ensure the correct molecular weight and properties of the resulting polymer (Ajioka et al., 1995).

In the ring-opening polymerization method, lactide is first produced by heating lactic acid under vacuum, and then the lactide is polymerized into PLA using a catalyst such as tin octoate or aluminum isopropoxide. The polymerization reaction takes place at lower temperatures (around 140–160 °C) and is carried out in the presence of an initiator, such as stannous octoate (Widjaja et al., 2023).

4.1.2 Properties of PLA

PLA has several advantages over traditional petroleum-based plastics. First, it is biodegradable and compostable, meaning that it can break down into natural compounds in the environment and does not contribute to the growing problem of plastic waste. Second, it is made from renewable resources, which reduces our dependence on fossil fuels. Third, it has a lower carbon footprint than traditional plastics, as it requires less energy to produce.

However, PLA also has some limitations. For example, it has relatively low heat resistance compared to traditional plastics, making it unsuitable for some applications such as hot food packaging. It also has a slower degradation rate under certain conditions, such as in landfills where there is limited oxygen, which can lead to the release of methane gas (Ali et al., 2023b).

4.1.3 Recent Advances in PLA Research

Despite its limitations, PLA is an attractive alternative to traditional plastics, and recent research has focused on improving its properties and finding new applications for the material.

One area of research has been to improve the heat resistance of PLA. Researchers have developed various approaches to increase the heat resistance of PLA, including blending it with other polymers, adding fillers such as clay or carbon nanotubes, and using chemical modifications to the polymer structure. For example, a study by Khammassi et al. (2022) investigated the use of graphene oxide as a filler in PLA to improve its thermal properties. The researchers found that the addition of graphene oxide improved the thermal stability and mechanical properties of the PLA.

Another area of research has been to improve the degradation rate of PLA under certain conditions. Researchers have investigated various methods to accelerate the degradation of PLA, including the use of enzymes, exposure to UV radiation, and modifying the polymer structure. For example, a study by Kim et al. (2022) investigated the use of cellulase enzymes to accelerate the degradation of PLA. The researchers found that the use of cellulase enzymes increased the rate of PLA degradation by up to 30 times.

Other research has focused on developing new applications for PLA. One area of interest is in the development of PLA-based composites for use in the automotive industry. A study investigated the use of PLA-based composites to produce lightweight, biodegradable automotive parts. The researchers found that the PLA-based composites had good mechanical properties and could be used to replace traditional petroleum-based plastics in automotive applications (Ilyas et al., 2021).

Another area of research is in the development of PLA-based textiles. A study by Jabbar et al. (2020) investigated the use of PLA fibers in the production of clothing. The researchers found that PLA fibers had good mechanical properties and could be used as a sustainable alternative to traditional textiles made from synthetic fibers.

4.2 Polyhydroxyalkanoates (PHA)

Polyhydroxyalkanoates (PHAs) are a group of biodegradable and biocompatible polymers that are produced by a wide range of microorganisms as intracellular carbon and energy storage compounds. PHAs are attracting significant attention as a sustainable alternative to conventional petroleum-based plastics due to their environmental benefits, such as renewability, biodegradability, and carbon neutrality (Tan et al., 2014).

4.2.1 Synthesis Process of PHA

The synthesis process of PHAs involves the fermentation of microorganisms, such as bacteria or yeast, in the presence of a carbon source, such as sugars, organic acids, or hydrocarbons. During the fermentation process, microorganisms convert the carbon source into a precursor molecule, such as β-hydroxyacyl-CoA, which is then polymerized by the action of PHA synthase enzymes into a linear or branched chain of PHA polymer. The type and properties of PHA produced depend on the microorganism and the carbon source used in the fermentation process.

PHAs can be synthesized using a wide range of microorganisms, including bacteria such as *Cupriavidus necator*, *Ralstonia eutropha*, and *Pseudomonas aeruginosa*, as well as yeast such as *Candida* and *Rhodotorula*. The choice of microorganism depends on factors such as the desired type of PHA, the availability and cost of the carbon source, and the ease of fermentation and downstream processing (Pederson et al., 2006; Chen, 2009).

The carbon source used in PHA fermentation can also vary widely, depending on factors such as cost, availability, and sustainability. Common carbon sources include glucose, fructose, sucrose, and other sugars, as well as organic acids such as propionic acid, butyric acid, and acetic acid. PHAs can also be produced from renewable resources, such as plant oils, lignocellulosic biomass, and waste streams from food and agricultural processing (Sehgal & Gupta, 2020).

The properties of PHAs can be tailored by adjusting the type and composition of monomers in the polymer, as well as by blending or copolymerizing PHAs with other biopolymers or synthetic polymers. For example, the mechanical properties of PHAs can be improved by blending them with natural polymers such as chitosan, or with synthetic polymers such as polycaprolactone. PHAs can also be copolymerized with other monomers to produce hybrid polymers with specific properties, such as improved thermal stability or biocompatibility.

4.2.2 Properties of PHA

- Biodegradability: PHAs are biodegradable under natural conditions, such as in soil, water, or compost, which can reduce the accumulation of plastic waste in the environment.
- Mechanical properties: PHAs exhibit a range of mechanical properties, from flexible to rigid, depending on their composition and structure. PHAs can also be processed into various forms, such as films, fibers, and 3D-printed objects, which makes them suitable for a variety of applications.
- Thermal properties: PHAs have a relatively low glass transition temperature, which makes them susceptible to deformation and melting at high temperatures. However, some types of PHAs, such as poly(3-hydroxybutyrate-co-3-hydroxyvalerate), exhibit improved thermal stability and can be processed at higher temperatures.

- Biocompatibility: PHAs are biocompatible and can be used in medical applications, such as tissue engineering, drug delivery, and implants. PHAs are also non-toxic and can be used in food packaging applications.

4.3 PBS (Polybutylene Succinate)

PBS is a biodegradable and renewable bioplastic that has gained attention as a potential alternative to traditional petroleum-based plastics. It is a thermoplastic polymer that is synthesized from succinic acid and 1,4-butanediol through a condensation polymerization process.

4.3.1 Synthesis Process of PBS

The synthesis process of PBS involves three main steps: production of succinic acid and 1,4-butanediol, polymerization of the monomers to form PBS, and purification of the synthesized PBS (Barletta et al., 2022).

4.3.2 Production of Succinic Acid and 1,4-Butanediol

Succinic acid is typically produced through the fermentation of renewable feedstocks such as corn or sugarcane. The fermentation process involves the use of microorganisms such as *Actinobacillus succinogenes* or *Escherichia coli* to convert sugars into succinic acid. The produced succinic acid is then purified through a series of steps such as filtration, distillation, and crystallization (Saxena et al., 2017).

1,4-Butanediol can also be produced through the fermentation of renewable feedstocks such as corn or sugar beet. The fermentation process involves the use of microorganisms such as *Klebsiella pneumoniae* or *Candida rugosa* to convert sugars into 1,4-butanediol. The produced 1,4-butanediol is then purified through a series of steps such as distillation and crystallization (Kumar et al., 2023).

4.3.3 Polymerization of Succinic Acid and 1,4-Butanediol to Form PBS

The polymerization process of succinic acid and 1,4-butanediol to form PBS can be carried out through various methods such as melt polymerization, solution polymerization, and solid-state polymerization. Melt polymerization involves heating the monomers in the presence of a catalyst at high temperature and pressure to form the polymer. Solution polymerization involves dissolving the monomers in a solvent and adding a catalyst to initiate the polymerization. Solid-state polymerization involves heating the monomers in the solid state in the presence of a catalyst to form the polymer (Minh et al., 2010; Kang et al., 2015; Vardon et al., 2017).

4.3.4 Purification of PBS

The synthesized PBS is then purified to remove any impurities such as unreacted monomers or catalysts. The purification process typically involves precipitation of the PBS from a solvent or non-solvent, followed by filtration and washing. The purity of the PBS is important for its properties and applications.

The synthesis process of PBS has several advantages over traditional petroleum-based plastics. First, it is produced from renewable resources, reducing our dependence on fossil fuels. Second, it is biodegradable and compostable, meaning that it can break down into natural compounds in the environment and does not contribute to the growing problem of plastic waste. Third, it has good mechanical and thermal properties, making it suitable for a range of applications such as packaging, textiles, and automotive parts (Platnieks et al., 2021).

Recent research has focused on improving the synthesis process of PBS to overcome its limitations and increase its potential applications. Researchers have investigated various methods to increase the yield and reduce the cost of PBS production, including optimizing fermentation conditions, using alternative feedstocks such as waste materials, and developing genetically modified microorganisms. Furthermore, researchers have developed various approaches to improve the properties of PBS, including blending it with other polymers, adding fillers such as natural fibers or nano clays, and modifying the polymer structure through copolymerization or grafting (Chen et al., 2023a; Narisetty et al., 2022).

Another approach to improving the properties of PBS is through copolymerization or grafting. Copolymerization involves incorporating other monomers into the polymerization process to form copolymers with different properties. Grafting involves attaching side chains or functional groups onto the polymer chain to modify its properties. These approaches can improve the thermal stability, mechanical properties, and processing characteristics of PBS (Rafiqah et al., 2021).

Recent research has also focused on improving the sustainability of the PBS synthesis process. One approach is to use waste materials such as lignocellulosic biomass as a feedstock to produce succinic acid and 1,4-butanediol. This not only reduces the cost of production but also provides an environmentally friendly way to dispose of waste materials. Furthermore, researchers have developed genetically modified microorganisms that can efficiently produce succinic acid and 1,4-butanediol from renewable feedstocks, reducing the reliance on fossil fuels.

4.4 *Polybutylene Adipate Terephthalate (PBAT)*

Polybutylene adipate terephthalate (PBAT) is a biodegradable and compostable bioplastic that has gained significant attention in recent years as a sustainable alternative to traditional petroleum-based plastics. PBAT is a copolyester made from adipic acid, 1,4-butanediol, and terephthalic acid, which are derived from renewable resources (Fukushima et al., 2012).

4.4.1 Synthesis Processes of PBAT

It is produced through a condensation polymerization process. This process involves the reaction of three monomers: adipic acid, 1,4-butanediol, and terephthalic acid, which are derived from renewable resources.

The synthesis process of PBAT begins with the reaction of adipic acid and 1,4-butanediol in the presence of a catalyst, typically tin(II) octoate, to form a diol intermediate. The catalyst initiates the reaction between the carboxylic acid group of the adipic acid and the hydroxyl group of the 1,4-butanediol, resulting in the formation of a diester. Water is produced as a byproduct of the reaction, and it is removed from the system through distillation. After the formation of the diol intermediate, terephthalic acid is added to the mixture, and the reaction is continued until the desired molecular weight is reached. The terephthalic acid reacts with the diol intermediate to form ester bonds, resulting in the formation of PBAT polymer chains (Mahata et al., 2021).

The reaction is carried out under controlled conditions, including temperature and pressure, to ensure the desired properties of the resulting polymer. The molecular weight of the polymer can be controlled by adjusting the ratio of the monomers and the reaction time (Jian et al., 2020).

The resulting PBAT polymer can be melt processed into various forms, such as films, fibers, and injection molded parts, making it suitable for a range of applications. The properties of PBAT can be tailored through the addition of fillers and modifiers, allowing it to be used in blends to achieve desired properties (Table 3).

Table 3 Bioplastics: technology and manufacturing perspectives

	PLA	PHA	PBS	PBAT
	Poly (lactic acid)	Poly (hydroxyal-kanoates)	Poly (butylene succinate)	Poly (butylene adipate terephthalate)
Raw materials	Bio-derived monomers—Lactic acid and lactides • Sugarcane to lactic acid by the fermentation process • Lactic acid to lactide by polymerization–depolymerization process	• Produced by microorganisms, including through bacterial fermentation of sugars or lipids	• Succinic acid • Butanediol • Tri-based catalyst	• Terephthalic acid • Adipic acid • Butanediol • Tri-based catalysts
Process	Ring-opening polymerization (ROP) of lactide for PLA (catalyst:Tin compound)	The polymer is obtained by extraction from microorganism	Melt polycondensation polymerization	Melt polycondensation polymerization
Manufacturers	• NatureWorks (joint venture between Cargill (US) and PTT (Thailand)-Ingeo biopolymer • Total-Corbion (joint-venture between Total (France) and Corbion (NL)	• Kaneka-Japan • Bio-on-Italy • Yield10 bioscience	• Mitsubishi chemical performance polymers (Japan)-BioPBS • Hexing chemical (China) • Xinfu pharmaceutical (China) • Showa high polymer-Bionolle	• BASF (Germany)-Ecoflex • Novamont (Italy) • Xinjiang blue ridge (China)-Tunhe • Lotte fine chemicals (S. Korea)-Enpol

Grades and applications	• Ingeo Biopolymer 6204D-NatureWorks Thermoplastic fiber grade resin • Ingeo Biopolymer 4032D-Natureworks For lamination and other packaging applications. It provides a barrier to flavor, grease and oil resistance • Ingeo Biopolymer 2500HP-NatureWorks FDA approved and can be used for packaging • PLA blend A-Total Corbion It possesses heat resistance like PP, PS, and ABS	• Ecomp 142 Krafrit group • Biodegradable PHA-based compound used for film applications • Ecomp 420-Krafit group • Biodegradable starch-based PHA compound. Used to produce twin wall sheets of 4 mm	• BioPBS FD92-Mitsubishi chemical • Paper coatings, sealants, hot beverage cups, boxes, and utensils for freshly cooked foods	• Ecovio F2331-BASF (ecoflex F&PLA) • This grade possesses high melt strength, good thermostability up to 230 °C, and good mechanical properties • Ecovio M2351 BASF (ecoflex F & PLA) • This is suitable for producing black, transparent, and colored mulch films • Ecovio T2308-BASF (ecoflex F & PLA) • It is a thermoformable version for food trays and cups • Ecovio IA1652 BASF (ecoflex F & PLA) • Mineral filler and PLA in high content. For injection molding. This grade is a printable, sealable, and easy to color compound

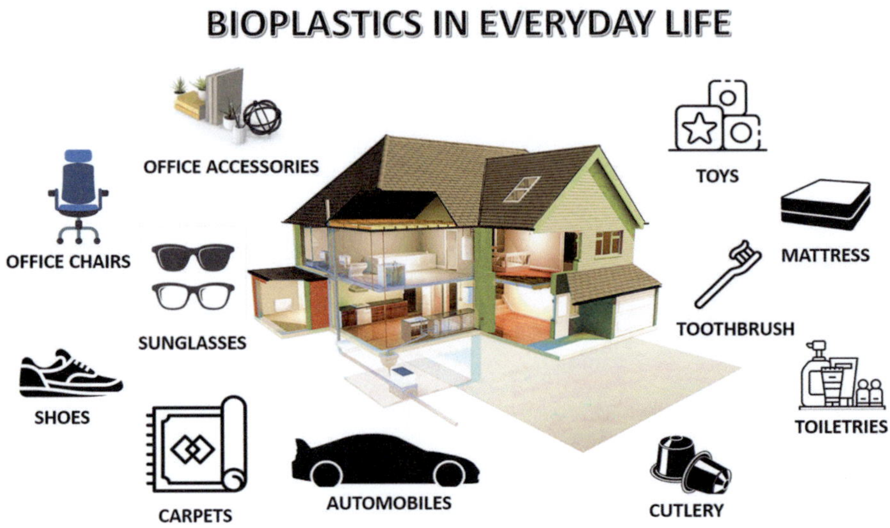

Fig. 3 Bioplastic in day-to-day life

5 Key Application Sectors

Bioplastics are used in a variety of applications (Fig. 3), with each type of bioplastic having its own unique set of properties that make it suitable for different uses. Here are some of the application sectors of bioplastics:

1. Packaging: This is the largest application sector for bioplastics, with food and beverage packaging accounting for most of the demand. Bioplastics offer several advantages for packaging, such as their biodegradability, compostability, and ability to reduce greenhouse gas emissions. PLA and PHA are two of the most used bioplastics in the packaging industry (Moshood et al., 2022; Rosenboom et al., 2022).
2. Agriculture: Bioplastics are used in the agriculture industry for applications such as mulch films, seedling trays, and irrigation pipes. Mulch films made from bioplastics can provide several benefits, such as weed control, moisture retention, and increased soil temperature. PBS and PBAT are two bioplastics commonly used in agriculture (Kasirajan & Ngouajio, 2012; Abbate et al., 2023).
3. Textiles: Bioplastics can be used in the textile industry to make sustainable fabrics and clothing. For example, PLA fibers can be used to make biodegradable textiles with properties like conventional polyester. Other bioplastics used in textiles include PHA and PBS (Kong et al., 2023).
4. Medical: Bioplastics are used in medical applications such as surgical sutures, drug delivery systems, and tissue engineering. PHB and PLA are two of the most used bioplastics in the medical industry. Biodegradable sutures made from PHB have several advantages over conventional sutures, such as reduced tissue trauma and faster healing (Naser et al., 2021).

5. Consumer goods: Bioplastics can be used in a variety of consumer goods, such as toys, electronics, and cosmetics packaging. PBAT and PLA are two of the most used bioplastics in consumer goods. Biodegradable toys made from PBAT can provide a sustainable alternative to conventional plastic toys (Upadhyaya & Sen Sarma, 2022).
6. Automotive industry: Bioplastics have been used in the automotive industry to reduce the weight of vehicles and increase fuel efficiency. They are used to create interior components such as dashboards, door panels, and seat cushions. For example, Mazda has used a bioplastic made from plant-derived materials in its MX-5 Miata roadster (Malkar et al., 2023; Rahman et al., 2023).

6 Comparative Study of Bioplastics and Conventional Plastics in Terms of Environmental Aspects

Bioplastics, sourced from renewable feedstocks, have shown potential in reducing greenhouse gas emissions compared to conventional plastics, predominantly derived from fossil fuels. Previous research, such as that conducted by Ingrao et al. (2015), indicates that bioplastics generally exhibit lower carbon footprints and contribute less to global warming potential. Moreover, studies by Degli Esposti et al. (2021) and Kawashima et al. (2019) have underscored the role of bioplastics in mitigating resource depletion, particularly in terms of reducing reliance on finite resources like crude oil (Kawashima et al., 2019; Degli Esposti et al., 2021).

Life cycle assessment (LCA) serves as a robust methodology for quantifying the environmental impacts of bioplastics and conventional plastics across their life cycles. Studies by Vinci et al. (2021) and Hobbs et al. (2021) have revealed substantial reductions in greenhouse gas emissions associated with bioplastics compared to conventional plastics (Vinci et al., 2021; Hobbs et al., 2021). For instance, Chen et al. reported a 30% reduction in carbon dioxide equivalents (CO_2e) emissions for bioplastic production compared to conventional plastics (Chen et al., 2023b). Additionally, Ali et al. found that bioplastics sourced from renewable feedstocks resulted in 40% lower energy consumption throughout their life cycle. These findings underscore the potential environmental benefits of bioplastics in mitigating climate change and reducing resource depletion (Ali et al., 2023a).

However, LCAs must consider various complexities to provide accurate assessments. Research by Espinoza-Orias et al. (2016) and Yutong et al. (2023) emphasized the significance of incorporating land use change and competition with food production into LCA frameworks for bioplastics derived from agricultural crops (Bishop et al., 2021). For example, Yutong et al. highlighted that bioplastics made from sugar beet feedstock could potentially lead to a 50% reduction in land use compared to conventional plastics derived from crude oil (Jin et al., 2023). Furthermore, studies by Bishop et al. (2021) have emphasized the importance of integrating factors such as water usage, soil quality, and biodiversity loss into LCA analyses to provide a comprehensive assessment of bioplastics' sustainability (Bishop et al., 2021).

Considering potential indirect environmental impacts is crucial in LCA analyses. Venancio et al. (2022) and Piemonte & Gironi (2011) demonstrated the importance of accounting for changes in land use patterns, carbon sequestration, and ecosystem services associated with bioplastics (Piemonte & Gironi, 2011; Venâncio et al., 2022). For instance, Escobar et al. quantified the carbon stock changes resulting from land use change for bioplastics production, highlighting the need to consider these indirect effects in LCA methodologies (Escobar & Britz, 2021).

7 Prospects of Bioplastics in Plastic Industries

The bioplastics in the plastic industry is gaining significant attention and momentum due to increasing environmental concerns, sustainability goals, and the desire to reduce the environmental impact of traditional plastics. Bioplastics represent a promising alternative, offering potential benefits in terms of resource efficiency, reduced carbon footprint, and decreased dependence on fossil fuels. Bioplastics are predicted to have a $20 billion market by 2026, having grown from a projected $9.2 billion in the current market every day (Research GV, 2021). Over 2.1 million tons of bioplastics were produced globally in 2020, and by 2025, that number is expected to rise to 2.9 million tons (EuropeanBioplastics, 2021). Asia continues to be the primary production hub, accounting for 46% of the world's total output of bioplastics. Europe, North America, and South America follow with 26%, 17%, and 10% of the worldwide production, respectively (EuropeanBioplastics, 2021).

Certain types of bioplastics are designed to be biodegradable and compostable, offering a solution to the problem of plastic pollution. Biodegradable plastics break down into natural compounds over time, reducing the environmental impact of plastic waste and minimizing harm to ecosystems. Currently, 0.8 million tons, or 40% of the world's bioplastics production volumes, come from drop-in bio-based polyethylene, bio-based polyethylene terephthalate, and bio-based polyamides (EuropeanBioplastics, 2021). The packaging business is acknowledged as using the most bioplastics nowadays, consuming close to one million tons, or 47% of all bioplastics produced. Although 4.8 billion hectares of land are used for agriculture worldwide, just 0.7 million hectares, or 0.015% of all arable land, are used to grow feedstocks that will eventually be used as precursors for bioplastics (EuropeanBioplastics, 2021). Thus, it is also expected that the manufacturing of bioplastics from agricultural crop leftovers will not compete with pasture, food, fodder, and arable land.

Growing awareness of environmental issues and increased consumer consciousness about the impact of plastic waste have led to a rising demand for sustainable alternatives. Many consumers and companies are actively seeking eco-friendly products, encouraging the plastic industry to adopt bioplastics as part of their commitment to corporate social responsibility. Governments and regulatory bodies are implementing measures to reduce the use of traditional plastics and promote the adoption of sustainable alternatives. Policies, such as bans on single-use plastics and incentives for bioplastics, are creating a supportive regulatory environment for the growth of the bioplastics sector.

While the prospect of bioplastics in the plastic industry is promising, challenges remain, including cost competitiveness, scalability of production, and addressing concerns about the competition between food and fuel crops for feedstock. However, ongoing research and development efforts, coupled with increasing awareness and regulatory support, indicate a positive trajectory for the integration of bioplastics into the mainstream plastic industry. As technology advances and economies of scale are achieved, bioplastics are likely to play a significant role in shaping a more sustainable future for the plastics industry.

8 Conclusion

In conclusion, this chapter provides a comprehensive overview of bioplastics, emphasizing their significance as a sustainable alternative to conventional plastics. Bioplastics, derived from renewable sources and possessing biodegradable or compostable properties, offer a promising solution to the environmental challenges posed by traditional plastics. The discussion covers various types of bioplastics, their synthesis processes, properties, applications, challenges, and recent research and development efforts. Despite facing challenges such as production costs, biodegradability requirements, and standardization issues, ongoing advancements in research aim to enhance the properties and reduce the production costs of bioplastics. The comparison between petro-based conventional plastics and bioplastics highlights the eco-friendliness, biodegradability, and diverse applications of the latter. The market analysis underscores the growing global demand for bioplastics, driven by increased consumer awareness, regulatory support, and technological advancements. The prospects of bioplastics in the plastic industry are promising, with the potential to revolutionize packaging, agriculture, textiles, medical applications, consumer goods, and automotive industries. As the demand for sustainable materials rises, and technology continues to progress, bioplastics are poised to play a crucial role in mitigating the environmental impact of plastic waste, fostering a more sustainable and circular economy.

References

Abbate, C., Scavo, A., Pesce, G. R., et al. (2023). Soil bioplastic mulches for agroecosystem sustainability: A comprehensive review. *Agriculture, 13*, 197. https://doi.org/10.3390/agriculture13010197

Abe, M. M., Martins, J. R., Sanvezzo, P. B., et al. (2021). Advantages and disadvantages of bioplastics production from starch and lignocellulosic components. *Polymers (Basel), 13*, 2484. https://doi.org/10.3390/polym13152484

Ahsan, W. A., Hussain, A., Lin, C., & Nguyen, M. K. (2023). Biodegradation of different types of bioplastics through composting—A recent trend in green recycling. *Catalysts, 13*, 294. https://doi.org/10.3390/catal13020294
Ajioka, M., Enomoto, K., Suzuki, K., & Yamaguchi, A. (1995). The basic properties of poly(lactic acid) produced by the direct condensation polymerization of lactic acid. *Journal of Environmental Polymer Degradation, 3*, 225–234. https://doi.org/10.1007/BF02068677
Ali, S. S., Abdelkarim, E. A., Elsamahy, T., et al. (2023a). Bioplastic production in terms of life cycle assessment: A state-of-the-art review. *Environmental Science and Ecotechnology, 15*, 100254. https://doi.org/10.1016/j.ese.2023.100254
Ali, W., Ali, H., Gillani, S., et al. (2023b). Polylactic acid synthesis, biodegradability, conversion to microplastics and toxicity: A review. *Environmental Chemistry Letters, 21*, 1761–1786. https://doi.org/10.1007/s10311-023-01564-8
Barletta, M., Aversa, C., Ayyoob, M., et al. (2022). Poly(butylene succinate) (PBS): Materials, processing, and industrial applications. *Progress in Polymer Science, 132*, 101579. https://doi.org/10.1016/j.progpolymsci.2022.101579
Barrino, F., De La Rosa-Ramírez, H., Schiraldi, C., et al. (2023). Preparation and characterization of new bioplastics based on polybutylene succinate (PBS). *Polymers (Basel), 15*, 1212. https://doi.org/10.3390/polym15051212
Bhattacharya, et al. Challenges and opportunities: plastic waste management in India, South Asian Born-Digital NGO Reports Collection Project, http://hdl.handle.net/2451/42242 (2018).
Bishop, G., Styles, D., & Lens, P. N. L. (2021). Environmental performance comparison of bioplastics and petrochemical plastics: A review of life cycle assessment (LCA) methodological decisions. *Resources, Conservation and Recycling, 168*, 105451. https://doi.org/10.1016/j.resconrec.2021.105451
Chen, G.-Q. (2009). A microbial polyhydroxyalkanoates (PHA) based bio- and materials industry. *Chemical Society Reviews, 38*, 2434. https://doi.org/10.1039/b812677c
Chen, C., Chen, X., Liu, L., et al. (2023a). Engineering microorganisms to produce bio-based monomers: Progress and challenges. *Fermentation, 9*, 137. https://doi.org/10.3390/fermentation9020137
Chen, G., Li, J., Sun, Y., et al. (2023b). Replacing traditional plastics with biodegradable plastics: Impact on carbon emissions. *Engineering, 32*, 152–162. https://doi.org/10.1016/j.eng.2023.10.002
Dammak, M., Fourati, Y., Tarrés, Q., et al. (2020). Blends of PBAT with plasticized starch for packaging applications: Mechanical properties, rheological behaviour and biodegradability. *Industrial Crops and Products, 144*, 112061. https://doi.org/10.1016/j.indcrop.2019.112061
Degli Esposti, M., Morselli, D., Fava, F., et al. (2021). The role of biotechnology in the transition from plastics to bioplastics: An opportunity to reconnect global growth with sustainability. *FEBS Open Bio, 11*, 967–983. https://doi.org/10.1002/2211-5463.13119
Escobar, N., & Britz, W. (2021). Metrics on the sustainability of region-specific bioplastics production, considering global land use change effects. *Resources, Conservation and Recycling, 167*, 105345. https://doi.org/10.1016/j.resconrec.2020.105345
EuropeanBioplastics. (2021). *Bioplastics market development update 2023*. Retrieved March 12, 2024, from https://www.european-bioplastics.org/market/
European Commission: Directorate-General for Environment, Turning the tide on single-use plastics, Publications Office, (2019). https://data.europa.eu/doi/10.2779/294711
Folino, A., Pangallo, D., & Calabrò, P. S. (2023). Assessing bioplastics biodegradability by standard and research methods: Current trends and open issues. *Journal of Environmental Chemical Engineering, 11*, 109424. https://doi.org/10.1016/j.jece.2023.109424
Fukushima, K., Wu, M.-H., Bocchini, S., et al. (2012). PBAT based nanocomposites for medical and industrial applications. *Materials Science and Engineering: C, 32*, 1331–1351. https://doi.org/10.1016/j.msec.2012.04.005
GlobalData. (n.d.). India Bioplastics Market Size, Segmentation by Category and Geography, Competitive Landscape and Forecast, 2017–2026.

Gong, L., Passari, A. K., Yin, C., et al. (2023). Sustainable utilization of fruit and vegetable waste bioresources for bioplastics production. *Critical Reviews in Biotechnology, 44*, 236–254. https://doi.org/10.1080/07388551.2022.2157241

Hobbs, S. R., Harris, T. M., Barr, W. J., & Landis, A. E. (2021). Life cycle assessment of bioplastics and food waste disposal methods. *Sustainability, 13*, 6894. https://doi.org/10.3390/su13126894

Ilyas, R. A., Sapuan, S. M., Harussani, M. M., et al. (2021). Polylactic acid (PLA) biocomposite: Processing, additive manufacturing and advanced applications. *Polymers (Basel), 13*, 1326. https://doi.org/10.3390/polym13081326

Ingrao, C., Tricase, C., Cholewa-Wójcik, A., et al. (2015). Polylactic acid trays for fresh-food packaging: A carbon footprint assessment. *The Science of the Total Environment, 537*, 385–398. https://doi.org/10.1016/j.scitotenv.2015.08.023

Jabbar, A., Tausif, M., Tahir, H. R., et al. (2020). Polylactic acid/lyocell fibre as an eco-friendly alternative to polyethylene terephthalate/cotton fibre blended yarns and knitted fabrics. *Journal of the Textile Institute, 111*, 129–138. https://doi.org/10.1080/00405000.2019.1624070

Jayakumar, A., Radoor, S., Siengchin, S., et al. (2023). Recent progress of bioplastics in their properties, standards, certifications and regulations: A review. *Science of the Total Environment, 878*, 163156. https://doi.org/10.1016/j.scitotenv.2023.163156

Jian, J., Xiangbin, Z., & Xianbo, H. (2020). An overview on synthesis, properties and applications of poly(butylene-adipate-co-terephthalate)–PBAT. *Advanced Industrial and Engineering Polymer Research, 3*, 19–26. https://doi.org/10.1016/j.aiepr.2020.01.001

Jin, Y., Lenzen, M., Montoya, A., et al. (2023). Greenhouse gas emissions, land use and employment in a future global bioplastics economy. *Resources, Conservation and Recycling, 193*, 106950. https://doi.org/10.1016/j.resconrec.2023.106950

Kang, K. H., Hong, U. G., Bang, Y., et al. (2015). Hydrogenation of succinic acid to 1,4-butanediol over re–Ru bimetallic catalysts supported on mesoporous carbon. *Applied Catalysis A: General, 490*, 153–162. https://doi.org/10.1016/j.apcata.2014.11.029

Kasirajan, S., & Ngouajio, M. (2012). Polyethylene and biodegradable mulches for agricultural applications: A review. *Agronomy for Sustainable Development, 32*, 501–529. https://doi.org/10.1007/s13593-011-0068-3

Kawashima, N., Yagi, T., & Kojima, K. (2019). How do bioplastics and fossil-based plastics play in a circular economy? *Macromolecular Materials and Engineering, 304*(9), 1900383. https://doi.org/10.1002/mame.201900383

Khammassi, S., Tarfaoui, M., Škrlová, K., et al. (2022). Poly(lactic acid) (PLA)-based nanocomposites: Impact of vermiculite, silver, and graphene oxide on thermal stability, isothermal crystallization, and local mechanical behavior. *Journal of Composites Science, 6*, 112. https://doi.org/10.3390/jcs6040112

Kim, T., Kim, D., & Park, Y. (2022). Recent progress in regenerated fibers for "green" textile products. *Journal of Cleaner Production, 376*, 134226. https://doi.org/10.1016/j.jclepro.2022.134226

Knowledge Sourcing Intelligence. (2023). India Bioplastics Market – Forecasts from 2022 to 2027. Knowledge Sourcing Intelligence. https://www.knowledge-sourcing.com/report/india-bioplastics-market.

Kong, U., Mohammad Rawi, N. F., & Tay, G. S. (2023). The potential applications of reinforced bioplastics in various industries: A review. *Polymers (Basel), 15*, 2399. https://doi.org/10.3390/polym15102399

Kumar, P., Park, H., Yuk, Y., et al. (2023). Developed and emerging 1,4-butanediol commercial production strategies: Forecasting the current status and future possibility. *Critical Reviews in Biotechnology, 44*, 530–546. https://doi.org/10.1080/07388551.2023.2176740

Mahata, D., Karthikeyan, S., Godse, R., & Gupta, V. K. (2021). Poly(butylene adipate-co-terephthalate) polyester synthesis process and product development. *Polymer Science, Series C, 63*, 102–111. https://doi.org/10.1134/S1811238221010045

Malkar, R., Kagale, S., Chavan, S., et al. (2023). Applications of bioplastics in sports and leisure. In *Handbook of bioplastics and biocomposites engineering applications* (pp. 299–315). Wiley.

Minh, D. P., Besson, M., Pinel, C., et al. (2010). Aqueous-phase hydrogenation of biomass-based succinic acid to 1,4-Butanediol over supported bimetallic catalysts. *Topics in Catalysis, 53*, 1270–1273. https://doi.org/10.1007/s11244-010-9580-y

More, N., Avhad, M., Utekar, S., & More, A. (2023). Polylactic acid (PLA) membrane—Significance, synthesis, and applications: A review. *Polymer Bulletin, 80*, 1117–1153. https://doi.org/10.1007/s00289-022-04135-z

Moshood, T. D., Nawanir, G., Mahmud, F., et al. (2022). Biodegradable plastic applications towards sustainability: A recent innovations in the green product. *Cleaner Engineering and Technology, 6*, 100404. https://doi.org/10.1016/j.clet.2022.100404

Narisetty, V., Okibe, M. C., Amulya, K., et al. (2022). Technological advancements in valorization of second generation (2G) feedstocks for bio-based succinic acid production. *Bioresource Technology, 360*, 127513. https://doi.org/10.1016/j.biortech.2022.127513

Naser, A. Z., Deiab, I., & Darras, B. M. (2021). Poly(lactic acid) (PLA) and polyhydroxyalkanoates (PHAs), green alternatives to petroleum-based plastics: A review. *RSC Advances, 11*, 17151–17196. https://doi.org/10.1039/D1RA02390J

Ncube, L. K., Ude, A. U., Ogunmuyiwa, E. N., et al. (2020). Environmental impact of food packaging materials: A review of contemporary development from conventional plastics to Polylactic acid based materials. *Materials (Basel), 13*, 4994. https://doi.org/10.3390/ma13214994

Pederson, E. N., McChalicher, C. W. J., & Srienc, F. (2006). Bacterial synthesis of PHA block copolymers. *Biomacromolecules, 7*, 1904–1911. https://doi.org/10.1021/bm0510101

Piemonte, V., & Gironi, F. (2011). Land-use change emissions: How green are the bioplastics? *Environmental Progress & Sustainable Energy, 30*, 685–691. https://doi.org/10.1002/ep.10518

Platnieks, O., Gaidukovs, S., Kumar Thakur, V., et al. (2021). Bio-based poly (butylene succinate): Recent progress, challenges and future opportunities. *European Polymer Journal, 161*, 110855. https://doi.org/10.1016/j.eurpolymj.2021.110855

Rafiqah, S. A., Khalina, A., Harmaen, A. S., et al. (2021). A review on properties and application of bio-based poly(butylene succinate). *Polymers (Basel), 13*, 1436. https://doi.org/10.3390/polym13091436

Rahman, M. Z., Rahman, M., Mahbub, T., et al. (2023). Advanced biopolymers for automobile and aviation engineering applications. *Journal of Polymer Research, 30*, 106. https://doi.org/10.1007/s10965-023-03440-z

Research GV. (2021). Bioplastics market size, share & trends analysis report by product (biodegradable, non-biodegradable), by application, by region, and segment forecasts, 2023–2030.

Rosenboom, J.-G., Langer, R., & Traverso, G. (2022). Bioplastics for a circular economy. *Nature Reviews Materials, 7*, 117–137. https://doi.org/10.1038/s41578-021-00407-8

Saxena, R. K., Saran, S., Isar, J., & Kaushik, R. (2017). Production and applications of succinic acid. In *Current developments in biotechnology and bioengineering* (pp. 601–630). Elsevier.

Sehgal, R., & Gupta, R. (2020). Polyhydroxyalkanoate and its efficient production: An eco-friendly approach towards development. *3 Biotech, 10*, 549. https://doi.org/10.1007/s13205-020-02550-5

Shafqat, A., Tahir, A., Mahmood, A., et al. (2020). A review on environmental significance carbon foot prints of starch based bio-plastic: A substitute of conventional plastics. *Biocatalysis and Agricultural Biotechnology, 27*, 101540. https://doi.org/10.1016/j.bcab.2020.101540

Tan, G.-Y., Chen, C.-L., Li, L., et al. (2014). Start a research on biopolymer Polyhydroxyalkanoate (PHA): A review. *Polymers (Basel), 6*, 706–754. https://doi.org/10.3390/polym6030706

Union, E. (2021). *Turning the tide on single-use plastics*. European Union.

Upadhyaya, S., & Sen Sarma, N. (2022). Biodegradable plastics as a substitute to traditional polythenes: A step toward a safer environment. In *Environmental sustainability and industries* (pp. 193–215). Elsevier.

Vardon, D. R., Settle, A. E., Vorotnikov, V., et al. (2017). Ru-Sn/AC for the aqueous-phase reduction of succinic acid to 1,4-Butanediol under continuous process conditions. *ACS Catalysis, 7*, 6207–6219. https://doi.org/10.1021/acscatal.7b02015

Vega-Baudrit, J. (2019). Polylactic acid (PLA) as a bioplastic and its possible applications in the food industry. *Journal of Food Science & Nutrition, 5*, 1–6. https://doi.org/10.24966/FSN-1076/100048

Venâncio, C., Lopes, I., & Oliveira, M. (2022). Bioplastics: Known effects and potential consequences to marine and estuarine ecosystem services. *Chemosphere, 309*, 136810. https://doi.org/10.1016/j.chemosphere.2022.136810

Vinci, G., Ruggieri, R., Billi, A., et al. (2021). Sustainable management of organic waste and recycling for bioplastics: A LCA approach for the Italian case study. *Sustainability, 13*, 6385. https://doi.org/10.3390/su13116385

Widjaja, T., Hendrianie, N., Nurkhamidah, S., et al. (2023). Poly lactic acid production using the ring opening polymerization (ROP) method using Lewis acid surfactant combined iron (Fe) catalyst (Fe(DS)3). *Heliyon, 9*, e17985. https://doi.org/10.1016/j.heliyon.2023.e17985

Yashavanth, P. R., Das, M., & Maiti, S. K. (2021). Recent progress and challenges in cyanobacterial autotrophic production of polyhydroxybutyrate (PHB), a bioplastic. *Journal of Environmental Chemical Engineering, 9*, 105379. https://doi.org/10.1016/j.jece.2021.105379

Conclusion and Way Forward

Suneel Pandey, Smita Mohanty, Girija K. Bharat, Hans N. Adam, Emmy Nøklebye, and Eirik H. Steindal

India's plastic waste management legislation framework has advanced significantly. In 2011, the Indian government's Ministry of Environment, Forests, and Climate Change (MoEFCC) released the plastic waste (management and handling) rules, which marked the beginning of extended producer responsibility (EPR) for plastic items in the country (CPCB, 2016). The present waste management system in India is the result of multiple revisions to the Plastic Waste Management (PWM) rules, 2016 and the Solid Waste Management (SWM) rules, 2016 (MoEFCC, 2016). These were reinforced by waste management standards, such as the CPCB standards for Segregation, Collection, and Disposal of Plastic Waste (CPCB, 2017a) and the Guidelines for Co-processing of Plastic Waste in Cement Kilns (CPCB, 2017b). The draft Uniform Framework for Extended Producer Responsibility (Under Plastic Waste Management Rules 2016) on June 26, 2020 is the producer's responsibility to facilitate a reverse collection mechanism and recycling of end-of-life, post-consumer waste. The EPR rules offer three options to producers:

(a) Pay a fee into a central corpus that would be spent toward managing the plastic waste;

S. Pandey
Environment and Waste Management Division, TERI, New Delhi, Delhi, India

S. Mohanty
Central Institute of Petrochemicals Engineering Technology, CIPET, Bhubaneswar, Odisha, India

G. K. Bharat (✉)
Mu Gamma Consultants, Gurugram, Haryana, India
e-mail: gbharat@mugammaconsultants.com

H. N. Adam · E. Nøklebye · E. H. Steindal
Norwegian Institute for Water Research (NIVA), Oslo, Norway

P. Chakraborty et al. (eds.), *Efficient and Enhanced Management of Plastic Waste in India*, https://doi.org/10.1007/978-3-031-89398-8_11

(b) Participate in and pay for establishing producer responsibility organizations (PROs) to collect and manage post-consumer plastic waste; and
(c) Buy credits from a system that would be established to balance the plastic waste they generate.

Although EPR has gained a foothold in the Indian waste management sector since the first introduction of the PWM Rules in 2011, its implementation mechanism was unclear to plastic waste producers, urban local bodies, and state regulators. All businesses that use plastic packaging must register as "producers" or "brand owners" in accordance with the PWM regulations of 2016 (Saahas Zero Waste, n.d.). Furthermore, if a manufacturer or brand owner operates in multiple states, PWM rules mandate that they register with SPCBs and CPCBs and submit an EPR strategy. According to the MoEFCC (2016), the PWM rules prohibit releasing goods onto the market without an EPR plan.

An effective EPR can be implemented through a partnership between policy-science and society. The recently published standard operating procedures (SOP) for EPR implementation have highlighted the need for a body with similar responsibilities (CPCB, 2021). The state government has been tasked by the SOP with this responsibility: to designate a nodal agency at the state level for the purpose of coordinating efforts to guarantee that the entire state, including town *panchayats*, cantonment boards, and rural local bodies, is covered by the EPR framework. Stakeholder cooperation and participation in the informal sector will need to be strengthened in addition to strict implementation of the EPR with set aims. Prioritization should be given to the following: better waste management infrastructure, scientific processing, high- and low-value plastic recycling, and the segregation and collection of plastic trash at the source. Low-value plastic waste typically seeps into the environment instead of being retrieved for recycling or energy recovery. Prioritizing training, capacity building, and behavior change initiatives is necessary to improve the collection and recycling of low-value plastics such as sachets, bags, and other items that aren't even collected.

To guarantee the effective collection and recycling of plastic waste, awareness among the many stakeholders—especially the users and generators of plastic waste—is crucial. Academic institutions and think tanks can assist the urban local bodies (ULBs) and *Gram Panchayats* in developing and implementing these kinds of initiatives. A better understanding of the phase-out of certain single-use plastic products (SUPPs) is needed for the effective implementation of the PWM Rules. Since public awareness can help policies to be implemented more effectively, creative public awareness campaigns are necessary to ensure that all policies are implemented as intended.

Following collection, transportation, and temporary storage, a growing amount of mixed and dry waste needs to be directed toward material recovery facilities (MRFs) for sorting and efficient recovery of different types of plastics. Since the generation of solid waste, especially plastics, is expected to increase in the near future, ULBs must immediately expand their MRF infrastructure and modify their operations to accommodate informal waste workers. Upskilling and capacity building are necessary for integrating the unorganized sector. Integration of technology with current capabilities is necessary to provide society with scalable solutions that adhere to environmental health and safety standards. At the city level, industry-producer responsibility organizations (PROs) and ULBs must collaborate to develop a shared responsibility model that will guarantee recycling funds are used efficiently and prevent duplication of effort. The shared responsibility model ought to keep an eye on how EPR is being implemented, since it continues to be a weak point in the execution of policies.

The States and urban local authorities have a duty to periodically assess and oversee the efficient execution of the policies. A long-term shift in behavior that supports the segregation, recycling, and reduction of low-value plastics—such as bags and sachets—as well as support for the overall management of plastics can result from placing a high priority on training and capacity building. Regular training courses should be prioritized by the government and industry.

Addressing the problem of plastic pollution in lakes and rivers as well as clean-up action plans and missions are among the national research needs. To be able to identify flaws in the current waste management system and plan for bridging the gaps, the States need to examine the sources of plastic waste leakage, hotspots, and paths of flows to water bodies and coastal water, as appropriate. The standardization of procedures for plastics sampling, identification, evaluations, analysis, and waste flow monitoring is another task that State PCBs must pursue. Research on upstream initiatives like recycling design, introduction of alternative packaging materials such as bioplastics, compostable plastics, and design for the environment (minimal hazardous materials) should also be funded by a national EPR fund, particularly as single-use plastics have been phased out starting on July 1, 2022. It is important to put regulatory measures in place in order to enhance the competitiveness of recycled products. Since community-led programs primarily focus on encouraging recycling, particularly for energy recovery and use in building materials, it is equally important to strengthen support for broader methods of reuse and recycling of plastic waste. Because of downcycling, energy recovery technologies are not the ideal choices for enhancing the resource's circular economy.

It is very crucial to support waste management practices such as segregation at source, as it enables the efficient functioning of Material Recovery Facilities (MRFs), where plastic waste can be sorted and recycled into value-added products`. This reduces the quantity of dry garbage and plastics dumped in landfills and waterways. Financial incentives make a huge difference to consumers for segregation. When it comes to low-value plastic garbage, such incentives can enhance collection (Types 5–7). As of now, only Types 1–4 are valuable, while Types 5–7 are not purchased. In the PWM industry, the idea of a minimum support price (MSP) may be taken into consideration. In order to improve waste management cycles in various cities and rural areas, it may be helpful to disseminate case studies that highlight the difficulties, successes, and lessons learned from zero landfill strategies, source segregation, the use of information communication and technology (ICT), enhanced stakeholder participation, improved communication techniques, and awareness generation. Through information sharing with stakeholders, training, and capacity building, a well-designed IEC program should continuously raise awareness. This will lead to behavioral changes, which in the absence of such changes may result in the failure of viable solutions and infrastructure.

To monitor, report, and intervene during plastic burning activities, Green Warriors from Gramme Panchayats and Urban Local Bodies may find it useful to receive this kind of training. Using ICT and enforcing sanctions for offenders could potentially lead to more efficient management of garbage-vulnerable areas. Focus should be placed on the health and safety of the informal waste pickers at work, giving them a voice in collective action, and providing financial support mechanisms like credit and health insurance. This would improve their standard of living and means of subsistence.

In order to achieve a long-term solution, industries and municipalities must work together in co-processing, a proven method of managing non-recyclable dry stream wastes. Cost-sharing and input waste quality are among the challenges that need to be addressed.

References

CPCB. (2016). *Plastic waste management rules*. Central Pollution Control Board (CPCB). https://cpcb.nic.in/rules-4/

CPCB. (2017a). *Consolidated guidelines for segregation, collection and disposal of plastic waste*. Central Pollution Control Board (CPCB). https://cpcb.nic.in/uploads/plasticwaste/Consolidate_Guidelines_for_disposal_of_PW.pdf

CPCB. (2017b). *Guidelines for co-processing of plastic waste in cement kilns (As per rule '5(b)' of plastic waste management rules, 2016)*. Central Pollution Control Board (CPCB). https://cpcb.nic.in/uploads/plasticwaste/co-processing_guidelines_final_23.05.17.pdf

CPCB. (2021). *Standard operating procedure for registration of producers, importers & brand-owners (PIBOs) under plastic waste management rules 2016 (as amended)*. Central Pollution Control Board (CPCB). https://cpcb.nic.in/uploads/plasticwaste/SOP_EPR_PWM_23-03-2021.pdf

MoEF&CC. (2016). *Solid waste management rules*. Ministry of Environment, Forest and Climate Change (MoEF&CC). https://cpcb.nic.in/uploads/MSW/SWM_2016.pdf

Saahas Zero Waste. (n.d.). https://www.businesscalltoaction.org/member/saahas-zero-waste

Index

P. Chakraborty et al. (eds.), *Efficient and Enhanced Management of Plastic Waste in India*, https://doi.org/10.1007/978-3-031-89398-8

MIX
Papier aus verantwortungsvollen Quellen
Paper from responsible sources
FSC® C105338

If you have any concerns about our products,
you can contact us on
ProductSafety@springernature.com

In case Publisher is established outside the EU,
the EU authorized representative is:
Springer Nature Customer Service Center GmbH
Europaplatz 3, 69115 Heidelberg, Germany

Printed by Libri Plureos GmbH
in Hamburg, Germany